Theory of Financial Risk and Derivative Pricing
From Statistical Physics to Risk Management, Second Edition

Risk control and derivative pricing have become of major concern to financial institutions, and there is a real need for adequate statistical tools to measure and anticipate the amplitude of the potential moves of the financial markets.

Summarising recent theoretical developments in the field, this second edition has been substantially expanded. New chapters now cover stochastic processes, Monte-Carlo methods, Black–Scholes theory, the theory of the yield curve, and Minority Game. There are discussions on aspects of data analysis, financial products, non-linear correlations, and herding, feedback and agent based models.

This book has become a classic reference for graduate students and researchers working in econophysics and mathematical finance, and for quantitative analysts working on risk management, derivative pricing and quantitative trading strategies.

JEAN-PHILIPPE BOUCHAUD was born in France in 1962. After studying at the French Lycée of London, he graduated from the Ecole Normale Supérieure in Paris, where he also obtained his Ph.D. in physics. He was then appointed by the CNRS until 1992, where he worked on diffusion in random media. After a year spent in the Cavendish Laboratory (Cambridge), Dr Bouchaud joined the Service de Physique de l'Etat Condensé (CEA-Saclay), where he worked on the dynamics of glassy systems and on granular media. He became interested in theoretical finance in 1991. He founded the company Science & Finance in 1994 with J.-P. Aguilar, which merged with Capital Fund Management (CFM) in 2000. His work in finance includes extreme risk control and alternative option pricing models. He teaches statistical mechanics and finance at Ecole Polytechnique and ENSAE, and supervises the research team at CFM with Marc Potters. He was awarded the IBM young scientist prize in 1990 and the C.N.R.S. Silver Medal in 1996.

MARC POTTERS is Head of Research and Chief Portfolio Manager at CFM, Paris, a systematic hedge fund trading equities, futures, options and bonds on worldwide electronic markets with over three billion dollars under management. With Jean-Philippe Bouchaud, he supervises the research of thirty physics Ph.D.s and the implementation of their results into automated trading strategies and risk control models. He has published numerous articles in the new field of statistical finance, in particular on Random Matrix Theory applied to portfolio management. Born in Belgium in 1969, Dr Potters grew up in Montréal. He obtained his Ph.D. in physics at Princeton University and was a post-doctoral fellow at the University of Rome La Sapienza. He joined Science & Finance (now CFM) in 1995.

Theory of Financial Risk and Derivative Pricing

From Statistical Physics to Risk Management

SECOND EDITION

Jean-Philippe Bouchaud and Marc Potters

CAMBRIDGE
UNIVERSITY PRESS

University Printing House, Cambridge CB2 8BS, United Kingdom

One Liberty Plaza, 20th Floor, New York, NY 10006, USA

477 Williamstown Road, Port Melbourne, VIC 3207, Australia

314-321, 3rd Floor, Plot 3, Splendor Forum, Jasola District Centre, New Delhi - 110025, India

79 Anson Road, #06-04/06, Singapore 079906

Cambridge University Press is part of the University of Cambridge.

It furthers the University's mission by disseminating knowledge in the pursuit of education, learning and research at the highest international levels of excellence.

www.cambridge.org
Information on this title: www.cambridge.org/9780521741866

First published 2000
This edition published 2003
Third printing 2006
Reprinted with corrections 2009
First paperback Edition 2009
Reprinted 2011

A catalogue record for this publication is available from the British Library

Library of Congress Cataloging in Publication data
Bouchaud, Jean–Philippe, 1962–.
Theory of financial risk and derivative pricing: from statistical physics to risk management / Jean–Philippe Bouchaud and Marc Potters. – 2ndedn.
p. cm.
Rev. edn of: Theory of financial risks. 2000.
Includes bibliographical references and index.
ISBN 0 521 819164 (hardback)
1. Finance. 2. Financial engineering. 3. Risk assessment. 4. Risk management.
I. Potters, Marc, 1969– II. Bouchaud, Jean–Philippe, 1962–Theory of financial risks. III. Title.
HG101.B68 2003
658.15′5–dc21 2003044037

ISBN 978-0-521-81916-9 Hardback
ISBN 978-0-521-74186-6 Paperback

Contents

Foreword

Since the 1980s an increasing number of physicists have been using ideas from statistical mechanics to examine financial data. This development was partially a consequence of the end of the cold war and the ensuing scarcity of funding for research in physics, but was mainly sustained by the exponential increase in the quantity of financial data being generated everyday in the world's financial markets.

Jean-Philippe Bouchaud and Marc Potters have been important contributors to this literature, and *Theory of Financial Risk and Derivative Pricing*, in this much revised second English-language edition, is an admirable summary of what has been achieved. The authors attain a remarkable balance between rigour and intuition that makes this book a pleasure to read.

To an economist, the most interesting contribution of this literature is a new way to look at the increasingly available high-frequency data. Although I do not share the authors' pessimism concerning long time scales, I agree that the methods used here are particularly appropriate for studying fluctuations that typically occur in frequencies of minutes to months, and that understanding these fluctuations is important for both scientific and pragmatic reasons. As most economists, Bouchaud and Potters believe that models in finance are never 'correct' – the specific models used in practice are often chosen for reasons of tractability. It is thus important to employ a variety of diagnostic tools to evaluate hypotheses and goodness of fit. The authors propose and implement a combination of formal estimation and statistical tests with less rigorous graphical techniques that help inform the data analyst. Though in some cases I wish they had provided conventional standard errors, I found many of their figures highly informative.

The first attempts at applying the methodology of statistical physics to finance dealt with individual assets. Financial economists have long emphasized the importance of correlations across assets returns. One important addition to this edition of *Theory of Financial Risk and Derivative Pricing* is the treatment of the joint behaviour of asset returns, including clustering, extreme correlations and the cross-sectional variation of returns, which is here named *variety*. This discussion plays an important role in risk management.

In this book, as in much of the finance literature inspired by physics, a *model* is typically a set of mathematical equations that 'fit' the data. However, in Chapter 20, Bouchaud and Potters study how such equations may result from the behaviour of economic agents. Much

of modern economic theory is occupied by questions of this kind, and here again physicists have much to contribute.

This text will be extremely useful for natural scientists and engineers who want to turn their attention to financial data. It will also be a good source for economists interested in getting acquainted with the very active research programme being pursued by the authors and other physicists working with financial data.

José A. Scheinkman
Theodore Wells '29 Professor of Economics, Princeton University

Preface

Je vais maintenant commencer à prendre toute la phynance. Après quoi je tuerai tout le monde et je m'en irai.

(A. Jarry, *Ubu roi.*)

Scope of the book

Finance is a rapidly expanding field of science, with a rather unique link to applications. Correspondingly, recent years have witnessed the growing role of financial engineering in market rooms. The possibility of easily accessing and processing huge quantities of data on financial markets opens the path to new methodologies, where systematic comparison between theories and real data not only becomes possible, but mandatory. This perspective has spurred the interest of the statistical physics community, with the hope that methods and ideas developed in the past decades to deal with complex systems could also be relevant in finance. Many holders of PhDs in physics are now taking jobs in banks or other financial institutions.

The existing literature roughly falls into two categories: either rather abstract books from the mathematical finance community, which are very difficult for people trained in natural sciences to read, or more professional books, where the scientific level is often quite poor.[†] Moreover, even in excellent books on the subject, such as the one by J. C. Hull, the point of view on derivatives is the traditional one of Black and Scholes, where the whole pricing methodology is based on the construction of *riskless strategies*. The idea of zero-risk is counter-intuitive and the reason for the existence of these riskless strategies in the Black–Scholes theory is buried in the premises of Ito's stochastic differential rules.

Recently, a handful of books written by physicists, including the present one,[‡] have tried to fill the gap by presenting the physicists' way of approaching scientific problems. The difference lies in priorities: the emphasis is less on rigour than on pragmatism, and no

[†] There are notable exceptions, such as the remarkable book by J. C. Hull, *Futures, Options and Other Derivatives*, Prentice Hall, 1997, or P. Wilmott, *Derivatives, The theory and practice of financial engineering*, John Wiley, 1998.

[‡] See 'Further reading' below.

theoretical model can ever supersede empirical data. Physicists insist on a detailed compar-
ison between 'theory' and 'experiments' (i.e. empirical results, whenever available), the art
of approximations and the systematic use of intuition and simplified arguments.

Indeed, it is our belief that a more intuitive understanding of standard mathematical
theories is needed for a better training of scientists and financial engineers in charge of
financial risks and derivative pricing. The models discussed in *Theory of Financial Risk
and Derivative Pricing* aim at accounting for real markets statistics where the construction of
riskless hedges is generally impossible and where the Black–Scholes model is inadequate.
The mathematical framework required to deal with these models is however not more
complicated, and has the advantage of making the issues at stake, in particular the problem
of risk, more transparent.

Much activity is presently devoted to create and develop new methods to measure and
control financial risks, to price derivatives and to devise decision aids for trading. We have
ourselves been involved in the construction of risk control and option pricing softwares for
major financial institutions, and in the implementation of statistical arbitrage strategies for
the company Capital Fund Management. This book has immensely benefited from the
constant interaction between theoretical models and practical issues. We hope that the
content of this book can be useful to all quants concerned with financial risk control and
derivative pricing, by discussing at length the advantages and limitations of various statistical
models and methods.

Finally, from a more academic perspective, the remarkable stability across markets and
epochs of the anomalous statistical features (fat tails, volatility clustering) revealed by the
analysis of financial time series begs for a simple, generic explanation in terms of agent
based models. This had led in the recent years to the development of the rich and interesting
models which we discuss. Although still in their infancy, these models will become, we
believe, increasingly important in the future as they might pave the way to more ambitious
models of collective human activities.

Style and organization of the book

Despite our efforts to remain simple, certain sections are still quite technical. We have used a
smaller font to develop the more advanced ideas, which are not crucial to the understanding
of the main ideas. Whole sections marked by a star (*) contain rather specialized material and
can be skipped at first reading. Conversely, crucial concepts and formulae are highlighted
by 'boxes', either in the main text or in the summary section at the end of each chapter.
Technical terms are set in boldface when they are first defined.

We have tried to be as precise as possible, but are sometimes deliberately sloppy and non-
rigorous. For example, the idea of probability is not axiomatized: its intuitive meaning is
more than enough for the purpose of this book. The notation $P(\cdot)$ means the probability dis-
tribution for the variable which appears between the parentheses, and not a well-determined
function of a dummy variable. The notation $x \to \infty$ does not necessarily mean that x tends
to infinity in a mathematical sense, but rather that x is 'large'. Instead of trying to derive

results which hold true in any circumstances, we often compare order of magnitudes of the different effects: small effects are neglected, or included perturbatively.[†]

Finally, we have not tried to be comprehensive, and have left out a number of important aspects of theoretical finance. For example, the problem of interest rate derivatives (swaps, caps, swaptions...) is not addressed. Correspondingly, we have not tried to give an exhaustive list of references, but rather, at the end of each chapter, a selection of books and specialized papers that are directly related to the material presented in the chapter. One of us (J.-P. B.) has been for several years an editor of the *International Journal of Theoretical and Applied Finance* and of the more recent *Quantitative Finance*, which might explain a certain bias in the choice of the specialized papers. Most papers that are still in e-print form can be downloaded from http://arXiv.org/ using their reference number.

This book is divided into twenty chapters. Chapters 1–4 deal with important results in probability theory and statistics (the central limit theorem and its limitations, the theory of extreme value statistics, the theory of stochastic processes, etc.). Chapter 5 presents some basic notions about financial markets and financial products. The statistical analysis of real data and empirical determination of statistical laws, are discussed in Chapters 6–8. Chapters 9–11 are concerned with the problems of inter-asset correlations (in particular in extreme market conditions) and the definition of risk, value-at-risk and expected shortfall. The theory of optimal portfolio, in particular in the case where the probability of extreme risks has to be minimized, is given in Chapter 12. The problem of forward contracts and options, their optimal hedge and the residual risk is discussed in detail in Chapters 13–15. The standard Black–Scholes point of view is given its share in Chapter 16. Some more advanced topics on options are introduced in Chapters 17 and 18 (such as exotic options, the role of transaction costs and Monte–Carlo methods). The problem of the yield curve, its theoretical description and some empirical properties are addressed in Chapter 19. Finally, a discussion of some of the recently proposed agent based models (in particular the now famous Minority Game) is given in Chapter 20. A short glossary of financial terms, an index and a list of symbols are given at the end of the book, allowing one to find easily where each symbol or word was used and defined for the first time.

Comparison with the previous editions

This book appeared in its first edition in French, under the title: *Théorie des Risques Financiers*, Aléa–Saclay–Eyrolles, Paris (1997). The second edition (or first English edition) was *Theory of Financial Risks*, Cambridge University Press (2000). Compared to this edition, the present version has been substantially reorganized and augmented – from five to twenty chapters, and from 220 pages to 400 pages. This results from the large amount of new material and ideas in the past four years, but also from the desire to make the book

[†] $a \simeq b$ means that a is of order b, $a \ll b$ means that a is smaller than, say, $b/10$. A computation neglecting terms of order $(a/b)^2$ is therefore accurate to 1%. Such a precision is usually enough in the financial context, where the uncertainty on the value of the parameters (such as the average return, the volatility, etc.), is often larger than 1%.

more self-contained and more accessible: we have split the book in shorter chapters focusing on specific topics; each of them ends with a summary of the most important points. We have tried to give explicitly many useful formulas (probability distributions, etc.) or practical clues (for example: how to generate random variables with a given distribution, in Appendix F.)

Most of the figures have been redrawn, often with updated data, and quite a number of new data analysis is presented. The discussion of many subtle points has been extended and, hopefully, clarified. We also added 'Derivative Pricing' in the title, since almost half of the book covers this topic.

More specifically, we have added the following important topics:

- A specific chapter on stochastic processes, continuous time and Ito calculus, and path integrals (see Chapter 3).
- A chapter discussing various aspects of data analysis and estimation techniques (see Chapter 4).
- A chapter describing financial products and financial markets (see Chapter 5).
- An extended description of non linear correlations in financial data (volatility clustering and the leverage effect) and some specific mathematical models, such as the multifractal Bacry–Muzy–Delour model or the Heston model (see Chapters 7 and 8).
- A detailed discussion of models of inter-asset correlations, multivariate statistics, clustering, extreme correlations and the notion of 'variety' (see Chapters 9 and 11).
- A detailed discussion of the influence of drift and correlations in the dynamics of the underlying on the pricing of options (see Chapter 15).
- A whole chapter on the Black-Scholes way, with an account of the standard formulae (see Chapter 16).
- A new chapter on Monte-Carlo methods for pricing and hedging options (see Chapter 18).
- A chapter on the theory of the yield curve, explaining in (hopefully) transparent terms the Vasicek and Heath–Jarrow–Morton models and comparing their predictions with empirical data. (See Chapter 19, which contains some material that was not previously published.)
- A whole chapter on herding, feedback and agent based models, most notably the minority game (see Chapter 20).

Many chapters now contain some new material that have never appeared in press before (in particular in Chapters 7, 9, 11 and 19). Several more minor topics have been included or developed, such as the theory of progressive dominance of extremes (Section 2.4), the anomalous time evolution of 'hypercumulants' (Section 7.2.1), the theory of optimal portfolios with non linear constraints (Section 12.2.2), the 'worst fluctuation' method to estimate the value-at-risk of complex portfolios (Section 12.4) and the theory of value-at-risk hedging (Section 14.5).

We hope that on the whole, this clarified and extended edition will be of interest both to newcomers and to those already acquainted with the previous edition of our book.

Acknowledgements[†]

This book owes much to discussions that we had with our colleagues and friends at Science and Finance/CFM: Jelle Boersma, Laurent Laloux, Andrew Matacz, and Philip Seager. We want to thank in particular Jean-Pierre Aguilar, who introduced us to the reality of financial markets, suggested many improvements, and supported us during the many years that this project took to complete.

We also had many fruitful exchanges over the years with Alain Arnéodo, Erik Aurell, Marco Avellaneda, Elie Ayache, Belal Baaquie, Emmanuel Bacry, François Bardou, Martin Baxter, Lisa Borland, Damien Challet, Pierre Cizeau, Rama Cont, Lorenzo Cornalba, Michel Dacorogna, Michael Dempster, Nicole El Karoui, J. Doyne Farmer, Xavier Gabaix, Stefano Galluccio, Irene Giardina, Parameswaran Gopikrishnan, Philippe Henrotte, Giulia Iori, David Jeammet, Paul Jefferies, Neil Johnson, Hagen Kleinert, Imre Kondor, Jean-Michel Lasry, Thomas Lux, Rosario Mantegna,[‡] Matteo Marsili, Marc Mézard, Martin Meyer, Aubry Miens, Jeff Miller, Jean-François Muzy, Vivienne Plerou, Benoît Pochart, Bernd Rosenow, Nicolas Sagna, José Scheinkman, Farhat Selmi, Dragan Šestović, Jim Sethna, Didier Sornette, Gene Stanley, Dietrich Stauffer, Ray Streater, Nassim Taleb, Robert Tompkins, Johaness Voït, Christian Walter, Mark Wexler, Paul Wilmott, Matthieu Wyart, Tom Wynter and Karol Życzkowski.

We thank Claude Godrèche, who was the editor for the French version of this book, and Simon Capelin, in charge of the two English editions at C.U.P., for their friendly advice and support, and José Scheinkman for kindly accepting to write a foreword. M. P. wishes to thank Karin Badt for not allowing any compromises in the search for higher truths. J.-P. B. wants to thank J. Hammann and all his colleagues from Saclay and elsewhere for providing such a free and stimulating scientific atmosphere, and Elisabeth Bouchaud for having shared so many far more important things.

This book is dedicated to our families and children and, more particularly, to the memory of Paul Potters.

Further reading

• ECONOPHYSICS AND 'PHYNANCE'
J. Baschnagel, W. Paul, *Stochastic Processes, From Physics to Finance*, Springer-Verlag, 2000.
J.-P. Bouchaud, K. Lauritsen, P. Alstrom (Edts), *Proceedings of "Applications of Physics in Financial Analysis"*, held in Dublin (1999), Int. J. Theo. Appl. Fin., **3**, (2000).
J.-P. Bouchaud, M. Marsili, B. Roehner, F. Slanina (Edts), *Proceedings of the Prague Conference on Application of Physics to Economic Modelling*, Physica A, **299**, (2001).
A. Bunde, H.-J. Schellnhuber, J. Kropp (Edts), *The Science of Disaster*, Springer-Verlag, 2002.
J. D. Farmer, *Physicists attempt to scale the ivory towers of finance*, in Computing in Science and Engineering, November 1999, reprinted in Int. J. Theo. Appl. Fin., **3**, 311 (2000).

[†] Funding for this work was made available in part by market inefficiencies.
[‡] Who kindly gave us the permission to reproduce three of his graphs.

M. Levy, H. Levy, S. Solomon, *Microscopic Simulation of Financial Markets*, Academic Press, San Diego, 2000.

R. Mantegna, H. E. Stanley, *An Introduction to Econophysics*, Cambridge University Press, Cambridge, 1999.

B. Roehner, *Patterns of Speculation: A Study in Observational Econophysics*, Cambridge University Press, 2002.

J. Voït, *The Statistical Mechanics of Financial Markets*, Springer-Verlag, 2001.

1

Probability theory: basic notions

All epistemological value of the theory of probability is based on this: that large scale random phenomena in their collective action create strict, non random regularity.
(Gnedenko and Kolmogorov, *Limit Distributions for Sums of Independent Random Variables*.)

1.1 Introduction

Randomness stems from our incomplete knowledge of reality, from the lack of information which forbids a perfect prediction of the future. Randomness arises from complexity, from the fact that causes are diverse, that tiny perturbations may result in large effects. For over a century now, Science has abandoned Laplace's deterministic vision, and has fully accepted the task of deciphering randomness and inventing adequate tools for its description. The surprise is that, after all, randomness has many facets and that there are many levels to uncertainty, but, above all, that a new form of predictability appears, which is no longer deterministic but *statistical*.

Financial markets offer an ideal testing ground for these statistical ideas. The fact that a large number of participants, with divergent anticipations and conflicting interests, are simultaneously present in these markets, leads to unpredictable behaviour. Moreover, financial markets are (sometimes strongly) affected by external news – which are, both in date and in nature, to a large degree unexpected. The statistical approach consists in drawing from past observations some information on the frequency of possible price changes. If one then assumes that these frequencies reflect some intimate mechanism of the markets themselves, then one may hope that these frequencies will remain stable in the course of time. For example, the mechanism underlying the roulette or the game of dice is obviously always the same, and one expects that the frequency of all possible outcomes will be invariant in time – although of course each individual outcome is random.

This 'bet' that probabilities are stable (or better, stationary) is very reasonable in the case of roulette or dice;[†] it is nevertheless much less justified in the case of financial markets – despite the large number of participants which confer to the system a certain

[†] The idea that science ultimately amounts to making the best possible guess of reality is due to R. P. Feynman (Seeking New Laws, in *The Character of Physical Laws*, MIT Press, Cambridge, MA, 1965).

regularity, at least in the sense of Gnedenko and Kolmogorov. It is clear, for example, that financial markets do not behave now as they did 30 years ago: many factors contribute to the evolution of the way markets behave (development of derivative markets, world-wide and computer-aided trading, etc.). As will be mentioned below, 'young' markets (such as emergent countries markets) and more mature markets (exchange rate markets, interest rate markets, etc.) behave quite differently. The statistical approach to financial markets is based on the idea that whatever evolution takes place, this happens sufficiently *slowly* (on the scale of several years) so that the observation of the recent past is useful to describe a not too distant future. However, even this 'weak stability' hypothesis is sometimes badly in error, in particular in the case of a crisis, which marks a sudden change of market behaviour. The recent example of some Asian currencies indexed to the dollar (such as the Korean won or the Thai baht) is interesting, since the observation of past fluctuations is clearly of no help to predict the amplitude of the sudden turmoil of 1997, see Figure 1.1.

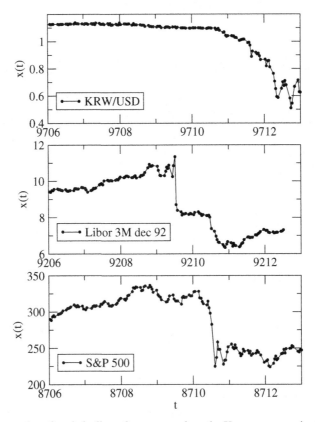

Fig. 1.1. Three examples of statistically unforeseen crashes: the Korean won against the dollar in 1997 (top), the British 3-month short-term interest rates futures in 1992 (middle), and the S&P 500 in 1987 (bottom). In the example of the Korean won, it is particularly clear that the distribution of price changes before the crisis was extremely narrow, and could not be extrapolated to anticipate what happened in the crisis period.

Hence, the statistical description of financial fluctuations is certainly imperfect. It is nevertheless extremely helpful: in practice, the 'weak stability' hypothesis is in most cases reasonable, at least to describe *risks*.[†]

In other words, the amplitude of the possible price changes (but not their sign!) is, to a certain extent, predictable. It is thus rather important to devise adequate tools, in order to *control* (if at all possible) financial risks. The goal of this first chapter is to present a certain number of basic notions in probability theory which we shall find useful in the following. Our presentation does not aim at mathematical rigour, but rather tries to present the key concepts in an intuitive way, in order to ease their empirical use in practical applications.

1.2 Probability distributions

Contrarily to the throw of a dice, which can only return an integer between 1 and 6, the variation of price of a financial asset[‡] can be arbitrary (we disregard the fact that price changes cannot actually be smaller than a certain quantity – a 'tick'). In order to describe a random process X for which the result is a real number, one uses a probability density $P(x)$, such that the probability that X is within a small interval of width dx around $X = x$ is equal to $P(x) dx$. In the following, we shall denote as $P(\cdot)$ the probability density for the variable appearing as the argument of the function. This is a potentially ambiguous, but very useful notation.

The probability that X is between a and b is given by the integral of $P(x)$ between a and b,

$$\mathcal{P}(a < X < b) = \int_a^b P(x)\, dx. \tag{1.1}$$

In the following, the notation $\mathcal{P}(\cdot)$ means the probability of a given event, defined by the content of the parentheses (\cdot).

The function $P(x)$ is a density; in this sense it depends on the units used to measure X. For example, if X is a length measured in centimetres, $P(x)$ is a probability density per unit length, i.e. per centimetre. The numerical value of $P(x)$ changes if X is measured in inches, but the probability that X lies between two specific values l_1 and l_2 is of course independent of the chosen unit. $P(x) dx$ is thus invariant upon a change of unit, i.e. under the change of variable $x \to \gamma x$. More generally, $P(x) dx$ is invariant upon any (monotonic) change of variable $x \to y(x)$: in this case, one has $P(x) dx = P(y) dy$.

In order to be a probability density in the usual sense, $P(x)$ must be non-negative ($P(x) \geq 0$ for all x) and must be normalized, that is that the integral of $P(x)$ over the whole range of possible values for X must be equal to one:

$$\int_{x_m}^{x_M} P(x)\, dx = 1, \tag{1.2}$$

[†] The prediction of *future returns* on the basis of past returns is however much less justified.
[‡] **Asset** is the generic name for a financial instrument which can be bought or sold, like stocks, currencies, gold, bonds, etc.

where x_m (resp. x_M) is the smallest value (resp. largest) which X can take. In the case where the possible values of X are not bounded from below, one takes $x_m = -\infty$, and similarly for x_M. One can actually always assume the bounds to be $\pm\infty$ by setting to zero $P(x)$ in the intervals $]-\infty, x_m]$ and $[x_M, \infty[$. Later in the text, we shall often use the symbol \int as a shorthand for $\int_{-\infty}^{+\infty}$.

An equivalent way of describing the distribution of X is to consider its cumulative distribution $\mathcal{P}_<(x)$, defined as:

$$\mathcal{P}_<(x) \equiv \mathcal{P}(X < x) = \int_{-\infty}^{x} P(x')\,dx'. \tag{1.3}$$

$\mathcal{P}_<(x)$ takes values between zero and one, and is monotonically increasing with x. Obviously, $\mathcal{P}_<(-\infty) = 0$ and $\mathcal{P}_<(+\infty) = 1$. Similarly, one defines $\mathcal{P}_>(x) = 1 - \mathcal{P}_<(x)$.

1.3 Typical values and deviations

It is quite natural to speak about 'typical' values of X. There are at least three mathematical definitions of this intuitive notion: the **most probable** value, the **median** and the **mean**. The most probable value x^* corresponds to the maximum of the function $P(x)$; x^* needs not be unique if $P(x)$ has several equivalent maxima. The median x_{med} is such that the probabilities that X be greater or less than this particular value are equal. In other words, $\mathcal{P}_<(x_{\mathrm{med}}) = \mathcal{P}_>(x_{\mathrm{med}}) = \frac{1}{2}$. The mean, or **expected value** of X, which we shall note as m or $\langle x \rangle$ in the following, is the average of all possible values of X, weighted by their corresponding probability:

$$m \equiv \langle x \rangle = \int x P(x)\,dx. \tag{1.4}$$

For a unimodal distribution (unique maximum), symmetrical around this maximum, these three definitions coincide. However, they are in general different, although often rather close to one another. Figure 1.2 shows an example of a non-symmetric distribution, and the relative position of the most probable value, the median and the mean.

One can then describe the fluctuations of the random variable X: if the random process is repeated several times, one expects the results to be scattered in a cloud of a certain 'width' in the region of typical values of X. This width can be described by the **mean absolute deviation** (MAD) E_{abs}, by the **root mean square** (RMS) σ (or, **standard deviation**), or by the 'full width at half maximum' $w_{1/2}$.

The mean absolute deviation from a given reference value is the average of the distance between the possible values of X and this reference value,[†]

$$E_{\mathrm{abs}} \equiv \int |x - x_{\mathrm{med}}| P(x)\,dx. \tag{1.5}$$

[†] One chooses as a reference value the median for the MAD and the mean for the RMS, because for a fixed distribution $P(x)$, these two quantities minimize, respectively, the MAD and the RMS.

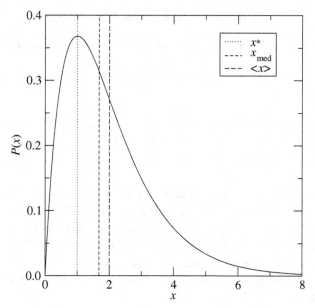

Fig. 1.2. The 'typical value' of a random variable X drawn according to a distribution density $P(x)$ can be defined in at least three different ways: through its mean value $\langle x \rangle$, its most probable value x^* or its median x_{med}. In the general case these three values are distinct.

Similarly, the **variance** (σ^2) is the mean distance squared to the reference value m,

$$\sigma^2 \equiv \langle (x - m)^2 \rangle = \int (x - m)^2 P(x) \, dx. \tag{1.6}$$

Since the variance has the dimension of x squared, its square root (the RMS, σ) gives the order of magnitude of the fluctuations around m.

Finally, the full width at half maximum $w_{1/2}$ is defined (for a distribution which is symmetrical around its unique maximum x^*) such that $P(x^* \pm (w_{1/2})/2) = P(x^*)/2$, which corresponds to the points where the probability density has dropped by a factor of two compared to its maximum value. One could actually define this width slightly differently, for example such that the total probability to find an event outside the interval $[(x^* - w/2), (x^* + w/2)]$ is equal to, say, 0.1. The corresponding value of w is called a quantile. This definition is important when the distribution has very fat tails, such that the variance or the mean absolute deviation are infinite.

The pair mean–variance is actually much more popular than the pair median–MAD. This comes from the fact that the absolute value is not an analytic function of its argument, and thus does not possess the nice properties of the variance, such as additivity under convolution, which we shall discuss in the next chapter. However, for the empirical study of fluctuations, it is sometimes preferable to use the MAD; it is more *robust* than the variance, that is, less sensitive to rare extreme events, which may be the source of large statistical errors.

1.4 Moments and characteristic function

More generally, one can define higher-order **moments** of the distribution $P(x)$ as the average of powers of X:

$$m_n \equiv \langle x^n \rangle = \int x^n P(x)\,\mathrm{d}x. \tag{1.7}$$

Accordingly, the mean m is the first moment ($n = 1$), and the variance is related to the second moment ($\sigma^2 = m_2 - m^2$). The above definition, Eq. (1.7), is only meaningful if the integral converges, which requires that $P(x)$ decreases sufficiently rapidly for large $|x|$ (see below).

From a theoretical point of view, the moments are interesting: if they exist, their knowledge is often equivalent to the knowledge of the distribution $P(x)$ itself.[†] In practice however, the high order moments are very hard to determine satisfactorily: as n grows, longer and longer time series are needed to keep a certain level of precision on m_n; these high moments are thus in general not adapted to describe empirical data.

For many computational purposes, it is convenient to introduce the **characteristic function** of $P(x)$, defined as its Fourier transform:

$$\hat{P}(z) \equiv \int \mathrm{e}^{\mathrm{i}zx} P(x)\,\mathrm{d}x. \tag{1.8}$$

The function $P(x)$ is itself related to its characteristic function through an inverse Fourier transform:

$$P(x) = \frac{1}{2\pi} \int \mathrm{e}^{-\mathrm{i}zx} \hat{P}(z)\,\mathrm{d}z. \tag{1.9}$$

Since $P(x)$ is normalized, one always has $\hat{P}(0) = 1$. The moments of $P(x)$ can be obtained through successive derivatives of the characteristic function at $z = 0$,

$$m_n = (-\mathrm{i})^n \left. \frac{\mathrm{d}^n}{\mathrm{d}z^n} \hat{P}(z) \right|_{z=0}. \tag{1.10}$$

One finally defines the **cumulants** c_n of a distribution as the successive derivatives of the logarithm of its characteristic function:

$$c_n = (-\mathrm{i})^n \left. \frac{\mathrm{d}^n}{\mathrm{d}z^n} \log \hat{P}(z) \right|_{z=0}. \tag{1.11}$$

The cumulant c_n is a polynomial combination of the moments m_p with $p \leq n$. For example $c_2 = m_2 - m^2 = \sigma^2$. It is often useful to normalize the cumulants by an appropriate power of the variance, such that the resulting quantities are dimensionless. One thus defines the **normalized cumulants** λ_n,

$$\lambda_n \equiv c_n / \sigma^n. \tag{1.12}$$

[†] This is not rigorously correct, since one can exhibit examples of different distribution densities which possess exactly the same moments, see Section 1.7 below.

One often uses the third and fourth normalized cumulants, called the **skewness** (ς) and **kurtosis** (κ),[†]

$$\varsigma \equiv \lambda_3 = \frac{\langle (x-m)^3 \rangle}{\sigma^3} \qquad \kappa \equiv \lambda_4 = \frac{\langle (x-m)^4 \rangle}{\sigma^4} - 3. \tag{1.13}$$

The above definition of cumulants may look arbitrary, but these quantities have remarkable properties. For example, as we shall show in Section 2.2, the cumulants simply add when one sums independent random variables. Moreover a Gaussian distribution (or the normal law of Laplace and Gauss) is characterized by the fact that all cumulants of order larger than two are identically zero. Hence the cumulants, in particular κ, can be interpreted as a measure of the distance between a given distribution $P(x)$ and a Gaussian.

1.5 Divergence of moments – asymptotic behaviour

The moments (or cumulants) of a given distribution do not always exist. A necessary condition for the nth moment (m_n) to exist is that the distribution density $P(x)$ should decay faster than $1/|x|^{n+1}$ for $|x|$ going towards infinity, or else the integral, Eq. (1.7), would diverge for $|x|$ large. If one only considers distribution densities that are behaving asymptotically as a power-law, with an exponent $1 + \mu$,

$$P(x) \sim \frac{\mu A_{\pm}^{\mu}}{|x|^{1+\mu}} \text{ for } x \to \pm\infty, \tag{1.14}$$

then all the moments such that $n \geq \mu$ are infinite. For example, such a distribution has no finite variance whenever $\mu \leq 2$. [Note that, for $P(x)$ to be a normalizable probability distribution, the integral, Eq. (1.2), must converge, which requires $\mu > 0$.]

> *The characteristic function of a distribution having an asymptotic power-law behaviour given by Eq. (1.14) is non-analytic around $z = 0$. The small z expansion contains regular terms of the form z^n for $n < \mu$ followed by a non-analytic term $|z|^{\mu}$ (possibly with logarithmic corrections such as $|z|^{\mu} \log z$ for integer μ). The derivatives of order larger or equal to μ of the characteristic function thus do not exist at the origin ($z = 0$).*

1.6 Gaussian distribution

The most commonly encountered distributions are the 'normal' laws of Laplace and Gauss, which we shall simply call **Gaussian** in the following. Gaussians are ubiquitous: for example, the number of *heads* in a sequence of a thousand coin tosses, the exact number of oxygen molecules in the room, the height (in inches) of a randomly selected individual,

[†] Note that it is sometimes $\kappa + 3$, rather than κ itself, which is called the kurtosis.

are all approximately described by a Gaussian distribution.[†] The ubiquity of the Gaussian can be in part traced to the central limit theorem (CLT) discussed at length in Chapter 2, which states that a phenomenon resulting from a large number of small independent causes is Gaussian. There exists however a large number of cases where the distribution describing a complex phenomenon is *not* Gaussian: for example, the amplitude of earthquakes, the velocity differences in a turbulent fluid, the stresses in granular materials, etc., and, as we shall discuss in Chapter 6, the price fluctuations of most financial assets.

A Gaussian of mean m and root mean square σ is defined as:

$$P_G(x) \equiv \frac{1}{\sqrt{2\pi\sigma^2}} \exp\left(-\frac{(x-m)^2}{2\sigma^2}\right). \tag{1.15}$$

The median and most probable value are in this case equal to m, whereas the MAD (or any other definition of the width) is proportional to the RMS (for example, $E_{abs} = \sigma\sqrt{2/\pi}$). For $m = 0$, all the odd moments are zero and the even moments are given by $m_{2n} = (2n-1)(2n-3)\ldots\sigma^{2n} = (2n-1)!!\,\sigma^{2n}$.

All the cumulants of order greater than two are zero for a Gaussian. This can be realized by examining its characteristic function:

$$\hat{P}_G(z) = \exp\left(-\frac{\sigma^2 z^2}{2} + imz\right). \tag{1.16}$$

Its logarithm is a second-order polynomial, for which all derivatives of order larger than two are zero. In particular, the kurtosis of a Gaussian variable is zero. As mentioned above, the kurtosis is often taken as a measure of the distance from a Gaussian distribution. When $\kappa > 0$ (**leptokurtic** distributions), the corresponding distribution density has a marked peak around the mean, and rather 'thick' tails. Conversely, when $\kappa < 0$, the distribution density has a flat top and very thin tails. For example, the uniform distribution over a certain interval (for which tails are absent) has a kurtosis $\kappa = -\frac{6}{5}$. Note that the kurtosis is bounded from below by the value -2, which corresponds to the case where the random variable can only take two values $-a$ and a with equal probability.

A Gaussian variable is peculiar because 'large deviations' are extremely rare. The quantity $\exp(-x^2/2\sigma^2)$ decays so fast for large x that deviations of a few times σ are nearly impossible. For example, a Gaussian variable departs from its most probable value by more than 2σ only 5% of the times, of more than 3σ in 0.2% of the times, whereas a fluctuation of 10σ has a probability of less than 2×10^{-23}; in other words, it never happens.

1.7 Log-normal distribution

Another very popular distribution in mathematical finance is the so-called **log-normal** law. That X is a log-normal random variable simply means that $\log X$ is normal, or Gaussian. Its use in finance comes from the assumption that the *rate of returns*, rather than the absolute

[†] Although, in the above three examples, the random variable cannot be negative. As we shall discuss later, the Gaussian description is generally only valid in a certain neighbourhood of the maximum of the distribution.

change of prices, are independent random variables. The increments of the logarithm of the price thus asymptotically sum to a Gaussian, according to the CLT detailed in Chapter 2. The log-normal distribution density is thus defined as:[†]

$$P_{LN}(x) \equiv \frac{1}{x\sqrt{2\pi\sigma^2}} \exp\left(-\frac{\log^2(x/x_0)}{2\sigma^2}\right), \tag{1.17}$$

the moments of which being: $m_n = x_0^n e^{n^2\sigma^2/2}$.

From these moments, one deduces the skewness, given by $\varsigma_3 = (e^{3\sigma^2} - 3e^{\sigma^2} + 2)/(e^{\sigma^2} - 1)^{3/2}$, ($\simeq 3\sigma$ for $\sigma \ll 1$), and the kurtosis $\kappa = (e^{6\sigma^2} - 4e^{3\sigma^2} + 6e^{\sigma^2} - 3)/(e^{\sigma^2} - 1)^2 - 3$, ($\simeq 19\sigma^2$ for $\sigma \ll 1$).

In the context of mathematical finance, one often prefers log-normal to Gaussian distributions for several reasons. As mentioned above, the existence of a random rate of return, or random interest rate, naturally leads to log-normal statistics. Furthermore, log-normals account for the following symmetry in the problem of exchange rates:[‡] if x is the rate of currency A in terms of currency B, then obviously, $1/x$ is the rate of currency B in terms of A. Under this transformation, $\log x$ becomes $-\log x$ and the description in terms of a log-normal distribution (or in terms of any other even function of $\log x$) is independent of the reference currency. One often hears the following argument in favour of log-normals: since the price of an asset cannot be negative, its statistics cannot be Gaussian since the latter admits in principle negative values, whereas a log-normal excludes them by construction. This is however a red-herring argument, since the description of the fluctuations of the price of a financial asset in terms of Gaussian or log-normal statistics is in any case an *approximation* which is only valid in a certain range. As we shall discuss at length later, these approximations are totally unadapted to describe extreme risks. Furthermore, even if a price drop of more than 100% is in principle possible for a Gaussian process,[§] the error caused by neglecting such an event is much smaller than that induced by the use of either of these two distributions (Gaussian or log-normal). In order to illustrate this point more clearly, consider the probability of observing n times 'heads' in a series of N coin tosses, which is exactly equal to $2^{-N} C_N^n$. It is also well known that in the neighbourhood of $N/2$, $2^{-N} C_N^n$ is very accurately approximated by a Gaussian of variance $N/4$; this is however not contradictory with the fact that $n \geq 0$ by construction!

Finally, let us note that for moderate volatilities (up to say 20%), the two distributions (Gaussian and log-normal) look rather alike, especially in the 'body' of the distribution (Fig. 1.3). As for the tails, we shall see later that Gaussians substantially underestimate their weight, whereas the log-normal predicts that large positive jumps are more frequent

[†] A log-normal distribution has the remarkable property that the knowledge of all its moments is not sufficient to characterize the corresponding distribution. One can indeed show that the following distribution: $\frac{1}{\sqrt{2\pi}} x^{-1} \exp[-\frac{1}{2}(\log x)^2][1 + a\sin(2\pi \log x)]$, for $|a| \leq 1$, has moments which are independent of the value of a, and thus coincide with those of a log-normal distribution, which corresponds to $a = 0$.

[‡] This symmetry is however not always obvious. The dollar, for example, plays a special role. This symmetry can only be expected between currencies of similar strength.

[§] In the rather extreme case of a 20% annual volatility and a zero annual return, the probability for the price to become negative after a year in a Gaussian description is less than one out of 3 million.

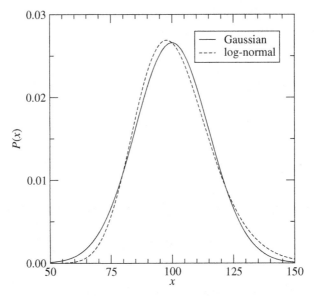

Fig. 1.3. Comparison between a Gaussian (thick line) and a log-normal (dashed line), with $m = x_0 = 100$ and σ equal to 15 and 15% respectively. The difference between the two curves shows up in the tails.

than large negative jumps. This is at variance with empirical observation: the distributions of absolute stock price changes are rather symmetrical; if anything, large negative draw-downs are more frequent than large positive draw-ups.

1.8 Lévy distributions and Paretian tails

Lévy distributions (noted $L_\mu(x)$ below) appear naturally in the context of the CLT (see Chapter 2), because of their stability property under addition (a property shared by Gaussians). The tails of Lévy distributions are however much 'fatter' than those of Gaussians, and are thus useful to describe multiscale phenomena (i.e. when both very large and very small values of a quantity can commonly be observed – such as personal income, size of pension funds, amplitude of earthquakes or other natural catastrophes, etc.). These distributions were introduced in the 1950s and 1960s by Mandelbrot (following Pareto) to describe personal income and the price changes of some financial assets, in particular the price of cotton. An important constitutive property of these Lévy distributions is their power-law behaviour for large arguments, often called **Pareto tails**:

$$L_\mu(x) \sim \frac{\mu A_\pm^\mu}{|x|^{1+\mu}} \text{ for } x \to \pm\infty, \tag{1.18}$$

where $0 < \mu < 2$ is a certain exponent (often called α), and A_\pm^μ two constants which we call **tail amplitudes**, or **scale parameters**: A_\pm indeed gives the order of magnitude of the

large (positive or negative) fluctuations of x. For instance, the probability to draw a number larger than x decreases as $\mathcal{P}_>(x) = (A_+/x)^\mu$ for large positive x.

One can of course in principle observe Pareto tails with $\mu \geq 2$; but, those tails do not correspond to the asymptotic behaviour of a Lévy distribution.

In full generality, Lévy distributions are characterized by an **asymmetry parameter** defined as $\beta \equiv (A_+^\mu - A_-^\mu)/(A_+^\mu + A_-^\mu)$, which measures the relative weight of the positive and negative tails. We shall mostly focus in the following on the symmetric case $\beta = 0$. The fully asymmetric case ($\beta = 1$) is also useful to describe strictly positive random variables, such as, for example, the time during which the price of an asset remains below a certain value, etc.

An important consequence of Eq. (1.14) with $\mu \leq 2$ is that the variance of a Lévy distribution is formally infinite: the probability density does not decay fast enough for the integral, Eq. (1.6), to converge. In the case $\mu \leq 1$, the distribution density decays so slowly that even the mean, or the MAD, fail to exist.[†] The scale of the fluctuations, defined by the width of the distribution, is always set by $A = A_+ = A_-$.

There is unfortunately no simple analytical expression for symmetric Lévy distributions $L_\mu(x)$, except for $\mu = 1$, which corresponds to a **Cauchy distribution** (or **Lorentzian**):

$$L_1(x) = \frac{A}{x^2 + \pi^2 A^2}. \tag{1.19}$$

However, the characteristic function of a symmetric Lévy distribution is rather simple, and reads:

$$\hat{L}_\mu(z) = \exp(-a_\mu |z|^\mu), \tag{1.20}$$

where a_μ is a constant proportional to the tail parameter A^μ:

$$A^\mu = \mu\Gamma(\mu - 1)\frac{\sin(\pi\mu/2)}{\pi}a_\mu \qquad 1 < \mu < 2, \tag{1.21}$$

and

$$A^\mu = (1 - \mu)\Gamma(\mu)\frac{\sin(\pi\mu/2)}{\pi\mu}a_\mu \qquad \mu < 1. \tag{1.22}$$

It is clear, from (1.20), that in the limit $\mu = 2$, one recovers the definition of a Gaussian. When μ decreases from 2, the distribution becomes more and more sharply peaked around the origin and fatter in its tails, while 'intermediate' events lose weight (Fig. 1.4). These distributions thus describe 'intermittent' phenomena, very often small, sometimes gigantic.

The moments of the symmetric Lévy distribution can be computed, when they exist. One finds:

$$\langle |x|^\nu \rangle = (a_\mu)^{\nu/\mu} \frac{\Gamma(-\nu/\mu)}{\mu\Gamma(-\nu)\cos(\pi\nu/2)}, \qquad -1 < \nu < \mu. \tag{1.23}$$

[†] The median and the most probable value however still exist. For a symmetric Lévy distribution, the most probable value defines the so-called **localization** parameter m.

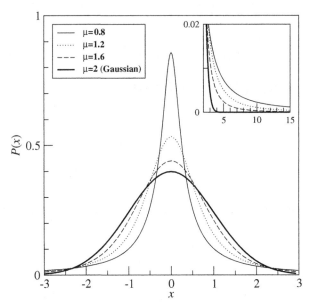

Fig. 1.4. Shape of the symmetric Lévy distributions with $\mu = 0.8$, 1.2, 1.6 and 2 (this last value actually corresponds to a Gaussian). The smaller μ, the sharper the 'body' of the distribution, and the fatter the tails, as illustrated in the inset.

Note finally that Eq. (1.20) does not define a probability distribution when $\mu > 2$, because its inverse Fourier transform is not everywhere positive.

In the case $\beta \neq 0$, one would have:

$$\hat{L}_\mu^\beta(z) = \exp\left[-a_\mu |z|^\mu \left(1 + i\beta \tan(\mu\pi/2)\frac{z}{|z|}\right)\right] \quad (\mu \neq 1). \tag{1.24}$$

It is important to notice that while the leading asymptotic term for large x is given by Eq. (1.18), there are subleading terms which can be important for finite x. The full asymptotic series actually reads:

$$L_\mu(x) = \sum_{n=1}^{\infty} \frac{(-)^{n+1}}{\pi n!} \frac{a_\mu^n}{x^{1+n\mu}} \Gamma(1+n\mu)\sin(\pi\mu n/2). \tag{1.25}$$

The presence of the subleading terms may lead to a bad empirical estimate of the exponent μ based on a fit of the tail of the distribution. In particular, the 'apparent' exponent which describes the function L_μ for finite x is larger than μ, and decreases towards μ for $x \to \infty$, but more and more slowly as μ gets nearer to the Gaussian value $\mu = 2$, for which the power-law tails no longer exist. Note however that one also often observes empirically the opposite behaviour, i.e. an apparent Pareto exponent which *grows* with x. This arises when the Pareto distribution, Eq. (1.18), is only valid in an intermediate regime $x \ll 1/\alpha$, beyond which the distribution decays exponentially, say as $\exp(-\alpha x)$. The Pareto tail is then 'truncated' for large values of x, and this leads to an effective μ which grows with x.

An interesting generalization of the symmetric Lévy distributions which accounts for this exponential cut-off is given by the **truncated Lévy distributions** (TLD), which will be of much use in the following. A simple way to alter the characteristic function Eq. (1.20) to account for an exponential cut-off for large arguments is to set:

$$\hat{L}_{\mu}^{(t)}(z) = \exp\left[-a_{\mu}\frac{(\alpha^2 + z^2)^{\frac{\mu}{2}}\cos\left(\mu\arctan(|z|/\alpha)\right) - \alpha^{\mu}}{\cos(\pi\mu/2)}\right], \tag{1.26}$$

for $1 \leq \mu \leq 2$. The above form reduces to Eq. (1.20) for $\alpha = 0$. Note that the argument in the exponential can also be written as:

$$\frac{a_{\mu}}{2\cos(\pi\mu/2)}[(\alpha + iz)^{\mu} + (\alpha - iz)^{\mu} - 2\alpha^{\mu}]. \tag{1.27}$$

The first cumulants of the distribution defined by Eq. (1.26) read, for $1 < \mu < 2$:

$$c_2 = \mu(\mu - 1)\frac{a_{\mu}}{|\cos\pi\mu/2|}\alpha^{\mu-2} \qquad c_3 = 0. \tag{1.28}$$

The kurtosis $\kappa = \lambda_4 = c_4/c_2^2$ is given by:

$$\lambda_4 = \frac{(3 - \mu)(2 - \mu)|\cos\pi\mu/2|}{\mu(\mu - 1)a_{\mu}\alpha^{\mu}}. \tag{1.29}$$

Note that the case $\mu = 2$ corresponds to the Gaussian, for which $\lambda_4 = 0$ as expected. On the other hand, when $\alpha \to 0$, one recovers a pure Lévy distribution, for which c_2 and c_4 are formally infinite. Finally, if $\alpha \to \infty$ with $a_{\mu}\alpha^{\mu-2}$ fixed, one also recovers the Gaussian.

As explained below in Section 3.1.3, the truncated Lévy distribution has the interesting property of being infinitely divisible for all values of α and μ (this includes the Gaussian distribution and the pure Lévy distributions).

Exponential tail: a limiting case

Very often in the following, we shall notice that in the formal limit $\mu \to \infty$, the power-law tail becomes an exponential tail, if the tail parameter is simultaneously scaled as $A^{\mu} = (\mu/\alpha)^{\mu}$. Qualitatively, this can be understood as follows: consider a probability distribution restricted to positive x, which decays as a power-law for large x, defined as:

$$\mathcal{P}_>(x) = \frac{A^{\mu}}{(A + x)^{\mu}}. \tag{1.30}$$

This shape is obviously compatible with Eq. (1.18), and is such that $\mathcal{P}_>(x = 0) = 1$. If $A = (\mu/\alpha)$, one then finds:

$$\mathcal{P}_>(x) = \frac{1}{[1 + (\alpha x/\mu)]^{\mu}} \xrightarrow[\mu\to\infty]{} \exp(-\alpha x). \tag{1.31}$$

1.9 Other distributions (*)

There are obviously a very large number of other statistical distributions useful to describe random phenomena. Let us cite a few, which often appear in a financial context:

- The discrete **Poisson distribution**: consider a set of points randomly scattered on the real axis, with a certain density ω (e.g. the times when the price of an asset changes). The number of points n in an arbitrary interval of length ℓ is distributed according to the Poisson distribution:

$$P(n) \equiv \frac{(\omega\ell)^n}{n!} \exp(-\omega\ell). \tag{1.32}$$

- The **hyperbolic distribution**, which interpolates between a Gaussian 'body' and exponential tails:

$$P_{\mathrm{H}}(x) \equiv \frac{1}{2x_0 K_1(\alpha x_0)} \exp -\left[\alpha\sqrt{x_0^2 + x^2}\right], \tag{1.33}$$

where the normalization $K_1(\alpha x_0)$ is a modified Bessel function of the second kind. For x small compared to x_0, $P_{\mathrm{H}}(x)$ behaves as a Gaussian although its asymptotic behaviour for $x \gg x_0$ is fatter and reads $\exp(-\alpha|x|)$.

From the characteristic function

$$\hat{P}_{\mathrm{H}}(z) = \frac{\alpha x_0 K_1(x_0\sqrt{1 + \alpha z})}{K_1(\alpha x_0)\sqrt{1 + \alpha z}}, \tag{1.34}$$

we can compute the variance

$$\sigma^2 = \frac{x_0 K_2(\alpha x_0)}{\alpha K_1(\alpha x_0)}, \tag{1.35}$$

and kurtosis

$$\kappa = 3\left(\frac{K_2(\alpha x_0)}{K_1(\alpha x_0)}\right)^2 + \frac{12}{\alpha x_0}\frac{K_2(\alpha x_0)}{K_1(\alpha x_0)} - 3. \tag{1.36}$$

Note that the kurtosis of the hyperbolic distribution is always between zero and three. (The skewness is zero since the distribution is even.)
In the case $x_0 = 0$, one finds the symmetric exponential distribution:

$$P_{\mathrm{E}}(x) = \frac{\alpha}{2} \exp(-\alpha|x|), \tag{1.37}$$

with even moments $m_{2n} = (2n)!\,\alpha^{-2n}$, which gives $\sigma^2 = 2\alpha^{-2}$ and $\kappa = 3$. Its characteristic function reads: $\hat{P}_{\mathrm{E}}(z) = \alpha^2/(\alpha^2 + z^2)$.

- The **Student distribution**, which also has power-law tails:

$$P_{\mathrm{S}}(x) \equiv \frac{1}{\sqrt{\pi}} \frac{\Gamma((1 + \mu)/2)}{\Gamma(\mu/2)} \frac{a^\mu}{(a^2 + x^2)^{(1+\mu)/2}}, \tag{1.38}$$

which coincides with the Cauchy distribution for $\mu = 1$, and tends towards a Gaussian in the limit $\mu \to \infty$, provided that a^2 is scaled as μ. This distribution is usually known as

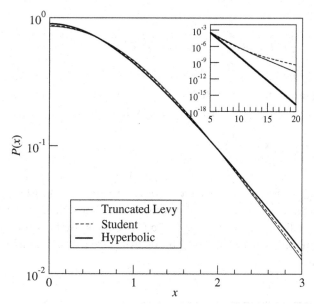

Fig. 1.5. Probability density for the truncated Lévy ($\mu = \frac{3}{2}$), Student and hyperbolic distributions. All three have two free parameters which were fixed to have unit variance and kurtosis. The inset shows a blow-up of the tails where one can see that the Student distribution has tails similar to (but slightly thicker than) those of the truncated Lévy.

Student's t-distribution with μ degrees of freedom, but we shall call it simply the Student distribution.

The even moments of the Student distribution read: $m_{2n} = (2n - 1)!!\Gamma(\mu/2 - n)/ \Gamma(\mu/2)(a^2/2)^n$, provided $2n < \mu$; and are infinite otherwise. One can check that in the limit $\mu \to \infty$, the above expression gives back the moments of a Gaussian: $m_{2n} = (2n - 1)!!\sigma^{2n}$. The kurtosis of the Student distribution is given by $\kappa = 6/(\mu - 4)$. Figure 1.5 shows a plot of the Student distribution with $\kappa = 1$, corresponding to $\mu = 10$.

Note that the characteristic function of Student distributions can also be explicitly computed, and reads:

$$\hat{P}_S(z) = \frac{2^{1-\mu/2}}{\Gamma(\mu/2)}(az)^{\mu/2}K_{\mu/2}(az), \tag{1.39}$$

where $K_{\mu/2}$ is the modified Bessel function. The cumulative distribution in the useful cases $\mu = 3$ and $\mu = 4$ with a chosen such that the variance is unity read:

$$\mathcal{P}_{S,>}(x) = \frac{1}{2} - \frac{1}{\pi}\left[\arctan x + \frac{x}{1 + x^2}\right] \qquad (\mu = 3, \, a = 1), \tag{1.40}$$

and

$$\mathcal{P}_{S,>}(x) = \frac{1}{2} - \frac{3}{4}u + \frac{1}{4}u^3, \qquad (\mu = 4, \ a = \sqrt{2}), \tag{1.41}$$

where $u = x/\sqrt{2 + x^2}$.

- The **inverse gamma** distribution, for positive quantities (such as, for example, volatilities, or waiting times), also has power-law tails. It is defined as:

$$P_\Gamma(x) = \frac{x_0^\mu}{\Gamma(\mu)x^{1+\mu}} \exp\left(-\frac{x_0}{x}\right). \tag{1.42}$$

Its moments of order $n < \mu$ are easily computed to give: $m_n = x_0^n \Gamma(\mu - n)/\Gamma(\mu)$. This distribution falls off very fast when $x \to 0$. As we shall see in Chapter 7, an inverse gamma distribution and a log-normal distribution can sometimes be hard to distinguish empirically. Finally, if the volatility σ^2 of a Gaussian is itself distributed as an inverse gamma distribution, the distribution of x becomes a Student distribution – see Section 9.2.5 for more details.

1.10 Summary

- The most probable value and the mean value are both estimates of the typical values of a random variable. Fluctuations around this value are measured by the root mean square deviation or the mean absolute deviation.
- For some distributions with very fat tails, the mean square deviation (or even the mean value) is infinite, and the typical values must be described using quantiles.
- The Gaussian, the log-normal and the Student distributions are some of the important probability distributions for financial applications.
- The way to generate numerically random variables with a given distribution (Gaussian, Lévy stable, Student, etc.) is discussed in Chapter 18, Appendix F.

- FURTHER READING

W. Feller, *An Introduction to Probability Theory and its Applications*, Wiley, New York, 1971.

P. Lévy, *Théorie de l'addition des variables aléatoires*, Gauthier Villars, Paris, 1937–1954.

B. V. Gnedenko, A. N. Kolmogorov, *Limit Distributions for Sums of Independent Random Variables*, Addison Wesley, Cambridge, MA, 1954.

G. Samorodnitsky, M. S. Taqqu, *Stable Non-Gaussian Random Processes*, Chapman & Hall, New York, 1994.

2

Maximum and addition of random variables

In the greatest part of our concernments, [God] has afforded us only the twilight, as I may so say, of probability; suitable, I presume, to that state of mediocrity and probationership he has been pleased to place us in here.

(John Locke, *An Essay Concerning Human Understanding*.)

2.1 Maximum of random variables – statistics of extremes

If one observes a series of N independent realizations of the same random phenomenon, a question which naturally arises, in particular when one is concerned about risk control, is to determine the order of magnitude of the *maximum* observed value of the random variable (which can be the price drop of a financial asset, or the water level of a flooding river, etc.). For example, in Chapter 10, the so-called 'value-at-risk' (VaR) on a typical time horizon will be defined as the possible maximum loss over that period (within a certain confidence level).

The law of large numbers tells us that an event which has a probability p of occurrence appears on average Np times on a series of N observations. One thus expects to observe events which have a probability of at least $1/N$. It would be surprising to encounter an event which has a probability much smaller than $1/N$. The order of magnitude of the largest event, Λ_{\max}, observed in a series of N independent identically distributed (iid) random variables is thus given by:

$$\mathcal{P}_>(\Lambda_{\max}) = 1/N. \tag{2.1}$$

More precisely, the full probability distribution of the maximum value $x_{\max} = \max_{i=1,N}\{x_i\}$, is relatively easy to characterize; this will justify the above simple criterion Eq. (2.1). The cumulative distribution $\mathcal{P}(x_{\max} < \Lambda)$ is obtained by noticing that if the maximum of all x_i's is smaller than Λ, all of the x_i's must be smaller than Λ. If the random variables are iid, one finds:

$$\mathcal{P}(x_{\max} < \Lambda) = [\mathcal{P}_<(\Lambda)]^N. \tag{2.2}$$

Note that this result is general, and does not rely on a specific choice for $P(x)$. When Λ is large, it is useful to use the following approximation:

$$\mathcal{P}(x_{\max} < \Lambda) = [1 - \mathcal{P}_>(\Lambda)]^N \simeq e^{-N\mathcal{P}_>(\Lambda)}. \tag{2.3}$$

Since we now have a simple formula for the distribution of x_{max}, one can invert it in order to obtain, for example, the median value of the maximum, noted Λ_{med}, such that $\mathcal{P}(x_{max} < \Lambda_{med}) = \frac{1}{2}$:

$$P_>(\Lambda_{med}) = 1 - \left(\frac{1}{2}\right)^{1/N} \simeq \frac{\log 2}{N}. \tag{2.4}$$

More generally, the value Λ_p which is greater than x_{max} with probability p is given by

$$P_>(\Lambda_p) \simeq -\frac{\log p}{N}. \tag{2.5}$$

The quantity Λ_{max} defined by Eq. (2.1) above is thus such that $p = 1/e \approx 0.37$. The probability that x_{max} is even *larger* than Λ_{max} is thus 63%. As we shall now show, Λ_{max} also corresponds, in many cases, to the *most probable value* of x_{max}.

Equation (2.5) will be very useful in Chapter 10 to estimate a maximal potential loss within a certain confidence level. For example, the largest daily loss Λ expected next year, with 95% confidence, is defined such that $\mathcal{P}_<(-\Lambda) = -\log(0.95)/250$, where $\mathcal{P}_<$ is the cumulative distribution of daily price changes, and 250 is the number of market days per year.

Interestingly, the distribution of x_{max} only depends, when N is large, on the asymptotic behaviour of the distribution of x, $P(x)$, when $x \to \infty$. For example, if $P(x)$ behaves as an exponential when $x \to \infty$, or more precisely if $\mathcal{P}_>(x) \sim \exp(-\alpha x)$, one finds:

$$\Lambda_{max} = \frac{\log N}{\alpha}, \tag{2.6}$$

which grows very slowly with N.[†] Setting $x_{max} = \Lambda_{max} + (u/\alpha)$, one finds that the deviation u around Λ_{max} is distributed according to the **Gumbel distribution**:

$$P(u) = e^{-e^{-u}} e^{-u}. \tag{2.7}$$

The most probable value of this distribution is $u = 0$.[‡] This shows that Λ_{max} is the most probable value of x_{max}. The result, Eq. (2.7), is actually much more general, and is valid as soon as $P(x)$ decreases more rapidly than any power-law for $x \to \infty$: the deviation between Λ_{max} (defined as Eq. (2.1)) and x_{max} is always distributed according to the Gumbel law, Eq. (2.7), up to a scaling factor in the definition of u.

The situation is radically different if $P(x)$ decreases as a power-law, cf. Eq. (1.14). In this case,

$$\mathcal{P}_>(x) \simeq \frac{A_+^\mu}{x^\mu}, \tag{2.8}$$

and the typical value of the maximum is given by:

$$\Lambda_{max} = A_+ N^{\frac{1}{\mu}}. \tag{2.9}$$

Numerically, for a distribution with $\mu = \frac{3}{2}$ and a scale factor $A_+ = 1$, the largest of $N =$

[†] For example, for a symmetric exponential distribution $P(x) = \exp(-|x|)/2$, the median value of the maximum of $N = 10\,000$ variables is only 6.3.

[‡] This distribution is discussed further in the context of financial risk control in Section 10.3.2, and drawn in Figure 10.1.

10 000 variables is on the order of 450, whereas for $\mu = \frac{1}{2}$ it is one hundred million! The complete distribution of the maximum, called the **Fréchet** distribution, is given by:

$$P(u) = \frac{\mu}{u^{1+\mu}} e^{-1/u^\mu} \qquad u = \frac{x_{\max}}{A_+ N^{\frac{1}{\mu}}}. \tag{2.10}$$

Its asymptotic behaviour for $u \to \infty$ is still a power-law of exponent $1 + \mu$. Said differently, both power-law tails and exponential tails *are stable with respect to the 'max' operation.*[†] The most probable value x_{\max} is now equal to $(\mu/1 + \mu)^{1/\mu} \Lambda_{\max}$. As mentioned above, the limit $\mu \to \infty$ formally corresponds to an exponential distribution. In this limit, one indeed recovers Λ_{\max} as the most probable value.

> *Equation (2.9) allows us to discuss intuitively the divergence of the mean value for $\mu \leq 1$ and of the variance for $\mu \leq 2$. If the mean value exists, the sum of N random variables is typically equal to Nm, where m is the mean (see also below). But when $\mu < 1$, the largest encountered value of X is on the order of $N^{1/\mu} \gg N$, and would thus be larger than the entire sum. Similarly, as discussed below, when the variance exists, the RMS of the sum is equal to $\sigma\sqrt{N}$. But for $\mu < 2$, x_{\max} grows faster than \sqrt{N}.*

More generally, one can rank the random variables x_i in decreasing order, and ask for an estimate of the nth encountered value, noted $\Lambda[n]$ below. (In particular, $\Lambda[1] = x_{\max}$.) The distribution P_n of $\Lambda[n]$ can be obtained in full generality as:

$$P_n(\Lambda[n]) = N C_{N-1}^{n-1} \, P(x = \Lambda[n]) \, (\mathcal{P}(x > \Lambda[n])^{n-1} (\mathcal{P}(x < \Lambda[n])^{N-n}. \tag{2.11}$$

The previous expression means that one has first to choose $\Lambda[n]$ among N variables (N ways), $n - 1$ variables among the $N - 1$ remaining as the $n - 1$ largest ones (C_{N-1}^{n-1} ways), and then assign the corresponding probabilities to the configuration where $n - 1$ of them are larger than $\Lambda[n]$ and $N - n$ are smaller than $\Lambda[n]$. One can study the position $\Lambda^*[n]$ of the maximum of P_n, and also the width of P_n, defined from the second derivative of $\log P_n$ calculated at $\Lambda^*[n]$. The calculation simplifies in the limit where $N \to \infty$, $n \to \infty$, with the ratio n/N fixed. In this limit, one finds a relation which generalizes Eq. (2.1):

$$\mathcal{P}_>(\Lambda^*[n]) = n/N. \tag{2.12}$$

The width w_n of the distribution is found to be given by:

$$w_n = \frac{1}{\sqrt{N}} \frac{\sqrt{(n/N) - (n/N)^2}}{P(x = \Lambda^*[n])}, \tag{2.13}$$

which shows that in the limit $N \to \infty$, the value of the nth variable is more and more sharply peaked around its most probable value $\Lambda^*[n]$, given by Eq. (2.12).

In the case of an exponential tail, one finds that $\Lambda^*[n] \simeq \log(N/n)/\alpha$; whereas in the case of power-law tails, one rather obtains:

$$\Lambda^*[n] \simeq A_+ \left(\frac{N}{n}\right)^{\frac{1}{\mu}}. \tag{2.14}$$

[†] A third class of laws, stable under 'max' concerns random variables, which are bounded from above – i.e. such that $P(x) = 0$ for $x > x_M$, with x_M finite. This leads to the **Weibull** distributions, which we will not consider further in this book. See Gumbel (1958) and Galambos (1987).

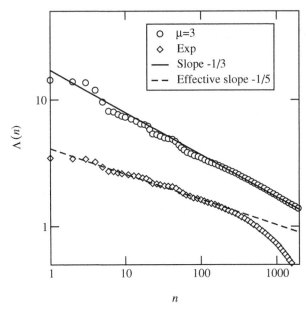

Fig. 2.1. Amplitude versus rank plots. One plots the value of the nth variable $\Lambda[n]$ as a function of its rank n. If $P(x)$ behaves asymptotically as a power-law, one obtains a straight line in log–log coordinates, with a slope equal to $-1/\mu$. For an exponential distribution, one observes an effective slope which is smaller and smaller as N/n tends to infinity. The points correspond to synthetic time series of length 5000, drawn according to a power-law with $\mu = 3$, or according to an exponential. Note that if the axes x and y are interchanged, then according to Eq. (2.12), one obtains an estimate of the cumulative distribution, $\mathcal{P}_>$.

This last equation shows that, for power-law variables, the encountered values are hierarchically organized: for example, the ratio of the largest value $x_{\max} \equiv \Lambda[1]$ to the second largest $\Lambda[2]$ is of the order of $2^{1/\mu}$, which becomes larger and larger as μ decreases, and conversely tends to one when $\mu \to \infty$.

The property, Eq. (2.14) is very useful in identifying empirically the nature of the tails of a probability distribution. One sorts in decreasing order the set of observed values $\{x_1, x_2, \ldots, x_N\}$ and one simply draws $\Lambda[n]$ as a function of n. If the variables are power-law distributed, this graph should be a straight line in log–log plot, with a slope $-1/\mu$, as given by Eq. (2.14) (Fig. 2.1). On the same figure, we have shown the result obtained for exponentially distributed variables. On this diagram, one observes an approximately straight line, but with an effective slope which varies with the total number of points N: the slope is less and less as N/n grows larger. In this sense, the formal remark made above, that an exponential distribution could be seen as a power-law with $\mu \to \infty$, becomes somewhat more concrete. Note that if the axes x and y of Figure 2.1 are interchanged, then according to Eq. (2.12), one obtains an estimate of the cumulative distribution, $\mathcal{P}_>$.

Let us finally note another property of power-laws, potentially interesting for their empirical determination. If one computes the average value of x conditioned to a certain

minimum value Λ:

$$\langle x \rangle_\Lambda = \frac{\int_\Lambda^\infty x P(x) \, dx}{\int_\Lambda^\infty P(x) \, dx}, \tag{2.15}$$

then, if $P(x)$ *decreases as in Eq. (1.14), one finds, for* $\Lambda \to \infty$,

$$\langle x \rangle_\Lambda = \frac{\mu}{\mu - 1} \Lambda, \tag{2.16}$$

independently of the tail amplitude A_\pm^μ.[†] *The average* $\langle x \rangle_\Lambda$ *is thus always of the same order as* Λ *itself, with a proportionality factor which diverges as* $\mu \to 1$.

2.2 Sums of random variables

In order to describe the statistics of future prices of a financial asset, one *a priori* needs a distribution density for all possible time intervals, corresponding to different trading time horizons. For example, the distribution of 5-min price fluctuations is different from the one describing daily fluctuations, itself different for the weekly, monthly, etc. variations. But in the case where the fluctuations are independent and identically distributed (iid), an assumption which is, however, usually not justified (see Chapter 7) it is possible to reconstruct the distributions corresponding to different time scales from the knowledge of that describing short time scales only. In this context, Gaussians and Lévy distributions play a special role, because they are stable: if the short time scale distribution is a stable law, then the fluctuations on all time scales are described by the same stable law – only the parameters of the stable law must be changed (in particular its width). More generally, if one sums iid variables, then, independently of the short time distribution, the law describing long times converges towards one of the stable laws: this is the content of the 'central limit theorem' (CLT). In practice, however, this convergence can be very slow and thus of limited interest, in particular if one is concerned about relatively short time scales.

2.2.1 Convolutions

What is the distribution of the sum of two independent random variables? This sum can, for example, represent the variation of price of an asset over the next two days which is the sum of the increment between today and tomorrow (δX_1) and between tomorrow and the day after tomorrow (δX_2), both assumed to be random and independent. (The notation δX_1 will throughout the book refer to *increments*.)

Let us thus consider $X = \delta X_1 + \delta X_2$ where δX_1 and δX_2 are two random variables, independent, and distributed according to $P_1(\delta x_1)$ and $P_2(\delta x_2)$, respectively. The probability that X is equal to x (within dx) is given by the sum over all possibilities of obtaining $X = x$ (that is all combinations of $\delta X_1 = \delta x_1$ and $\delta X_2 = \delta x_2$ such that $\delta x_1 + \delta x_2 = x$), weighted by their respective probabilities. The variables δX_1 and δX_2 being independent, the joint probability that $\delta X_1 = \delta x_1$ and $\delta X_2 = x - \delta x_1$ is equal to $P_1(\delta x_1)P_2(x - \delta x_1)$, from which one

[†] This means that μ can be determined by a one parameter fit only.

obtains:

$$P(x, N = 2) = \int P_1(x')P_2(x - x')\,dx'. \tag{2.17}$$

This equation defines the **convolution** between P_1 and P_2, which we shall write $P = P_1 \star P_2$. The generalization to the sum of N independent random variables is immediate. If $X = \delta X_1 + \delta X_2 + \cdots + \delta X_N$ with δX_i distributed according to $P_i(\delta x_i)$, the distribution of X is obtained as:

$$P(x, N) = \int P_1(x_1') \ldots P_{N-1}(x_{N-1}')P_N(x - x_1' - \cdots - x_{N-1}') \prod_{i=1}^{N-1} dx_i'. \tag{2.18}$$

One thus understands how powerful is the hypothesis that the increments are iid, i.e. that $P_1 = P_2 = \cdots = P_N$. Indeed, according to this hypothesis, one only needs to know the distribution of increments over a unit time interval to reconstruct that of increments over an interval of length N: it is simply obtained by convoluting the elementary distribution N times with itself.

The analytical or numerical manipulations of Eqs. (2.17) and (2.18) are much eased by the use of Fourier transforms, for which convolutions become simple products. The equation $P(x, N = 2) = [P_1 \star P_2](x)$, reads in Fourier space:

$$\hat{P}(z, N = 2) = \int e^{iz(x-x'+x')} \int P_1(x')P_2(x - x')\,dx'\,dx \equiv \hat{P}_1(z)\hat{P}_2(z). \tag{2.19}$$

In order to obtain the Nth convolution of a function with itself, one should raise its characteristic function to the power N, and then take its inverse Fourier transform.

2.2.2 Additivity of cumulants and of tail amplitudes

It is clear that the mean of the sum of two random variables (independent or not) is equal to the sum of the individual means. The mean is thus additive under convolution. Similarly, if the random variables are independent, one can show that their variances (when they both exist) are also additive. More generally, all the cumulants (c_n) of two independent distributions simply add. This follows from the fact that since the characteristic functions multiply, their logarithm add. The additivity of cumulants is then a simple consequence of the linearity of derivation.

The cumulants of a given law convoluted N times with itself thus follow the simple rule $c_{n,N} = N c_{n,1}$ where the $\{c_{n,1}\}$ are the cumulants of the elementary distribution P_1. Since the cumulant c_n has the dimension of X to the power n, its relative importance is best measured in terms of the normalized cumulants:

$$\lambda_n^N \equiv \frac{c_{n,N}}{(c_{2,N})^{\frac{n}{2}}} = \frac{c_{n,1}}{(c_{2,1})^{\frac{n}{2}}} N^{1-n/2}. \tag{2.20}$$

The normalized cumulants thus decay with N for $n > 2$; the higher the cumulant, the faster the decay: $\lambda_n^N \propto N^{1-n/2}$. The kurtosis κ, defined above as the fourth normalized cumulant, thus decreases as $1/N$.[†] This is basically the content of the CLT: when N is very large, the cumulants of order > 2 become negligible. Therefore, the distribution of the sum is only characterized by its first two cumulants (mean and variance): it is a Gaussian.

Let us now turn to the case where the elementary distribution $P_1(\delta x_1)$ decreases as a power-law for large arguments x_1 (cf. Eq. (1.14)), with a certain exponent μ. The cumulants of order higher than μ are thus divergent. By studying the small z singular expansion of the Fourier transform of $P(x, N)$, one finds that the above additivity property of cumulants is bequeathed to the tail amplitudes A_\pm^μ: the asymptotic behaviour of the distribution of the sum $P(x, N)$ *still behaves as a power-law* (which is thus conserved by addition for all values of μ, provided one takes the limit $x \to \infty$ *before* $N \to \infty$ – see the discussion in Section 2.3.3), with a tail amplitude given by:

$$A_{\pm,N}^\mu \equiv N A_\pm^\mu. \tag{2.21}$$

The tail parameter thus plays the role, for power-law variables, of a generalized cumulant.

2.2.3 Stable distributions and self-similarity

If one adds random variables distributed according to an arbitrary law $P_1(\delta x_1)$, one constructs a random variable which has, in general, a different probability distribution ($P(x, N) = [P_1]^{\star N}(x)$). However, for certain special distributions, the law of the sum has exactly the same shape as the elementary distribution – these are called *stable* laws. The fact that two distributions have the 'same shape' means that one can find a (N-dependent) translation and dilation of x such that the two laws coincide:

$$P(x, N)\,dx = P_1(x_1)\,dx_1 \quad \text{where } x = a_N x_1 + b_N. \tag{2.22}$$

The distribution of increments on a certain time scale (week, month, year) is thus **scale invariant**, provided the variable X is properly rescaled. In this case, the chart giving the evolution of the price of a financial asset as a function of time has the same statistical structure, independently of the chosen elementary time scale – only the average slope and the amplitude of the fluctuations are different. These charts are then called **self-similar**, or, using a better terminology introduced by Mandelbrot, **self-affine** (Figs. 2.2 and 2.3).

The family of all possible stable laws coincide with the Lévy distributions defined above, which include Gaussians as the special case $\mu = 2$. This is easily seen in Fourier space, using the explicit shape of the characteristic function of the Lévy distributions. We shall specialize here for simplicity to the case of symmetric distributions $P_1(-x_1) = P_1(x_1)$, for which the

[†] We will later (Section 7.1.4) study the case of sums of correlated random variables for which the kurtosis still decays to zero, but more slowly, as $N^{-\nu}$ with $\nu < 1$.

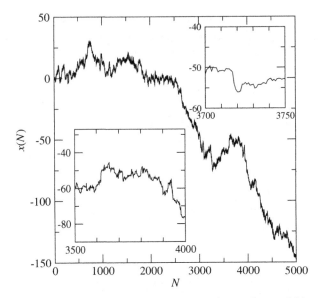

Fig. 2.2. Example of a self-affine function, obtained by summing random variables. One plots the sum x as a function of the number of terms N in the sum, for a Gaussian elementary distribution $P_1(\delta x_1)$. Several successive 'zooms' reveal the self-similar nature of the function, here with $a_N = N^{1/2}$. Note however that for high resolutions, one observes the breakdown of self similarity, related to the existence of a discretization scale. Note the absence of large scale jumps (present in Fig. 2.3); they are discussed in Chapter 3.

translation factor is zero ($b_N \equiv 0$). The scale parameter is then given by $a_N = N^{1/\mu}$,[†] and one finds, for $\mu < 2$:

$$\langle |x|^q \rangle^{\frac{1}{q}} \propto A N^{\frac{1}{\mu}} \qquad q < \mu \tag{2.23}$$

where $A = A_+ = A_-$. In words, the above equation means that the order of magnitude of the fluctuations on 'time' scale N is a factor $N^{1/\mu}$ larger than the fluctuations on the elementary time scale. However, once this factor is taken into account, the probability distributions are identical. One should notice the smaller the value of μ, the faster the growth of fluctuations with time.

2.3 Central limit theorem

We have thus seen that the stable laws (Gaussian and Lévy distributions) are **fixed points** of the convolution operation. These fixed points are actually also **attractors**, in the sense that any distribution convoluted with itself a large number of times finally converges towards a stable law (apart from some very pathological cases). Said differently, the limit distribution of the sum of a large number of random variables is a stable law. The precise formulation of this result is known as the **central limit theorem** (CLT).

[†] The case $\mu = 1$ is special and involves extra logarithmic factors.

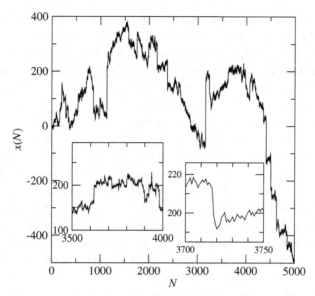

Fig. 2.3. In this case, the elementary distribution $P_1(\delta x_1)$ decreases as a power-law with an exponent $\mu = 1.5$. The scale factor is now given by $a_N = N^{2/3}$. Note that, contrarily to the previous graph, one clearly observes the presence of sudden 'jumps', which reflect the existence of very large values of the elementary increment δx_1. Again, for high resolutions, one observes the breakdown of self similarity, related to the existence of a discretization scale.

2.3.1 Convergence to a Gaussian

The classical formulation of the CLT deals with sums of iid random variables of *finite variance* σ^2 towards a Gaussian. In a more precise way, the result is then the following:

$$\lim_{N \to \infty} \mathcal{P}\left(u_1 \leq \frac{x - mN}{\sigma\sqrt{N}} \leq u_2\right) = \int_{u_1}^{u_2} \frac{1}{\sqrt{2\pi}} e^{-u^2/2}\, du, \tag{2.24}$$

for all *finite* u_1, u_2. Note however that for finite N, the distribution of the sum $X = \delta X_1 + \cdots + \delta X_N$ in the tails (corresponding to extreme events) can be very different from the Gaussian prediction; but the weight of these non-Gaussian regions tends to zero when N goes to infinity. The CLT only concerns the *central* region, which keeps a finite weight for N large: we shall come back in detail to this point below.

The main hypotheses ensuring the validity of the Gaussian CLT are the following:

- The δX_i must be independent random variables, or at least not 'too' correlated (the correlation function $\langle \delta x_i \delta x_j \rangle - m^2$ must decay sufficiently fast when $|i - j|$ becomes large, see Section 2.5 below). For example, in the extreme case where all the δX_i are perfectly correlated (i.e. they are all equal), the distribution of X is the same as that of the individual δX_i (once the factor N has been properly taken into account).

- The random variables δX_i need not necessarily be identically distributed. One must however require that the variance of all these distributions are not too dissimilar, so that no one of the variances dominates over all the others (as would be the case, for example, if the variances were themselves distributed as a power-law with an exponent $\mu < 1$). In this case, the variance of the Gaussian limit distribution is the average of the individual variances. This also allows one to deal with sums of the type $X = p_1 \delta X_1 + p_2 \delta X_2 + \cdots + p_N \delta X_N$, where the p_i are arbitrary coefficients; this case is relevant in many circumstances, in particular in portfolio theory (cf. Chapter 10).[†]

- Formally, the CLT only applies in the limit where N is infinite. In practice, N must be large enough for a Gaussian to be a good approximation of the distribution of the sum. The minimum required value of N (called N^* below) depends on the elementary distribution $P_1(\delta x_1)$ and its distance from a Gaussian. Also, N^* depends on how far in the tails one requires a Gaussian to be a good approximation, which takes us to the next point.

- As mentioned above, the CLT does not tell us anything about the tails of the distribution of X; only the central part of the distribution is well described by a Gaussian. The 'central' region means a region of width at least of the order of $\sqrt{N}\sigma$ around the mean value of X. The actual width of the region where the Gaussian turns out to be a good approximation for large finite N crucially depends on the elementary distribution $P_1(\delta x_1)$. This problem will be explored in Section 2.3.3. Roughly speaking, this region is of width $\sim N^{3/4}\sigma$ for 'narrow' symmetric elementary distributions, such that all even moments are finite. This region is however sometimes of much smaller extension: for example, if $P_1(\delta x_1)$ has power-law tails with $\mu > 2$ (such that σ is finite), the Gaussian 'realm' grows barely faster than \sqrt{N} (as $\sim \sqrt{N \log N}$).

> The above formulation of the CLT requires the existence of a finite variance. This condition can be somewhat weakened to include some 'marginal' distributions such as a power-law with $\mu = 2$. In this case the scale factor is not $a_N = \sqrt{N}$ but rather $a_N = \sqrt{N \log N}$. However, as we shall discuss in the next section, elementary distributions which decay more slowly than $|x|^{-3}$ do not belong the the Gaussian basin of attraction. More precisely, the necessary and sufficient condition for $P_1(\delta x_1)$ to belong to this basin is that:
>
> $$\lim_{u \to \infty} u^2 \frac{\mathcal{P}_{1<}(-u) + \mathcal{P}_{1>}(u)}{\int_{|u'|<u} u'^2 P_1(u') \, du'} = 0. \tag{2.25}$$
>
> This condition is always satisfied if the variance is finite, but allows one to include the marginal cases such as a power-law with $\mu = 2$.

> ### The central limit theorem and information theory
>
> It is interesting to notice that the Gaussian is the law of maximum entropy – or minimum information – such that its variance is fixed. The missing information quantity \mathcal{I} (or

[†] If the p_i's are themselves random variables with a Pareto tail exponent $\mu < 1$, then the distribution of X for a given realization of the p_i's does not converge to any fixed shape in the limit $N \to \infty$, except if the δX_i are Gaussian variables, in which case the distribution of X is Gaussian but with a variance that depends on p_i.

entropy) associated with a probability distribution P is defined as:[†]

$$\mathcal{I}[P] \equiv -\int P(x) \log P(x) \, dx. \tag{2.26}$$

The distribution maximizing $\mathcal{I}[P]$ for a given value of the variance is obtained by taking a functional derivative with respect to $P(x)$:

$$\frac{\partial}{\partial P(x)} \left[\mathcal{I}[P] - \zeta \int x'^2 P(x') \, dx' - \zeta' \int P(x') \, dx' \right] = 0, \tag{2.27}$$

where ζ is fixed by the condition $\int x^2 P(x) \, dx = \sigma^2$ and ζ' by the normalization of $P(x)$. It is immediately seen that the solution to Eq. (2.27) is indeed the Gaussian. The numerical value of its entropy is:

$$\mathcal{I}_G = \frac{1}{2} + \frac{1}{2} \log(2\pi) + \log(\sigma) \approx 1.419 + \log(\sigma). \tag{2.28}$$

For comparison, one can compute the entropy of the symmetric exponential distribution, which is:

$$\mathcal{I}_E = 1 + \frac{\log 2}{2} + \log(\sigma) \approx 1.346 + \log(\sigma). \tag{2.29}$$

It is important to realize that the convolution operation is 'information burning', since all the details of the elementary distribution $P_1(x_1)$ progressively disappear while the Gaussian distribution emerges.

Note that generalized 'non additive' entropies can also be constructed, as (see Tsallis (1988)):

$$\mathcal{I}_q[P] \equiv \int P^q(x) \, dx. \tag{2.30}$$

Formally, the usual entropy is recovered as $\mathcal{I}[P] = \lim_{q \to 1}(1 - \mathcal{I}_q[P])/(q - 1)$. The maximization of $\mathcal{I}_q[P]$ with a fixed variance and normalization gives, for $q < 1$, a Student distribution of index $\mu = 2/(1 - q) - 1$. In the limit $q \to 1$, $\mu \to \infty$ and one recovers the Gaussian (see Section 1.9).

2.3.2 Convergence to a Lévy distribution

Let us now turn to the case of the sum of a large number N of iid random variables, asymptotically distributed as a power-law with $\mu < 2$, and with a tail amplitude $A^\mu = A_+^\mu = A_-^\mu$ (cf. Eq. (1.14)). The variance of the distribution is thus infinite. The limit distribution for large N is then a stable Lévy distribution of exponent μ and with a tail amplitude $N A^\mu$. If the positive and negative tails of the elementary distribution $P_1(\delta x_1)$ are characterized by different amplitudes (A_-^μ and A_+^μ) one then obtains an asymmetric Lévy distribution with parameter $\beta = (A_+^\mu - A_-^\mu)/(A_+^\mu + A_-^\mu)$. If the 'left' exponent is different from the 'right' exponent ($\mu_- \neq \mu_+$), then the smallest of the two wins and one finally obtains a totally asymmetric Lévy distribution ($\beta = -1$ or $\beta = 1$) with exponent $\mu = \min(\mu_-, \mu_+)$. The CLT generalized to Lévy distributions applies with the same precautions as in the Gaussian case above.

[†] Note that entropy is defined up to an additive constant. It is common to add 1 to the above definition.

A very important difference between the Gaussian CLT and the generalized 'Lévy' CLT, which appears visually when comparing Figs. (2.2, 2.3) is the following: in the Gaussian case, the order of magnitude of the sum of N terms is \sqrt{N} and is much larger (in the large N limit) than the largest of these N terms. In other words, no single term contributes to a finite fraction of the whole sum. Conversely, in the case $\mu < 2$, the whole sum and the largest term have the same order of magnitude ($N^{1/\mu}$). The largest term in the sum contributes now to a finite fraction $0 < f < 1$ of the sum, even when $N \to \infty$. Note that f does not converge to a fixed number, but has a non trivial distribution over the interval $[0, 1]$.[†]

Technically, a distribution $P_1(\delta x_1)$ belongs to the basin of attraction of the Lévy distribution $L_{\mu,\beta}$ if and only if:

$$\lim_{u \to \infty} \frac{\mathcal{P}_{1<}(-u)}{\mathcal{P}_{1>}(u)} = \frac{1 - \beta}{1 + \beta}; \tag{2.31}$$

and for all r,

$$\lim_{u \to \infty} \frac{\mathcal{P}_{1<}(-u) + \mathcal{P}_{1>}(u)}{\mathcal{P}_{1<}(-ru) + \mathcal{P}_{1>}(ru)} = r^\mu. \tag{2.32}$$

A distribution with an asymptotic tail given by Eq. (1.14) is such that,

$$\mathcal{P}_{1<}(u) \underset{u \to -\infty}{\simeq} \frac{A_-^\mu}{|u|^\mu} \text{ and } \mathcal{P}_{1>}(u) \underset{u \to \infty}{\simeq} \frac{A_+^\mu}{u^\mu}, \tag{2.33}$$

and thus belongs to the attraction basin of the Lévy distribution of exponent μ and asymmetry parameter $\beta = (A_+^\mu - A_-^\mu)/(A_+^\mu + A_-^\mu)$.

2.3.3 Large deviations

The CLT teaches us that the Gaussian approximation is justified to describe the 'central' part of the distribution of the sum of a large number of random variables (of finite variance). However, the definition of the *centre* has remained rather vague up to now. The CLT only states that the probability of finding an event in the *tails* goes to zero for large N. In the present section, we characterize more precisely the region where the Gaussian approximation is valid.

If X is the sum of N iid random variables of mean m and variance σ^2, one defines a 'rescaled variable' U as:

$$U = \frac{X - Nm}{\sigma\sqrt{N}}, \tag{2.34}$$

which according to the CLT tends towards a Gaussian variable of zero mean and unit variance. Hence, for any *fixed* u, one has:

$$\lim_{N \to \infty} \mathcal{P}_>(u) = \mathcal{P}_{G>}(u), \tag{2.35}$$

[†] On this point, see: Derrida (1997).

where $\mathcal{P}_{G>}(u)$ is the related to the error function, and describes the weight contained in the tails of the Gaussian:

$$\mathcal{P}_{G>}(u) = \int_u^\infty \frac{1}{\sqrt{2\pi}} \exp(-v^2/2) \, dv \equiv \tfrac{1}{2}\mathrm{erfc}\left(\frac{u}{\sqrt{2}}\right). \tag{2.36}$$

However, the above convergence is *not uniform*. The value of N such that the approximation $\mathcal{P}_>(u) \simeq \mathcal{P}_{G>}(u)$ becomes valid depends on u. Conversely, for fixed N, this approximation is only valid for u not too large: $|u| \ll u_0(N)$.

One can estimate $u_0(N)$ in the case where the elementary distribution $P_1(x_1)$ is 'narrow', that is, decreasing faster than any power-law when $|x_1| \to \infty$, such that all the moments are finite. In this case, all the cumulants of P_1 are finite and one can obtain a systematic expansion in powers of $N^{-1/2}$ of the difference $\Delta\mathcal{P}_>(u) \equiv \mathcal{P}_>(u) - \mathcal{P}_{G>}(u)$,

$$\Delta\mathcal{P}_>(u) \simeq \frac{\exp(-u^2/2)}{\sqrt{2\pi}} \left(\frac{Q_1(u)}{N^{1/2}} + \frac{Q_2(u)}{N} + \cdots + \frac{Q_k(u)}{N^{k/2}} + \cdots\right), \tag{2.37}$$

where the $Q_k(u)$ are polynomials functions which can be expressed in terms of the normalized cumulants λ_n (cf. Eq. (1.12)) of the elementary distribution. More explicitly, the first two terms are given by:

$$Q_1(u) = \tfrac{1}{6}\lambda_3(u^2 - 1) \tag{2.38}$$

and

$$Q_2(u) = \tfrac{1}{72}\lambda_3^2 u^5 + \tfrac{1}{8}\left(\tfrac{1}{3}\lambda_4 - \tfrac{10}{9}\lambda_3^2\right) u^3 + \left(\tfrac{5}{24}\lambda_3^2 - \tfrac{1}{8}\lambda_4\right) u. \tag{2.39}$$

One recovers the fact that if all the cumulants of $P_1(\delta x_1)$ of order larger than two are zero, all the Q_k are also identically zero and so is the difference between $P(x, N)$ and the Gaussian.

For a general asymmetric elementary distribution P_1, the skewness $\varsigma \equiv \lambda_3$ is non-zero. The leading term in the above expansion when N is large is thus $Q_1(u)$. For the Gaussian approximation to be meaningful, one must at least require that this term is small in the central region where u is of order one, which corresponds to $x - mN \sim \sigma\sqrt{N}$. This thus imposes that $N \gg N^* = \lambda_3^2$. The Gaussian approximation remains valid whenever the relative error is small compared to 1. For large u (which will be justified for large N), the relative error is obtained by dividing Eq. (2.37) by $\mathcal{P}_{G>}(u) \simeq \exp(-u^2/2)/(u\sqrt{2\pi})$. One then obtains the following condition:[†]

$$\lambda_3 u^3 \ll N^{1/2} \quad \text{i.e.} \quad |x - Nm| \ll \sigma\sqrt{N}\left(\frac{N}{N^*}\right)^{1/6}. \tag{2.40}$$

This shows that the central region has an extension growing as $N^{2/3}$.

A symmetric elementary distribution is such that $\lambda_3 \equiv 0$; it is then the kurtosis $\kappa = \lambda_4$ that fixes the first correction to the Gaussian when N is large, and thus the extension of the

[†] The above arguments can actually be made fully rigorous, see Feller (1971).

central region. The conditions now read: $N \gg N^* = \lambda_4$ and

$$\lambda_4 u^4 \ll N \quad \text{i.e. } |x - Nm| \ll \sigma \sqrt{N} \left(\frac{N}{N^*} \right)^{1/4}. \tag{2.41}$$

The central region now extends over a region of width $N^{3/4}$.

As an example of application, let us compute the MAD of a symmetric distribution to first order in kurtosis. For a Gaussian distribution of standard deviation $\sigma = 1$, one has $\langle |u| \rangle = \sqrt{2/\pi}$. Using Eq. (2.37) with $\lambda_3 = 0$ at the kurtosis order, one finds that the correction is given, after integration by parts, by:

$$\langle |u| \rangle = \sqrt{2/\pi} + \frac{\kappa_N}{12\sqrt{2\pi}} \int_0^\infty (u^3 - 3u) \exp(-u^2/2) \mathrm{d}u = \sqrt{2/\pi} \left(1 - \frac{\kappa_N}{24} \right). \tag{2.42}$$

Interestingly, this simple formula is quite accurate even for κ of order one. For example, the MAD of a symmetric exponential distribution (for which $\kappa = 3$) is $\sigma/\sqrt{2}$, whereas the above formula gives an approximate MAD of 0.698σ, off by only 1%. More generally, one can compute the relative correction of the qth absolute moment compared to the Gaussian value, where q is an arbitrary real positive number. This quantity can be called a hypercumulant of order q and will be noted Λ_q. To the kurtosis order one finds a very simple formula:

$$\Lambda_q = \frac{\Delta \langle |u|^q \rangle}{\langle |u|^q \rangle_G} = \frac{q(q-2)}{24} \kappa_N. \tag{2.43}$$

For $q = 1$, one recovers the previous result $-\kappa_N/24$, whereas for $q = 2$ the correction vanishes as it should, since the variance is fixed. This formula will be used in Chapter 7, as an illuminating test of the convergence of price returns towards Gaussian statistics.

The results of the present section do not directly apply if the elementary distribution $P_1(\delta x_1)$ decreases as a power-law ('broad distribution'). In this case, some of the cumulants are infinite and the above cumulant expansion, Eq. (2.37), is meaningless. In the next section, we shall see that in this case the 'central' region is much more restricted than in the case of 'narrow' distributions. We shall then describe in Section 2.3.6, the case of 'truncated' power-law distributions, where the above conditions become asymptotically relevant. These laws however may have a very large kurtosis, which depends on the point where the truncation becomes noticeable, and the above condition $N \gg \lambda_4$ can be hard to satisfy.

2.3.4 Steepest descent method and Cramèr function (*)

More generally, when N is large, one can write the distribution of the sum of N iid random variables as:[†]

$$P(x, N) \underset{N \to \infty}{\simeq} \exp \left[-NS \left(\frac{x}{N} \right) \right], \tag{2.44}$$

where S is the so-called **Cramèr function**, which gives some information about the probability of X even outside the 'central' region. When the variance is finite, S grows as

[†] We assume that their mean is zero, which can always be achieved through a suitable shift of δx_1.

$S(u) \propto u^2$ for small u's, which again leads to a Gaussian central region. For finite u, S can be computed using the saddle point (or steepest descent) method, valid for N large: see e.g. Hinch (1991). This very useful method allows one to obtain an asymptotic expansion, for large N, of integrals of the type:

$$\mathcal{J}_N = \int_C f(z) \exp[Ng(z)] \, dz, \qquad (2.45)$$

where f and g are slowly varying, analytic function of the complex variable z and the integral is over a certain contour C in the complex plane. The method makes use of two important properties of analytic functions, which allows one to prove that for large N, \mathcal{J}_N is dominated by the vicinity of the stationary point z^* such that $g'(z^*) = 0$, that we assume for simplicity to be unique. The two properties are (i) that one can deform the integration contour such as to go through z^* without changing the value of the integral; (ii) the contour levels of the real part of $g(z)$ are everywhere orthogonal to the contour levels of the imaginary part of $g(z)$. This second property means that along the steepest descent (or ascent) lines of the real part of $g(z)$, the imaginary part of $g(z)$ is constant. Therefore, if one deforms the contour of integration such as to cross z^* in the direction of steepest descent, the phase of $\exp[Ng(z)]$ is locally constant, and the integral does not averge to zero. The final result reads, for N large:

$$\mathcal{J}_N \simeq \sqrt{\frac{2\pi}{-Ng''(z^*)}} \, f(z^*) \exp[Ng(z^*)], \qquad (2.46)$$

up to N^{-1} corrections in the prexponential term. This result also holds when $f(z)$ and $g(z)$ are real function of the real variable z, in which case z^* is simply the point where $g(z)$ reaches its maximum value.

Let us now apply this method to find the Cramèr function. By definition,

$$P(x, N) = \frac{1}{2\pi} \int \exp N \left(-iz\frac{x}{N} + \log[\hat{P}_1(z)] \right) dz. \qquad (2.47)$$

When N is large, the above integral is therefore dominated by the neighbourhood of the point z^* where the term in the exponential is stationary. The results can be written as:

$$P(x, N) \simeq \exp\left[-NS\left(\frac{x}{N}\right) \right], \qquad (2.48)$$

with $S(u)$ given by:

$$\left.\frac{d\log[\hat{P}_1(z)]}{dz}\right|_{z=z^*} = iu \qquad S(u) = -iz^*u + \log[\hat{P}_1(z^*)], \qquad (2.49)$$

which, in principle, allows one to estimate $P(x, N)$ even outside the central region. Note that if $S(u)$ is finite for finite u, the corresponding probability is exponentially small in N.

2.3.5 The CLT at work on simple cases

It is helpful to give some flesh to the above general statements, by working out explicitly the convergence towards the Gaussian in two exactly soluble cases. On these examples, one clearly sees the domain of validity of the CLT as well as its limitations.

Let us first study the case of positive random variables distributed according to the exponential distribution:

$$P_1(x_1) = \Theta(x_1)\alpha e^{-\alpha x_1}, \tag{2.50}$$

where $\Theta(x_1)$ is the function equal to 1 for $x_1 \geq 0$ and to 0 otherwise. A simple computation shows that the above distribution is correctly normalized, has a mean given by $m = \alpha^{-1}$ and a variance given by $\sigma^2 = \alpha^{-2}$. Furthermore, the exponential distribution is asymmetrical; its third moment is given by $c_3 = \langle (x - m)^3 \rangle = 2\alpha^{-3}$, or $\varsigma = 2$.

The sum of N such variables is distributed according to the Nth convolution of the exponential distribution. According to the CLT this distribution should approach a Gaussian of mean mN and of variance $N\sigma^2$. The Nth convolution of the exponential distribution can be computed exactly. The result is:[†]

$$P(x, N) = \Theta(x)\alpha^N \frac{x^{N-1}e^{-\alpha x}}{(N-1)!}, \tag{2.51}$$

which is called a **Gamma distribution** of index N. At first sight, this distribution does not look very much like a Gaussian! For example, its asymptotic behaviour is very far from that of a Gaussian: the 'left' side is strictly zero for negative x, while the 'right' tail is exponential, and thus much fatter than the Gaussian. It is thus very clear that the CLT does not apply for values of x too far from the mean value. However, the central region around $Nm = N\alpha^{-1}$ is well described by a Gaussian. The most probable value (x^*) is defined as:

$$\frac{d}{dx}x^{N-1}e^{-\alpha x}\bigg|_{x^*} = 0, \tag{2.52}$$

or $x^* = (N - 1)m$. An expansion in $x - x^*$ of $P(x, N)$ then gives us:

$$\log P(x, N) = -K(N - 1) - \log m - \frac{\alpha^2(x - x^*)^2}{2(N - 1)}$$
$$+ \frac{\alpha^3(x - x^*)^3}{3(N - 1)^2} + O(x - x^*)^4, \tag{2.53}$$

where

$$K(N) \equiv \log N! + N - N \log N \underset{N \to \infty}{\simeq} \tfrac{1}{2}\log(2\pi N). \tag{2.54}$$

Hence, to second order in $x - x^*$, $P(x, N)$ is given by a Gaussian of mean $(N - 1)m$ and variance $(N - 1)\sigma^2$. The relative difference between N and $N - 1$ goes to zero for large N. Hence, for the Gaussian approximation to be valid, one requires not only that N be

[†] This result can be shown by induction using the definition (Eq. (2.17)).

large compared to one, but also that the higher-order terms in $(x - x^*)$ be negligible. The cubic correction is small compared to 1 as long as $\alpha |x - x^*| \ll N^{2/3}$, in agreement with the above general statement, Eq. (2.40), for an elementary distribution with a non-zero third cumulant. Note also that for $x \to \infty$, the exponential behaviour of the Gamma function coincides (up to subleading terms in x^{N-1}) with the asymptotic behaviour of the elementary distribution $P_1(x_1)$.

Another very instructive example is provided by a distribution which behaves as a power-law for large arguments, but at the same time has a finite variance to ensure the validity of the CLT. Consider the following explicit example of a Student distribution with $\mu = 3$:

$$P_1(x_1) = \frac{2a^3}{\pi \left(x_1^2 + a^2\right)^2}, \tag{2.55}$$

where a is a positive constant. This symmetric distribution behaves as a power-law with $\mu = 3$ (cf. Eq. (1.14)); all its cumulants of order larger than or equal to three are infinite. However, its variance is finite and equal to a^2.

It is useful to compute the characteristic function of this distribution,

$$\hat{P}_1(z) = (1 + a|z|)e^{-a|z|}, \tag{2.56}$$

and the first terms of its small z expansion, which read:

$$\hat{P}_1(z) \simeq 1 - \frac{z^2 a^2}{2} + \frac{|z|^3 a^3}{3} + O(z^4). \tag{2.57}$$

The first singular term in this expansion is thus $|z|^3$, as expected from the asymptotic behaviour of $P_1(x_1)$ in x_1^{-4}, and the divergence of the moments of order larger than three.

The Nth convolution of $P_1(\delta x_1)$ thus has the following characteristic function:

$$\hat{P}_1^{\,N}(z) = (1 + a|z|)^N e^{-aN|z|}, \tag{2.58}$$

which, expanded around $z = 0$, gives:

$$\hat{P}_1^{\,N}(k) \simeq 1 - \frac{N z^2 a^2}{2} + \frac{N |z|^3 a^3}{3} + O(z^4). \tag{2.59}$$

Note that the $|z|^3$ singularity (which signals the divergence of the moments m_n for $n \geq 3$) does not disappear under convolution, even if at the same time $P(x, N)$ converges towards the Gaussian. The resolution of this apparent paradox is again that the convergence towards the Gaussian only concerns the centre of the distribution, whereas the tail in x^{-4} survives for ever (as was mentioned in Section 2.2.3).

As follows from the CLT, the centre of $P(x, N)$ is well approximated, for N large, by a Gaussian of zero mean and variance $N a^2$:

$$P(x, N) \simeq \frac{1}{\sqrt{2\pi N}a} \exp\left(-\frac{x^2}{2N a^2}\right). \tag{2.60}$$

On the other hand, since the power-law behaviour is conserved upon addition and that the tail amplitudes simply add (cf. Eq. (1.14)), one also has, for large x's:

$$P(x, N) \underset{x \to \infty}{\simeq} \frac{2Na^3}{\pi x^4}. \tag{2.61}$$

The above two expressions, Eqs. (2.60) and (2.61), are not incompatible, since these describe two very different regions of the distribution $P(x, N)$. For fixed N, there is a characteristic value $x_0(N)$ beyond which the Gaussian approximation for $P(x, N)$ is no longer accurate, and the distribution is described by its asymptotic power-law regime. The order of magnitude of $x_0(N)$ is fixed by looking at the point where the two regimes match to one another:

$$\frac{1}{\sqrt{2\pi N a}} \exp\left(-\frac{x_0^2}{2Na^2}\right) \simeq \frac{2Na^3}{\pi x_0^4}. \tag{2.62}$$

One thus finds,

$$x_0(N) \simeq a\sqrt{N \log N}, \tag{2.63}$$

(neglecting subleading corrections for large N).

This means that the rescaled variable $U = X/(a\sqrt{N})$ becomes for large N a Gaussian variable of unit variance, but this description ceases to be valid as soon as $u \sim \sqrt{\log N}$, which grows very slowly with N. For example, for N equal to a million, the Gaussian approximation is only acceptable for fluctuations of u of less than three or four RMS!

Finally, the CLT states that the weight of the regions where $P(x, N)$ substantially differs from the Gaussian goes to zero when N becomes large. For our example, one finds that the probability that X falls in the tail region rather than in the central region is given by:

$$\mathcal{P}_<(x_0) + \mathcal{P}_>(x_0) \simeq 2 \int_{a\sqrt{N \log N}}^{\infty} \frac{2a^3 N}{\pi x^4} \, \mathrm{d}x \propto \frac{1}{\sqrt{N} \log^{3/2} N}, \tag{2.64}$$

which indeed goes to zero for large N.

Since the kurtosis is infinite, one cannot expect the cumulant expansion given by Eq. (2.37) to hold. From the exact expression of the characteristic function $\hat{P}_1^N(z)$ given above, one can work out the leading correction to the Gaussian for large N. The trick is to set $z = \hat{z}/\sqrt{N}$ and let N tend to infinity for fixed \hat{z}: this has the virtue of selecting the region where the sum is of order \sqrt{N}, that is $u = x/\sqrt{N}$ of order unity. In this limit, the characteristic function writes:

$$\hat{P}_1^N(z) = \left(1 + \frac{1}{3\sqrt{N}} |\hat{z}a|^3\right) e^{-\hat{z}^2 a^2/2}. \tag{2.65}$$

The corresponding correction to the Gaussian distribution reads:

$$\Delta P(u) = \frac{1}{3\pi\sqrt{N}} \int_0^\infty (\hat{z}a)^3 \cos(\hat{z}u) e^{-\hat{z}^2 a^2/2} \mathrm{d}\hat{z}, \tag{2.66}$$

which tends to zero as $N^{-1/2}$, much more slowly than if the kurtosis was finite. The shape of this correction is shown in Fig. 2.4. From the knowledge of $\Delta P(u)$, one can compute,

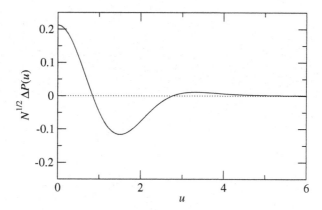

Fig. 2.4. Plot of $\sqrt{N}\,\Delta P(u)$, Eq. (2.66), as a function of u.

for example, the correction to the MAD for large N and find:

$$\langle |u| \rangle = \sqrt{\frac{2}{\pi}}\left(1 - \frac{1}{6\sqrt{2\pi N}}\right), \tag{2.67}$$

that can be compared to Eq. (2.42) above.

The above arguments are not special to the case $\mu = 3$ and in fact apply more generally, as long as $\mu > 2$, i.e. when the variance is finite. In the general case, one finds that the CLT is valid in the region $|x| \ll x_0 \propto \sqrt{N \log N}$, and that the weight of the non-Gaussian tails is given by:

$$\mathcal{P}_<(x_0) + \mathcal{P}_>(x_0) \propto \frac{1}{N^{\mu/2-1} \log^{\mu/2} N}, \tag{2.68}$$

which tends to zero for large N. However, one should notice that as μ approaches the 'dangerous' value $\mu = 2$, the weight of the tails becomes more and more important. The correction to the Gaussian (for example the MAD) only decays as $N^{1-\mu/2}$. For $\mu < 2$, the whole argument collapses since the weight of the tails would grow with N. In this case, however, the convergence is no longer towards the Gaussian, but towards the Lévy distribution of exponent μ.

2.3.6 Truncated Lévy distributions

An interesting case is when the elementary distribution $P_1(\delta x_1)$ is a truncated Lévy distribution (TLD) as defined in Section 1.8. If one considers the sum of N random variables distributed according to a TLD, the condition for the CLT to be valid reads (for $\mu < 2$):[†]

$$N \gg N^* = \lambda_4 \Longrightarrow (N a_\mu)^{\frac{1}{\mu}} \gg \alpha^{-1}. \tag{2.69}$$

[†] One can see by inspection that the other conditions, concerning higher-order cumulants in Eq. (2.37), which read $N^{k-1}\lambda_{2k} \gg 1$, are actually equivalent to the one written here.

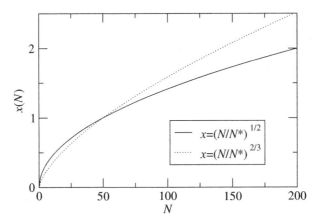

Fig. 2.5. Behaviour of the typical value of X as a function of N for TLD variables. When $N \ll N^*$, x grows as $N^{1/\mu}$ (dotted line). When $N \sim N^*$, x reaches the value α^{-1} and the exponential cut-off starts being relevant. When $N \gg N^*$, the behaviour predicted by the CLT sets in, and one recovers $x \propto \sqrt{N}$ (plain line).

This condition has a very simple intuitive meaning. A TLD behaves very much like a pure Lévy distribution as long as $x \ll \alpha^{-1}$. In particular, it behaves as a power-law of exponent μ and tail amplitude $A^\mu \propto a_\mu$ in the region where x is large but still much smaller than α^{-1} (we thus also assume that α is very small). If N is not too large, most values of x fall in the Lévy-like region. The largest value of x encountered is thus of order $x_{\max} \simeq A N^{1/\mu}$ (cf. Eq. (2.9)). If x_{\max} is very small compared to α^{-1}, it is consistent to forget the exponential cut-off and think of the elementary distribution as a pure Lévy distribution. One thus observes a first regime in N where the typical value of X grows as $N^{1/\mu}$, as if α was zero.[†] However, as illustrated in Figure 2.5, this regime ends when x_{\max} reaches the cut-off value α^{-1}: this happens precisely when N is of the order of N^* defined above. For $N > N^*$, the variable X progressively converges towards a Gaussian variable of width \sqrt{N}, at least in the region where $|x| \ll \sigma N^{3/4}/N^{*1/4}$. The typical amplitude of X thus behaves (as a function of N) as sketched in Figure 2.5. Notice that the asymptotic part of the distribution of X (outside the central region) decays as an exponential for all values of N.

2.3.7 Conclusion: survival and vanishing of tails

The CLT thus teaches us that if the number of terms in a sum is large, the sum becomes (nearly) a Gaussian variable. This sum can represent the temporal aggregation of the daily fluctuations of a financial asset, or the aggregation, in a portfolio, of different stocks. The Gaussian (or non-Gaussian) nature of this sum is thus of crucial importance for risk control, since the extreme tails of the distribution correspond to the most 'dangerous' fluctuations. As we have discussed above, fluctuations are never Gaussian in the far-tails: one can explicitly

[†] Note however that the variance of X grows like N for all N. However, the variance is dominated by the cut-off and, in the region $N \ll N^*$, grossly overestimates the typical values of X, see Section 2.3.2.

show that if the elementary distribution decays as a power-law (or as an exponential, which formally corresponds to $\mu = \infty$), the distribution of the sum decays in the very same manner outside the central region, i.e. much more slowly than the Gaussian. The CLT simply ensures that these tail regions are expelled more and more towards large values of X when N grows, and their associated probability is smaller and smaller. When confronted with a concrete problem, one must decide whether N is large enough to be satisfied with a Gaussian description of the risks. In particular, if N is less than the characteristic value N^* defined above, the Gaussian approximation is very bad.

2.4 From sum to max: progressive dominance of extremes (*)

Although taking the sum of N iid random variables or extracting the largest value may look completely unrelated and lead to different limit distributions, one may actually find a formulation where the two problems are continuously related. Let us consider the following generalized sum:

$$S_q = \sum_{i=1}^{N} \delta x_i^q, \tag{2.70}$$

where q is a real number and δX are restricted to be positive random variables. It is clear that when $q = 1$, one recovers the simple sum considered in the above section. Provided that all the moments of X are finite, the Gaussian CLT applies to S_q for all finite q in the limit $N \to \infty$ since the random variables δx_i^q are still iid. However, if one chooses $q \to \infty$ and N finite, the largest term in the sum completely dominates over all the others, in such a way that, in that limit, $S_q \simeq \delta x_{\max}^q$. Is there a way to take simultaneously the limit where q *and* N tend to infinity such that one escapes from the standard CLT?

Let us consider the specific case $\delta X = \exp Y$ where the random variable Y is Gaussian, of zero mean and variance σ^2. In financial applications, S_q would correspond to the value of a portfolio containing N assets with random returns Y after time q, such that the initial value of each asset is unity. In physics, S_q is a partition function if one identifies Y with an energy and q as an inverse temperature. The problem defined by Eq. (2.70) is known as the **random energy model**, and is relevant for the understanding of a surprisingly large number of complex systems, from amorphous, glassy materials to error correcting codes or other computational science problems (see Derrida (1981), Montanari (2001)).

Interestingly, one finds that the statistics of the variable S_q defined by Eq. (2.70) has three different regimes as q increases, reflecting precisely the three different regimes of the CLT discussed above. For small q, the statistics of S_q is Gaussian, for larger q it becomes a fully asymmetric ($\beta = 1$) Lévy stable distribution with a finite mean but infinite variance ($1 < \mu < 2$) and finally, for still larger q's, a fully asymmetric Lévy stable distribution with an infinite mean ($\mu < 1$). The precise relation between μ and q reads $\mu = \sqrt{2 \log N}/q\sigma$ (see Bovier et al. (2002)). The non trivial limit where one does not reach the CLT nor the extreme value statistics is therefore the regime where q diverges as $\sqrt{\log N}$. For 'small' enough q, all the terms contributing to S_q are of similar magnitude and the usual Gaussian

CLT applies. As q becomes larger, the largest terms in the sum S_q are more and more dominant, until a point where only a finite number of terms actually contribute to S_q, even in the $N \to \infty$ limit: this corresponds to the case $\mu < 1$. Eventually, for $q \to \infty$ or $\mu \to 0$, only one term contributes to S_q and one recovers the extreme value statistics detailed in Section 2.1. Note that these results can be extended to the case where Y is not Gaussian, but has tails decaying faster than exponentially, for example as $\exp - y^c$, $c > 1$. Only the precise relation between μ and q is affected.

Coming back to our financial example, the above results mean that for a given number of assets in the portfolio, a correct diversification (corresponding to a Gaussian distribution of the value of the portfolio) is only achieved up to 'time' $q^* = \sqrt{\log N}/\sqrt{2}\sigma$. For longer times, most of the investor's wealth will be concentrated in only a few assets, unless the portfolio is frequently rebalanced.

2.5 Linear correlations and fractional Brownian motion

Let us assume that the correlation function $C_{i,j}$ of the random variable δX (defined as $\langle \delta x_i \delta x_j \rangle - m^2$) is non-zero for $i \neq j$. We also assume that the process is *stationary*, i.e. that $C_{i,j}$ only depends on $|i - j|$: $C_{i,j} = C(|i - j|)$, with $C(\infty) = 0$. The variance of the sum can be expressed in terms of the matrix C as:[†]

$$\langle x^2 \rangle = \sum_{i,j=1}^{N} C_{i,j} = N\sigma^2 + 2N \sum_{\ell=1}^{N} \left(1 - \frac{\ell}{N}\right) C(\ell), \tag{2.71}$$

where $\sigma^2 \equiv C(0)$. From this expression, it is readily seen that if $C(\ell)$ decays faster than $1/\ell$ for large ℓ, the sum over ℓ tends to a constant for large N, and thus the variance of the sum still grows as N, as for the usual CLT. If however $C(\ell)$ decays for large ℓ as a power-law $\ell^{-\nu}$, with $\nu < 1$, then the variance grows faster than N, as $N^{2-\nu}$ – correlations thus enhance fluctuations. Hence, when $\nu < 1$, the standard CLT certainly has to be amended. The problem of the limit distribution in these cases is however not solved in general. For example, if the δX_i are correlated Gaussian variables, it is easy to show that the resulting sum is also Gaussian of width $\propto N^{1-\nu/2}$, whatever the value of $\nu < 1$: a linear combination of Gaussian variables remains Gaussian.

Another solvable but non trivial case is when the δX_i are correlated Gaussian variables, but one takes the sum of the squares of the δX_i's. This sum converges towards a Gaussian of width \sqrt{N} whenever $\nu > \frac{1}{2}$, but towards a non-trivial limit distribution of a new kind (i.e. neither Gaussian nor Lévy stable) when $\nu < \frac{1}{2}$. In this last case, the proper rescaling factor must be chosen as $N^{1-\nu}$.

One can also construct *anti-correlated* random variables, the sum of which grows *slower* than \sqrt{N}. In the case of power-law correlated or anti-correlated Gaussian random variables, one speaks of **fractional Brownian motion**.[‡] Let us briefly explain how such processes can

[†] We again assume in the following, without loss of generality, that the mean m is zero.
[‡] See Mandelbrot and Van Ness (1968).

be constructed. We will first consider the set of random variables x_k defined as:

$$x_k = \frac{s_0}{\sqrt{M}} \sum_{m=1}^{M/2} \left(\frac{2\pi m}{M}\right)^{\frac{\nu-3}{2}} \left(\xi_m e^{\frac{2i\pi mk}{M}} + \xi_m^* e^{-\frac{2i\pi mk}{N}}\right),$$ (2.72)

where the ξ_m's are $M/2$ independent complex Gaussian variables of zero mean and unit variance, i.e. both the real and imaginary part are independent and of variance equal to $1/2$. Therefore, one has:

$$\langle \xi_m^2 \rangle = \langle \xi_m^{*2} \rangle = 0 \qquad \langle \xi_m \xi_m^* \rangle = 1.$$ (2.73)

From this representation, one may compute the variance of the lagged difference $x_{k+N} - x_k$. In the limit where $1 \ll N \ll M$, one finds:

$$\langle (x_{k+N} - x_k)^2 \rangle \simeq 2s_0^2 \Gamma(\nu - 2) \cos\frac{\pi\nu}{2} N^{2-\nu}.$$ (2.74)

When $0 < \nu < 1$, the variance grows faster than N and one speaks about a **persistent** random walk. A plot of x_k versus k for $\nu = 2/5$ is given in Fig. 2.6, where 'trends' can be observed and reflect the correlations present in the process (see below). Note that again this plot is self affine. When $2 > \nu > 1$, on the other hand, the variance grows slower than N and one now speaks about an **anti-persistent** random walk. A plot of x_k versus k for $\nu = 3/2$ is given in Fig. 2.6, where now the walk is seen to curl back on itself, reflecting the anticorrelations built in the process. The case $\nu = 1$ corresponds to the usual random walk.

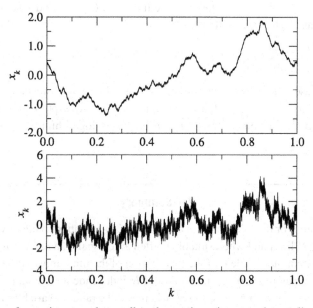

Fig. 2.6. Example of a persistent random walk and an anti-persistent random walk, constructed using Eq. (2.72) with $M = 50\,000$. The exponent ν is chosen to be, respectively, $\nu = 2/5 < 1$ and $\nu = 3/2 > 1$. Note that is both cases, the process is self-affine. The case $\nu = 1$ corresponds to the usual random walk – see Fig. 2.2.

A way to see this is to compute the correlations of the increments $\delta x_k = x_{k+1} - x_k$. Using the representation (2.72), one finds that the correlation function $C(\ell) = \langle \delta x_k \delta x_{k+\ell} \rangle$ is given, for $M \gg 1$, by:

$$C(\ell) = 4s_0^2 \int_0^{2\pi} \frac{\cos z\ell}{z^{3-\nu}} (1 - \cos z)\, \mathrm{d}z. \tag{2.75}$$

For $0 < \nu < 1$, this integral is found to behave, for large ℓ, as ℓ^ν, corresponding to long range correlations. For $1 < \nu < 2$, on the other hand, the integral oscillates in sign and integrates to zero, leading to an antipersistent behaviour.

Another, perhaps more direct way to construct long ranged correlated Gaussian random variables is to define the increments δx_k themselves as sums:

$$\delta x_k = \sum_{\ell=0}^{\infty} K(\ell)\xi_{k-\ell}, \tag{2.76}$$

where the ξ's are now real independent Gaussian random variables of unit variance and $K(\ell)$ a certain kernel. If k represents the time, δx_k can be thought of as resulting of a superposition of past influences, with a certain memory kernel $K(\ell)$. The computation of $\langle \delta x_i \delta x_j \rangle$ is straightforward and leads to:

$$C(|i - j|) = \sum_{\ell=0}^{\infty} K(\ell)K(\ell + |i - j|). \tag{2.77}$$

Choosing $K(\ell) \sim K_0/\ell^{(1+\nu)/2}$ for large ℓ, one finds that $C(|i - j|)$ decays asymptotically, for $0 < \nu < 1$, as $C_0/|i - j|^\nu$ with:

$$C_0 = K_0^2 \Gamma(\nu) \frac{\Gamma(1/2 - \nu/2)}{\Gamma(1/2 + \nu/2)}. \tag{2.78}$$

An important case, which we will discuss later, is when $\nu = 0$. In this case, one needs to introduce a large scale cut-off, for example, $K(\ell) = K_0 \exp(-\alpha\ell)/\sqrt{\ell}$. In the limit $1 \ll |i - j| \ll \alpha^{-1}$, one finds that $C(|i - j|)$ behaves logarithmically, as $C(|i - j|) \simeq -K_0^2 \log(\alpha|i - j|)$.

2.6 Summary

- The maximum (or the minimum) in a series of N independent random variables has, in the large N limit and when suitably shifted and rescaled, a *universal* distribution, to a large extent independent of the distribution of the random variables themselves. The order of magnitude of the maximum of N independent Gaussian variables only grows very slowly with N, as $\sqrt{\log N}$, whereas if the random variables have a Pareto tail with an exponent μ, the largest term grows much faster, as $N^{1/\mu}$. The former case corresponds to an asymptotic *Gumbel* distribution, the latter to an asymptotic *Fréchet* distribution.

- The sum of a series of N independent random variables with zero mean also has, in the large N limit and when suitably rescaled, a *universal* distribution, again to a large extent independent of the distribution of the random variables themselves (central limit theorem). If the variance of these variables is finite, this distribution is Gaussian, with a width growing as \sqrt{N}. If the variance is infinite, as is the case for Pareto variables with an exponent $\mu < 2$, the limit distribution is a Lévy stable law, with a width growing as $N^{1/\mu}$. In the latter case, the largest term in the series remains of the same order of magnitude as the whole sum in the limit $N \to \infty$.
- The above universal *limit distributions* only apply in the scaling regime, but do not describe the far tails, which retain some of the characteristics of the initial distribution. For example, the sum of Pareto variables with an exponent $\mu > 2$ converges towards a Gaussian random variable in a region of width $\sqrt{N \log N}$, outside which the initial Pareto power-law tail is recovered.

- FURTHER READING

J. Baschnagel, W. Paul, *Stochastic Processes, From Physics to Finance*, Springer-Verlag, 2000.

F. Bardou, J.-P. Bouchaud, A. Aspect, C. Cohen-Tannoudji, *Lévy statistics and Laser cooling*, Cambridge University Press, 2002.

J. Beran, *Statistics for Long-Memory Processes*, Chapman & Hall, 1994.

J.-P. Bouchaud, A. Georges, *Anomalous diffusion in disordered media: statistical mechanisms, models and physical applications*, Phys. Rep., **195**, 127 (1990).

P. Embrechts, C. Klüppelberg, T. Mikosch, *Modelling Extremal Events for Insurance and Finance*, Springer-Verlag, Berlin, 1997.

W. Feller, *An Introduction to Probability Theory and its Applications*, Wiley, New York, 1971.

J. Galambos, *The Asymptotic Theory of Extreme Order Statistics*, R. E. Krieger Publishing Co., Malabar, Florida, 1987.

B. V. Gnedenko, A. N. Kolmogorov, *Limit Distributions for Sums of Independent Random Variables*, Addison Wesley, Cambridge, MA, 1954.

E. J. Gumbel, *Statistics of Extremes*, Columbia University Press, New York, 1958.

E. J. Hinch, *Perturbation methods*. Cambridge University Press, 1991.

P. Lévy, *Théorie de l'addition des variables aléatoires*, Gauthier Villars, Paris, 1937–1954.

B. B. Mandelbrot, *The Fractal Geometry of Nature*, Freeman, San Francisco, 1982.

B. B. Mandelbrot, *Fractals and Scaling in Finance*, Springer, New York, 1997.

R. Mantegna, H. E. Stanley, *An Introduction to Econophysics*, Cambridge University Press, Cambridge, 1999.

E. Montroll, M. Shlesinger, The Wonderful World of Random Walks, in *Nonequilibrium Phenomena II, From Stochastics to Hydrodynamics*, Studies in Statistical Mechanics XI (J. L. Lebowitz, E. W. Montroll, eds) North-Holland, Amsterdam, 1984.

G. Samorodnitsky, M. S. Taqqu, *Stable Non-Gaussian Random Processes*, Chapman & Hall, New York, 1994.

J. Voït, *The Statistical Mechanics of Financial Markets*, Springer-Verlag, 2001.

N. Wiener, *Extrapolation, Interpolation and Smoothing of Time Series*, Wiley, New York, 1949.

G. Zaslavsky, U. Frisch, M. Shlesinger, *Lévy Flights and Related Topics in Physics*, Lecture Notes in Physics **450**, Springer, New York, 1995.

• SOME SPECIALIZED PAPERS

J.-P. Bouchaud, M. Mézard, *Universality classes for extreme value statistics*, J. Phys. A., **30**, 7997 (1997).

A. Bovier, I. Kurkova, M. Loewe, *Fluctuations of the free energy in the REM and the p-spin SK model*, Ann. Probab., **30**, 605 (2002).

B. Derrida, *The Random Energy Model*, Phys. Rev. B, **24**, 2613 (1981).

B. Derrida, *Non-Self averaging effects in sum of random variables*, Physica **D**, **107**, 186 (1997).

I. Koponen, *Analytic approach to the problem of convergence to truncated Lévy flights towards the Gaussian stochastic process*, Physical Review, **E 52**, 1197 (1995).

B. B. Mandelbrot, J. W. Van Ness, *Fractional Brownian Motion, fractional noises and applications*, SIAM Review, **10**, 422 (1968).

A. Montanari, *The glassy phase of Gallager codes*, Eur. Phys. J., **B 23**, 121 (2001).

M. Romeo et al., *Broad distribution effects in sums of lognormal random variables*, Eur. Phys. J., **B 32**, 513 (2003).

C. Tsallis, *Possible generalization of Boltzmann–Gibbs statistics*, J. Stat. Phys., **52**, 479 (1988).

• SOME RECENT READS

G. Ben Arous, L. V. Bogachev, S. A. Molchanov, *Limit theorems for sums of random exponentials*, Probability Theory and Related Fields, **132**, 579 (2005).

Y. Malevergne, D. Sornette, *Extreme Financial Risks: From Dependence to Risk Management* (Springer, 2005).

3

Continuous time limit, Ito calculus and path integrals

Comment oser parler des lois du hasard ? Le hasard n'est-il pas l'antithèse de toute loi ?[†]
(Joseph Bertrand, *Calcul des probabilités*.)

3.1 Divisibility and the continuous time limit

3.1.1 Divisibility

We have discussed in the previous chapter how the sum of two iid random variables can be computed if the distribution of these two variables is known. One can ask the opposite question: knowing the probability distribution of a variable X, is it possible to find two iid random variables such that $X = \delta X_1 + \delta X_2$, or more precisely such that the distribution of X can be written as the convolution of two identical distributions. If this is possible, the variable X is said to be **divisible**.

We already know cases where this is possible. For example, if X has a Gaussian distribution of variance σ^2, one can choose X_1 and X_2 to be have Gaussian distribution of variance $\sigma^2/2$. More generally, if X is a Lévy stable random variable of parameter a_μ (see Eq. (1.20)), then X_1 and X_2 are also Lévy stable random variables with parameter $a_\mu/2$. However, all random variables are not divisible: as we show below, if X has a uniform distribution in the interval $[a, b]$, it cannot be written as $X = \delta X_1 + \delta X_2$ with $\delta X_1, \delta X_2$ iid.

More formally, this problem translates into the properties of characteristic functions $\hat{P}(z)$. Since the convolution amounts to a simple product for characteristic functions, the question is whether the square root of a characteristic function is still the Fourier transform of a *bona fide* probability distribution, i.e. a non negative function of integral equal to one. This imposes certain conditions on the characteristic function, for example on the cumulants when they exist. For example, the kurtosis κ of any probability distribution is bounded from below by -2. This comes from the inequality $\langle x^4 \rangle \geq \langle x^2 \rangle^2$, which itself is a consequence of the positivity of the distribution $P(x)$. This is enough to show that a uniform variable is not divisible. The kurtosis of the uniform distribution is equal to $\kappa = -6/5$. From the additivity of the cumulants under convolution, the kurtosis κ_1 of the putative distribution P_1 such that $P = [P_1]^{*2}$ has to be twice that of P, i.e. $\kappa_1 = -12/5 < -2$. Therefore P_1 cannot be everywhere positive, and does not correspond to a probability distribution.

[†] How dare we speak of the laws of chance? Is not chance the antithesis of all law?

3.1.2 Infinite divisibility

One can obviously generalize the above question to the case of N-divisibility: can one decompose a random variable X as a sum of N iid random variables? Said differently, when is $\hat{P}(z)^{1/N}$ the Fourier transform of a probability distribution? Again, the case of Gaussian variables is simple. Consider a Gaussian random variable of variance $\sigma^2 T$ and zero mean.[†] Its characteristic function is given by $\hat{P}(z) = \exp{-\sigma^2 z^2 T}$. Taking the Nth root we find $\hat{P}(z)^{1/N} = \exp{-\sigma^2 z^2 T/N}$, i.e. the characteristic function of a Gaussian distribution of variance $\sigma^2 T/N$. One can take the $N \to \infty$ limit, and find that a Gaussian random variable of variance $\sigma^2 T$ can be thought of as an infinite sum of random Gaussian increments δX_i of vanishing variance:

$$X_N = \sum_{i=0}^{N-1} \delta X_i \longrightarrow X(T) = \int_0^T dX(t), \qquad t = T\frac{i}{N}. \tag{3.1}$$

In the limit $N \to \infty$, the quantity t becomes continuous and can be thought of as 'time'. The resulting 'trajectory' $X(t)$ is then called the continuous time Brownian motion and has many very interesting properties. For example, $X(\lambda t)$ is a Gaussian variable of variance equal to that of $X(t)$ multiplied by a factor λ. One can also show that $X(t)$ is *continuous* in the following sense. The probability $\mathcal{P}_>(\epsilon)$ that any one of the N increments exceeds in size a certain small but finite value ϵ is given by:

$$\mathcal{P}_>(\epsilon) = 1 - \left(1 - 2 \int_\epsilon^\infty \frac{\exp{-Nu^2/2\sigma^2 T}}{\sqrt{2\pi\sigma^2 T/N}} \, du \right)^N. \tag{3.2}$$

In the limit $N \to \infty$, ϵ fixed, one can use an asymptotic estimate of the above integral to find:

$$\mathcal{P}_>(\epsilon) \simeq 1 - \left(1 - \sqrt{\frac{2T}{\pi N}} \frac{\sigma}{\epsilon} \exp{-\frac{N\epsilon^2}{2\sigma^2 T}} \right)^N \simeq \sqrt{\frac{2NT}{\pi}} \frac{\sigma}{\epsilon} \exp{-\frac{N\epsilon^2}{2\sigma^2 T}}, \tag{3.3}$$

which tends to zero when $N \to \infty$. Therefore, in the continuous time limit, almost surely, the resulting process does not jump.

Another simple example of infinite divisibility is when X is a Lévy stable random variable of parameter a_μ, then the increments δX are also Lévy stable random variables with parameter a_μ/N. However, this process is not continuous in the large N limit, since in this case one has, for a symmetric distribution:

$$\mathcal{P}_>(\epsilon) \simeq 1 - \left(1 - 2 \int_\epsilon^\infty \frac{A^\mu}{N\mu u^{1+\mu}} \, du \right)^N \longrightarrow 1 - \exp[-2(A/\epsilon)^\mu], \tag{3.4}$$

which is finite. Therefore there is finite probability to observe a discontinuity of amplitude $> \epsilon$ in the 'trajectory' $X(t)$.

[†] We introduce a factor T that will prove useful when taking the continuum limit. The symbol σ^2 is a variance per unit time, or a diffusion constant and its squared-root (σ) is called the **volatility**.

A graphical representation of these two processes was shown in Figures 2.2 and 2.3. The Lévy process shown in Figure 2.3 has a few clear finite discontinuities, while the Gaussian one (Fig. 2.2) looks continuous to the eye. Strictly speaking, the processes shown are discrete (hence discontinuous), but the largest scale is indistinguishable (to the eye) from a true continuum limit.

Due to the additivity of the cumulants (when they exist), one can assert in the general case that the iid increments needed to reconstruct the variable X have a variance which is a factor N smaller than the variance of X, but a kurtosis a factor N *larger* than that of X. The only way to obtain a random variable with a finite kurtosis in the limit $N \to \infty$ is to start with increments that have an infinite kurtosis. More generally, the normalized cumulants of the increments λ_n are a factor $N^{n/2-1}$ larger than those of the variable X.

3.1.3 Poisson jump processes

Let us consider a random sequence of N increments δX_i constructed as follows: for each i, the value of δX_i is chosen to be either 0 with probability $1 - p/N$, or to a finite value ('jump') distributed according to a certain law $P_J(\delta x)$, with probability p/N, where p is finite. The quantity p is therefore the average number of jumps in the trajectories. It is straightforward to write down the probability density of individual increments,

$$P_1(\delta x) = \left(1 - \frac{p}{N}\right) \delta(\delta x - 0) + \frac{p}{N} \hat{P}_J(\delta x), \tag{3.5}$$

where $\delta(.)$ is the Dirac function corresponding to the cases where $\delta X = 0$. The variance of the increments is equal to $p\langle \delta x^2 \rangle_J / N$, whereas the kurtosis is given by:

$$\kappa = \frac{N \langle \delta x^4 \rangle_J}{p \langle \delta x^2 \rangle_J^2} - 3, \tag{3.6}$$

where $\langle \ldots \rangle_J$ refers to the moments of $P_J(\delta x)$. In the limit where p is fixed and $N \to \infty$, the increments have a vanishing variance and a diverging kurtosis.

Introducing the characteristic functions and noting that the Fourier transform of $\delta(.)$ is unity, allows one to get the following expression:

$$\hat{P}(z) = \left(1 - \frac{p}{N} + \frac{p}{N} \hat{P}_J(z)\right)^N. \tag{3.7}$$

It can be useful to realize that since the increments are independent and the number of jumps n has a binomial distribution, the distribution P of the sum of these increments can also be written as:

$$\hat{P}(z) = \sum_{n=0}^{N} C_N^n \left(\frac{p}{N}\right)^n \left(1 - \frac{p}{N}\right)^{N-n} [\hat{P}_J(z)]^n [1]^{N-n}, \tag{3.8}$$

where [1] in the above expression comes from the Dirac peak at zero.

In the large N limit, Eq. (3.7) still simplifies to:

$$\hat{P}(z) = \exp -p(1 - \hat{P}_J(z)),$$ (3.9)

and defines a **Poisson jump process** of intensity p. By construction, this Poisson jump process is infinitely divisible, as is obvious from the above form of $\hat{P}(z)$: the $1/N$th power of $\hat{P}(z)$ corresponds to the same jump process with a reduced intensity p/N.

One can generalize the above construction by allowing the increments to be, with probability $1 - p/N$, Gaussian with variance σ^2/N instead of being zero. The process is then, in the limit $N \to \infty$, a mixture of a continuous Brownian motion interrupted by random jumps of finite amplitude. The average number of jumps is equal to $N \times p/N = p$. The corresponding characteristic function reads:

$$\hat{P}(z) = \exp -\frac{1}{2}\sigma^2 z^2 - p(1 - \hat{P}_J(z)).$$ (3.10)

Any infinitely divisible process can be written as a mixture of a continuous Brownian motion and a Poisson jump process. Therefore, any non Gaussian infinitely divisible process contains jumps and cannot be continuous in the continuous time limit.

Lévy and truncated Lévy processes

As an illustration, it is interesting to consider the case where jumps are power law distributed, as:

$$P_J(\delta x) = \frac{A^\mu}{2\mu |\delta x|^{1+\mu}}\,\Theta(|\delta x| - A),$$ (3.11)

where Θ is the Heaviside function, and $\mu < 2$. The quantity $1 - \hat{P}_J(z)$ reads in this case:

$$1 - \hat{P}_J(z) = \frac{A^\mu}{\mu}\int_A^\infty \frac{1 - \cos zu}{u^{1+\mu}}\,du.$$ (3.12)

Note that when $\mu < 2$, the above integral converges in the limit $A \to 0$, and is equal to $c_\mu |z|^\mu$, where c_μ is a certain constant. If we take the limit $A \to 0$ and $p \to \infty$ in Eq. (3.9) such as to keep the product pA^μ fixed, we discover that $\hat{P}(z)$ is precisely the characteristic function of a symmetric Lévy stable distribution. This is expected since the average number of jumps p tends to infinity when $A \to 0$, and therefore the variable X is an infinite sum of random increments of diverging variance.

In order to construct truncated Lévy processes, one simply introduces an exponential cut-off in the distribution of jumps given by Eq. (3.11), as:

$$P_J(\delta x) \simeq \frac{A^\mu}{2\mu |\delta x|^{1+\mu}}\exp(-\alpha|\delta x|)\Theta(|\delta x| - A),$$ (3.13)

where the normalization is only correct in the limit $\alpha A \to 0$. The quantity $1 - \hat{P}_J(z)$ reads in this case:

$$1 - \hat{P}_J(z) \simeq \frac{A^\mu}{\mu}\int_A^\infty \frac{(1 - \cos zu)e^{-\alpha u}}{u^{1+\mu}}\,du,$$ (3.14)

which, in the limit $A \to 0$, exactly reproduces Eq. (1.26). Therefore, the truncated Lévy process is by construction infinitely divisible, and has jumps in the continuous time limit.[†]

3.2 Functions of the Brownian motion and Ito calculus

3.2.1 Ito's lemma

Let us return to the case of Gaussian increments which, as shown above, leads to a continuous process $X(t)$ in the continuous time limit. Since in this limit, increments are vanishingly small, one can expect that some type of differential calculus could be devised. Consider a certain function $F(X, t)$, possibly explicitly time dependent, of the process X. How does $F(X, t)$ vary between t and $t + dt$? The usual rule of differential calculus is to write:

$$dF = \frac{\partial F}{\partial t} dt + \frac{\partial F}{\partial X} dX(t). \tag{3.15}$$

However, one can see that this formula is at best ambiguous, for the following reason. Take for example $F(X, t) = X^2$, and average the value of F over all realizations of the Gaussian process. By definition, one obtains the variance of the process, which is equal to $\sigma^2 t$. Therefore, $d \langle F \rangle \equiv \sigma^2 dt$. On the other hand, if we average Eq. (3.15) and note that for our particular case $\partial F / \partial t \equiv 0$, we obtain:

$$\langle dF \rangle = 2 \langle X(t) dX(t) \rangle = 2 \lim_{N \to \infty} \langle X_i \delta X_i \rangle, \tag{3.16}$$

where the last equality refers to the discrete construction of the Brownian motion. This construction assumes that each increment δX_i is an independent random variable, in particular independent of X_i since the latter is the sum of increments *up to* $i - 1$ but excluding i. Since we assume that δX_i has a zero mean, one would then finds $\langle dF \rangle = 2 \langle X_i \delta X_i \rangle = 0$, in contradiction of the direct result $d \langle F \rangle \equiv \sigma^2 dt$.

This apparent contradiction comes from the fact that the notation $dX(t)$ is misleading, because the order of magnitude of $dX(t)$ is *not* dt, as would be in ordinary differential calculus, but of order \sqrt{dt}, which is much larger. A way to see this is again to refer to the discrete case where the variance of the increments is σ^2 / N. Since dt is the limit when N is large of $1/N$, the order of magnitude of the elementary increments which become dX in the continuous time limit is indeed $\sigma / \sqrt{N} = \sigma \sqrt{dt}$.[‡] Therefore, although the Brownian motion is continuous, it is not differentiable since $dX/dt \to \infty$ when $dt \to 0$.

This has a very important consequence for the differential calculus rules since terms of order dX^2 are of first order in dt, and should be carefully retained. Let us illustrate this in details in the above case $F(X, t) = F(X) = X^2$, by first concentrating on the discrete case. Since $F(x_i) = x_i^2$, one has:

$$F(x_{i+1}) - F(x_i) = (x_{i+1} + x_i)(x_{i+1} - x_i) = (2x_i + \delta x_i)\delta x_i = 2x_i \delta x_i + \delta x_i^2. \tag{3.17}$$

[†] See Koponen (1995).
[‡] In the following, we choose $T = 1$.

The first term is the $F'(x)dx$ expected from differential calculus. The second is a random variable of mean equal to $\langle \delta x_i^2 \rangle = \sigma^2/N$ and variance $\langle \delta x_i^4 \rangle - \langle \delta x_i^2 \rangle^2 = 2\sigma^4/N^2$ (recall that δx is Gaussian). Therefore, one can write:

$$\delta x_i^2 = \frac{\sigma^2}{N} + \frac{\sqrt{2}\sigma^2}{N}\xi_i, \tag{3.18}$$

where ξ is a random variable of zero mean and unit variance. The point now is that the effect of ξ is in fact negligible in the limit $N \to \infty$, and that this last term can be dropped. Indeed, if we sum Eq. (3.17) from $i = 0$ to $i = N - 1$, we find:

$$\sum_{i=0}^{N-1} F(x_{i+1}) - F(x_i) = \sum_{i=0}^{N-1} 2x_i\delta x_i + \sigma^2 + \frac{\sqrt{2}\sigma^2}{N}\sum_{i=0}^{N-1}\xi_i. \tag{3.19}$$

But the last term, using the CLT, is of order $\sqrt{2N}\sigma^2/N$ and vanishes in the large N limit. It therefore does not contribute, in that limit, to the evolution of F. One can see this at work on Fig. 3.1, where we have plotted $\sum_{i=1}^{n} \delta x_i^2$ as a function of n for $N = 100$, 1000 and 10000. One observes very clearly that for large N, $\sum_{i=1}^{n} \delta x_i^2$ is very well approximated by $\sigma^2 n/N$. This is in fact yet another illustration of the CLT.

Coming back to Eq. (3.17), we conclude that in the continuum limit where the random term containing ξ can be discarded in Eq. (3.18), one can write:

$$dF = 2XdX + \sigma^2 dt. \tag{3.20}$$

One now finds the correct result for $\langle dF \rangle$, namely $\sigma^2 dt$.

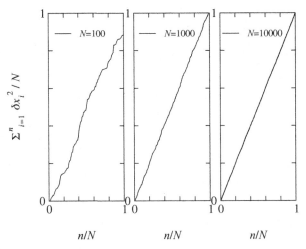

Fig. 3.1. Illustration of the Ito lemma at work. In the continuum limit the random variable $\int_0^t dX^2$ becomes equal to $\sigma^2 t$ with certainty. In this example, we have shown $\sum_{i=1}^{n} \delta x_i^2$, for increments δx_i with variance $1/N$. As N goes to infinity we approach the continuum limit, where the fluctuating sum becomes a straight line.

Turning now to the general case where F is arbitrary, one finds:

$$F(x_{i+1}) = F(x_i + \delta x_i) = F(x_i) + F'(x_i)\delta x_i + \frac{1}{2}F''(x_i)\delta x_i^2 + \cdots \tag{3.21}$$

where the terms neglected are of order δx^3, that is $(dt)^{3/2}$ in the continuous time limit, and therefore much smaller than dt. The same argument as above suggests that one can neglect the random part in δx_i^2. This comes from the fact that one can, if F is sufficiently regular, apply the CLT to the sum $\sum_{i=0}^{N-1} F''(x_i)\xi_i$, and find, as above, that the contribution of this term vanishes as $N^{-1/2}$. A more rigorous proof constitutes the famous **Ito lemma**, which makes precise the so-called **stochastic differential** rule for a general function $F(X, t)$:

$$dF = \frac{\partial F}{\partial t}dt + \frac{\partial F}{\partial X}dX(t) + \frac{\sigma^2}{2}\frac{\partial^2 F}{\partial x^2}dt, \tag{3.22}$$

where the last term is known as the Ito correction which appears because dX is of order \sqrt{dt} and not dt. The above Ito formula is at the heart of many applications, in particular in mathematical finance. Its use for option pricing will be discussed in Chapter 16.

Note that Ito's formula (3.22) is not time reversal invariant. Assume again that F does not explicitly depend on time. Reversing the arrow of time changes dX into $-dX$, but keeps the Ito correction unchanged, so that dF does not become $-dF$, as one might have expected. The reason for this is the explicit choice of an arrow of time when one assumes that δX_i is chosen *after* the point X_i is known.

One can extend in a straightforward way the above formula to the case where F depends on several, possibly correlated Brownian motions $X_\alpha(t)$, with $\alpha = 1, \ldots, M$:

$$dF = \frac{\partial F}{\partial t}dt + \sum_{\alpha=1}^{M}\frac{\partial F}{\partial X_\alpha}dX_\alpha(t) + \sum_{\alpha,\beta=1}^{M}\frac{C_{\alpha\beta}}{2}\frac{\partial^2 F}{\partial X_\alpha \partial X_\beta}dt, \tag{3.23}$$

with $C_{\alpha\beta} = \langle dX_\alpha dX_\beta \rangle$.

3.2.2 Novikov's formula

One can derive an important formula for an arbitrary function $G(\xi_1, \xi_2, \ldots, \xi_M)$ of correlated Gaussian variables $\xi_1, \xi_2, \ldots, \xi_M$, which reads:

$$\langle \xi_i G \rangle = \sum_{j=1}^{M}\langle \xi_i \xi_j \rangle \left\langle \frac{\partial G}{\partial x_j} \right\rangle. \tag{3.24}$$

This formula can easily be obtained by using the multivariate Gaussian distribution:

$$P(\xi_1, \xi_2, \ldots, \xi_M) = \frac{1}{Z}\exp{-\frac{1}{2}\sum_{k,\ell=1}^{M}\xi_k C_{k\ell}^{-1}\xi_\ell}, \tag{3.25}$$

where the C is the correlation matrix $C_{k\ell} \equiv \langle \xi_k \xi_\ell \rangle$, and Z a normalization factor equal to $(2\pi)^{M/2}\sqrt{\det \mathbf{C}}$. Therefore, the following identity is true:

$$\xi_i P \equiv -\sum_{j=1}^{M} C_{ij} \frac{\partial P}{\partial \xi_j}, \tag{3.26}$$

as can be seen by inspection, using the fact that $\sum_j C_{ij} C_{j\ell}^{-1} \equiv \delta_{i\ell}$. Inserting this formula in the integral defining $\langle \xi_i G \rangle$, and integrating by parts over all variables ξ_j, one finally ends up with Eq. (3.24).

3.2.3 Stratonovich's prescription

As an application of the Novikov's formula, we consider another construction of the continuous time Brownian motion, where the increments dX have a small but non zero correlation time τ that we will take infinitely small after the continuous time limit $dt \to 0$ has been taken. If $\tau \neq 0$, then X becomes differentiable and we choose:

$$\left\langle \frac{dX(t)}{dt} \frac{dX(t')}{dt'} \right\rangle = \frac{\sigma^2}{\tau} g\left(\frac{|t - t'|}{\tau}\right), \tag{3.27}$$

where g is a certain decaying function that we normalize as

$$2 \int_0^\infty g(u) du = 1. \tag{3.28}$$

The reason of this normalization is the following: since $X(t) = \int_0^t dX(t')$, one has:

$$\langle X(t)^2 \rangle = \left\langle \left(\int_0^t \frac{dX(t')}{dt'} dt' \right)^2 \right\rangle = \int_0^t \int_0^t \left\langle \frac{dX(t')}{dt'} \frac{dX(t'')}{dt''} \right\rangle dt' dt''. \tag{3.29}$$

Using (3.27) one then has:

$$\langle X(t)^2 \rangle = \frac{\sigma^2}{\tau} \int_0^t \left(\int_{-t'}^{t-t'} g\left(\frac{|t'' - t'|}{\tau}\right) dt'' \right) dt'. \tag{3.30}$$

Changing variable to $u = |t'' - t'|/\tau$ and assuming that τ is very small, one finally finds:

$$\langle X(t)^2 \rangle = \sigma^2 t \int_{-\infty}^\infty g(|u|) du, \tag{3.31}$$

which we want to identify with $\sigma^2 t$ in order to keep the same variance as for the above construction of the Brownian motion.

 Since X is now differentiable (because τ is finite), one can apply the usual differential rule Eq. (3.15) to any function $F(X, t)$. But now it is no longer true, in our above example where $F(X, t) = X^2$, that $X(t)$ and $dX(t)$ are uncorrelated, which was the origin of the contradiction discussed above. The average of the product $F'(X)dX(t)$ can actually be computed using the continuous time version of Novikov's formula, where the role of the ξ's is played by the dX/dt's. Noting that $\partial X(t)/\partial dX(t') = 1$ when $t > t'$ and 0

Table 3.1. Differences between Ito and Stratonovich prescriptions.

Ito	Stratonovich
$\langle dX^2 \rangle = \sigma^2 dt$	$\langle dX^2 \rangle = \sigma^2 dt$
$\langle X dX \rangle = 0$	$\langle X dX \rangle = \sigma^2 dt/2$
$dF(X,t) = \frac{\partial F}{\partial X}dX + \frac{\partial F}{\partial t}dt + \frac{\sigma^2}{2}\frac{\partial^2 F}{\partial X^2}dt$	$dF(X,t) = \frac{\partial F}{\partial X}dX + \frac{\partial F}{\partial t}dt$
functions are evaluated before the increments	functions are evaluated simultaneously to the increments
process dF not time reversal invariant	process dF time reversal invariant
standard in finance	standard in physics

otherwise, one has, using Eq. (3.24):

$$\left\langle F'(X(t),t)\frac{dX(t)}{dt} \right\rangle = \int_0^t \left\langle \frac{dX(t)}{dt}\frac{dX(t')}{dt'} \right\rangle \langle F''(X(t),t)\rangle dt', \qquad (3.32)$$

or, using again (3.27) and the change of variable $u = |t - t'|/\tau$,

$$\left\langle F'(X(t),t)\frac{dX(t)}{dt} \right\rangle = \langle F''(X(t),t)\rangle \, \sigma^2 \int_0^\infty g(u)du = \frac{\sigma^2}{2}\langle F''(X(t),t)\rangle, \qquad (3.33)$$

because the integral over u is only half of what is needed to obtain the full normalization of g. Note that the last term is exactly the average of the Ito correction term. So, in this case, the average of Eq. (3.15) where the Ito term is absent is identical to the average of the Ito formula, which assumes that $dX(t)$ is independent of the past. This point of view (no correction term but a small correlation time) is called the **Stratonovich pre-scription**. *A summary of the differences between the Ito and Stratonovich prescriptions is given in Table 3.1.*

3.3 Other techniques

3.3.1 Path integrals

Let us now discuss another important aspect of random processes, concerning the probability to observe a whole trajectory, and not only its 'end point' X. Again, it is more convenient to think first in discrete time. The probability to observe a certain trajectory $\mathcal{T} = \{x_1, x_2, \ldots, x_N\}$, knowing the starting point x_0 is, for iid increments, given by:

$$\mathcal{P}(\mathcal{T}) = \prod_{i=0}^{N-1} P_1(x_{i+1} - x_i), \qquad (3.34)$$

where P_1 is the elementary distribution of the increments. It often happens that one has to compute the average over all trajectories of a quantity \mathcal{O} that depends on the whole trajectory. In simple cases, this quantity is expressed only in terms of the different positions

x_i along the trajectory, for example:

$$\mathcal{O} = \exp \sum_{i=0}^{N-1} V(x_i), \tag{3.35}$$

where $V(x)$ is a certain function of x. An example of this form arises in the context of the interest rate curve and will be given in Chapter 19. The average is thus expressed as:

$$\langle \mathcal{O} \rangle_N = \int \prod_{i=0}^{N-1} \left[P_1(x_{i+1} - x_i) e^{V(x_i)} dx_i \right]. \tag{3.36}$$

It is interesting to introduce a restricted average, where only trajectories *ending* at a given point x are taken into account:

$$O(x, N) \equiv \int \delta(x_N - x) \prod_{i=0}^{N-1} \left[P_1(x_{i+1} - x_i) e^{V(x_i)} dx_i \right], \tag{3.37}$$

with, obviously, $\langle \mathcal{O} \rangle_N = \int O(x, N) dx$. From the above definitions, one can write the following recursion rule to relate the quantities at 'time' $N + 1$ and N:

$$O(x, N + 1) = e^{V(x)} \int P_1(x - x') O(x', N). \tag{3.38}$$

This equation is exact. It is interesting to consider its continuous time limit, which can be obtained by assuming that V is of order dt (i.e. $V = \hat{V} dt$), and that P_1 is very sharply peaked around its mean $m\,dt$, with a variance equal to $\sigma^2 dt$, with dt very small. Now, we can Taylor expand $O(x', N)$ around x in powers of $x' - x$. If all the rescaled moments $\hat{m}_n = m_n / \sigma^n$ of P_1 are finite, one finds:

$$O(x, N + 1) = e^{dt \hat{V}(x)} \Bigg(O(x - mdt, N)$$
$$+ \sum_{n=2}^{\infty} \frac{(-1)^n \hat{m}_n \sigma^n}{n!} \frac{\partial^n O(x - mdt, N)}{\partial x^n} (dt)^{n/2} \Bigg). \tag{3.39}$$

Expanding this last formula to order dt and neglecting terms of order $(dt)^{3/2}$ and higher, we find:

$$\lim_{dt \to 0} \frac{O(x, N + 1) - O(x, N)}{dt} \equiv \frac{\partial O(x, T)}{\partial T}$$
$$= -m \frac{\partial O(x, T)}{\partial x} + \frac{\sigma^2}{2} \frac{\partial^2 O(x, T)}{\partial x^2} + \hat{V}(x) O(x, T). \tag{3.40}$$

Therefore, the quantity $O(x, t)$ obeys a partial differential 'diffusion' equation, with a source term equal to $\hat{V}(x) O(x, t)$. One can thus interpret $O(x, t)$ as a density of Brownian walkers which 'proliferate' or disappear with an x-dependent rate $\hat{V}(x)$, i.e. each walker has a probability $\hat{V}(x) dt$ to give one offspring if $\hat{V}(x) > 0$ or to disappear if $\hat{V}(x) < 0$.

We have thus shown a very interesting result: the quantity defined as a path integral, Eq. (3.37) obeys a certain partial differential equation. Conversely, it can be very useful to represent the solution of a partial differential equation of the type (3.40) as a path integral, which can easily be simulated using Monte-Carlo methods, using for example the language of Brownian walkers alluded to above. The continuous limit of Eq. (3.37) can be constructed when P_1 is Gaussian, in which case:

$$\prod_{i=0}^{N-1} P_1(x_{i+1} - x_i)e^{V(x_i)} = \prod_{i=0}^{N-1} \exp\left(-\frac{1}{2\sigma^2 dt}(x_{i+1} - x_i - m dt)^2 + dt \hat{V}(x_i)\right)$$

$$\xrightarrow{dt \to 0} \exp\left(-\int_0^T \left[\frac{1}{2\sigma^2}\left(\frac{dx}{dt} - m\right)^2 - \hat{V}(x)\right] dt\right). \qquad (3.41)$$

Therefore, the quantity $O(x, T)$, defined as:

$$O(x, T) = \int_{x(0)=x_0}^{x(T)=x} \exp\left(-\int_0^T \left[\frac{1}{2\sigma^2}\left(\frac{dx}{dt} - m\right)^2 - \hat{V}(x)\right] dt\right) \mathcal{D}x, \qquad (3.42)$$

where $\mathcal{D}x$ is the differential measure over paths and where the paths are constrained to start from point x_0 and arrive at point x, is the solution of the partial differential equation (3.40) with initial condition $O(x, T = 0) = \delta(x - x_0)$. This is the Feynmann–Kac path integral representation.

Note that one can generalize this construction to the case where the mean increment m and the variance σ^2 depend both on x and t.

3.3.2 Girsanov's formula and the Martin–Siggia–Rose trick (*)

Let us end this rather technical chapter on two variations on path integral representations. Suppose that one wants to compute the average of a certain quantity \mathcal{O} over biased Brownian motion trajectories, with a certain local bias $m(x, t)$:

$$\langle \mathcal{O} \rangle_m \equiv \int \prod_t \mathcal{O}(x, t) \exp\left(\int_0^T \frac{1}{2\sigma^2}\left(\frac{dx}{dt} - m(x, t)\right)^2 dt\right) \mathcal{D}x. \qquad (3.43)$$

Expanding the square, we see that this average can be seen as an average over an ensemble of unbiased Brownian motion trajectories of a modified observable $\hat{\mathcal{O}}$ that reads:

$$\hat{\mathcal{O}}(x, t) = \mathcal{O}(x, t) \exp\left[\frac{m(x, t)}{\sigma^2}\frac{dx}{dt} - \frac{m(x, t)^2}{2\sigma^2}\right], \qquad (3.44)$$

such that

$$\langle \mathcal{O} \rangle_m = \langle \hat{\mathcal{O}} \rangle_0. \qquad (3.45)$$

This is called the **Girsanov formula** (see e.g. Baxter and Rennie (1996)).

Another 'trick', that can be useful in some circumstances, is to introduce an auxiliary dynamical variable $\hat{x}(t)$ and to rewrite:

$$\exp\left(\int_0^T -\frac{1}{2\sigma^2}\left(\frac{dx}{dt}-m(x,t)\right)^2 dt\right)$$

$$\propto \int \exp\left(\int_0^T \left[-\frac{\sigma^2}{2}\hat{x}(t)^2 + i\left(\frac{dx}{dt}-m(x,t)\right)\hat{x}(t)\right]dt\right)\mathcal{D}\hat{x}, \tag{3.46}$$

where $i^2 = -1$ in the above formula. This transformation, plugged in the general path integral (3.42), is usually called the Martin-Siggia-Rose representation (see e.g. Zinn–Justin (1989)).

3.4 Summary

- An infinitely divisible process is one that can be decomposed as an infinite sum of iid increments. The only infinitely divisible continuous process is continuous time Brownian motion. Other infinitely divisible processes, such as Lévy and truncated Lévy processes, contain discontinuities (jumps).
- The time evolution of a function of continuous time Brownian motion is described by Ito's stochastic calculus rules, Eq. (3.22). Compared to ordinary differential calculus there appears an extra second order differential term.

- FURTHER READING

M. Baxter, A. Rennie, *Financial Calculus, An Introduction to Derivative Pricing*, Cambridge University Press, Cambridge, 1996.

N. Ikeda, S. Watanabe, M. Fukushima, H. Kunita, *Ito's stochastic calculus and probability theory*, Tokyo, Springer, 1996.

P. Lévy, *Théorie de l'addition des variables aléatoires*, Gauthier Villars, Paris, 1937–1954.

A. Papoulis, *Probability, Random Variables, and Stochastic Processes*, 3rd edn, McGraw-Hill, New York, 1991.

N. G. van Kampen, *Stochastic processes in physics and chemistry*, North Holland, 1981.

J. Zinn-Justin, *Quantum field theory and critical phenomena*, Oxford University Press, 1989.

- SPECIALIZED PAPER

I. Koponen, *Analytic approach to the problem of convergence to truncated Lévy flights towards the Gaussian stochastic process*, Physical Review, **E 52**, 1197 (1995).

4

Analysis of empirical data

Les lois générales de la nature sont un ensemble d'exceptions non exceptionnelles, et,
par conséquent, sans aucun intérêt; l'exception exceptionnelle seule ayant un intérêt.[†]

(Alfred Jarry)

4.1 Estimating probability distributions

4.1.1 Cumulative distribution and densities – rank histogram

Imagine that one observes a series of N realizations of the random iid variable X,
$\{x_1, x_2, \ldots, x_N\}$, drawn from an unknown source of which one would like to determine
either its distribution density or its cumulative distribution. To determine the distribution
density using empirical data, one has to choose a certain width w for the bins to con-
struct frequency histograms. Another possibility is to smooth the data using, for example,
a Gaussian with a certain width w. More precisely, an estimate of the probability density
$P(x)$ can be obtained by computing:

$$P(x) = \frac{1}{N\sqrt{2\pi w^2}} \sum_{k=1}^{N} \exp\left(-\frac{(x_k - x)^2}{2w^2}\right).$$
(4.1)

(Note that by construction $\int P(x)\mathrm{d}x = 1$.) Even when the width w is carefully chosen,
part of the information is lost. It is furthermore difficult to characterize the tails of the
distribution, corresponding to rare events, since most bins in this region are empty – one
should in principle choose a position dependent bin width $w(x)$ to account for the rarefaction
of events in the tails.

On the other hand, the construction of the cumulative distribution does not require to
choose a bin width. The trick is to order the observed data according to their values, for
example in decreasing order. The value x_k of the kth variable (out of N) is then such that:

$$\mathcal{P}_>(x_k) = \frac{k}{N + 1}.$$
(4.2)

[†] The laws of nature are a collection of non exceptional exceptions, and therefore devoid of interest. Only
exceptional exceptions are genuinely interesting.

This result comes from the following observation: if one were to draw an $(N + 1)$th random variable from the same distribution, this new variable would have a probability $1/(N + 1)$ of being the largest, a probability $1/(N + 1)$ of being the second largest, and so forth. In other words, there is an *a priori* equal probability $1/(N + 1)$ that it would fall within any of the $N + 1$ intervals defined by the previously drawn variables. The probability that it would fall above the kth one, x_k is therefore equal to the number of intervals beyond x_k, which is equal to k, times $1/N + 1$. This is also equal, by definition, to $\mathcal{P}_>(x_k)$, hence Eq. (4.2). Unfortunately in this construction, we do not have an empirical value for $\mathcal{P}_>(x)$ when x is not one of the N observed points, we must therefore interpolate the distribution on the intermediate points.

Another approach is to make the hypothesis that the observed values were equally probable and that the unobserved values have zero probability, hence

$$P(x) = \frac{1}{N} \sum_{k=1}^{N} \delta(x_k - x), \tag{4.3}$$

which can be though as the limit $w \to 0$ of Eq. (4.1). In this construction $\mathcal{P}_>(x)$ is ill-defined for the points $\{x_k\}$ and is piecewise constant everywhere else,

$$\mathcal{P}_>(x) = \frac{k}{N} \qquad \text{for} \quad x_k < x < x_{k+1}. \tag{4.4}$$

Yet another, slightly different way to state this is to notice that, for large N, the most probable value $\Lambda^*[k]$ of x_k is such that $\mathcal{P}_>(\Lambda^*[k]) = k/N$: see also the discussion in Section 2.1, and Eq. (2.12).

Since the rare event part of the distribution is of particular interest, it is convenient to choose a logarithmic scale for the probabilities. Furthermore, in order to check visually the symmetry of the probability distributions, we have in the following systematically used $\mathcal{P}_<(-\delta x)$ for the negative increments, and $\mathcal{P}_>(\delta x)$ for positive δx.

4.1.2 Kolmogorov–Smirnov test

Once the empirical cumulative distribution is determined, one usually tries to find a mathematical distribution which reproduces as accurately as possible the empirical data. An interesting measure of the goodness of the fit is provided by the **Kolmogorov-Smirnov test**. Let $\mathcal{P}_{>,e}(x)$ be the empirical distribution obtained using the second rank ordering technique above, Eq. (4.4). This function jumps by an amount $1/N$ every time a data point is encountered, and is constant otherwise. Now, let $\mathcal{P}_{>,th}(x)$ be the theoretical cumulative distribution to be tested. Kolmogorov and Smirnov define the distance D between the two distributions as the maximum value of their absolute difference:

$$D = \max_x |\mathcal{P}_{>,th}(x) - \mathcal{P}_{>,e}(x)|. \tag{4.5}$$

Since $\mathcal{P}_{>,th}(x)$ is monotone, this maximum is reached for a value of $x = x_k$ belonging to the data set. Imagine that the data points are indeed chosen according to the theoretical

distribution $\mathcal{P}_{>,th}(x)$. The distance D is then a random quantity that depends on the actual series of x_k, but the distribution of D is exactly known when N is large and is, remarkably, totally independent of the shape of $\mathcal{P}_{>,th}(x)$! The cumulative distribution of D is given by:

$$\mathcal{P}_>(D) = Q_{KS}(\sqrt{N}D) \qquad Q_{KS}(u) = 2\sum_{n=1}^{\infty}(-1)^{n-1}\exp(-2n^2u^2). \tag{4.6}$$

This formula means that *if* $\mathcal{P}_{>,th}(x)$ is the correct distribution, then the distance D between the empirical cumulative distribution and the theoretical distribution should decay for large N as u/\sqrt{N}, where the random variable u has a known distribution, Q_{KS}. In practice, one computes from the data the quantity D from which one deduces, using Eq. (4.6), the probability $\mathcal{P}_>(D)$. If this number is found to be small, say 0.01, then the hypothesis that the two distributions are the same is very unlikely: the hypothesis can be rejected with 99% confidence.

Note that one can also compare two empirical distributions using the Kolmogorov-Smirnov test, without knowing the true distribution of any of them. This is very interesting for financial applications. One can ask whether the distribution of returns in a given period of time (say during the year 1999) is or is not the same as the distribution of returns the following year (2000). In this case, one constructs the distance D as:

$$D = \max_{x}|\mathcal{P}_{>,e1}(x) - \mathcal{P}_{>,e2}(x)|, \tag{4.7}$$

where the index 1, 2 refers to the two samples, of sizes N_1 and N_2. The cumulative distribution of D is now given by:

$$\mathcal{P}_>(D) = Q_{KS}\left(\sqrt{\frac{N_1 N_2}{N_1 + N_2}}D\right). \tag{4.8}$$

The Kolmogorov-Smirnov test is interesting because it gives *universal* answers, independent of the shape of the tested distributions. However, this test, as any other, has its weaknesses. In particular, it is clear that by construction, $|\mathcal{P}_{>,th}(x) - \mathcal{P}_{>,e}(x)|$ is zero for $x \to \pm\infty$ since in these cases, both quantities tend to 0 or 1. Therefore, the value of x that maximizes this difference is generically around the median, and the Kolmogorov-Smirnov test is not very sensitive to the quality of the fit in the tails – a potentially crucial region. It is thus interesting to look for other goodness of fit criteria.

4.1.3 Maximum likelihood

Suppose that $P_\lambda(x)$ denotes a family of probability distribution parametrized by one or several parameters, collectively noted λ. The *a priori* probability to observe the particular series $\{x_1, x_2, \ldots, x_N\}$ chosen according to $P_\lambda(x)$ is proportional to:

$$P_\lambda(x_1)P_\lambda(x_2)\ldots P_\lambda(x_N). \tag{4.9}$$

The **most likely** value λ^* of λ is such that this *a priori* probability is maximized. Taking for example:

- $P_\lambda(x)$ to be a Gaussian distribution of mean m and variance σ^2, one finds that the most likely values of m and σ^2 are, respectively, the empirical mean and variance of the data points. Indeed, the likelihood reads:

$$P_\lambda(x_1)P_\lambda(x_2)\dots P_\lambda(x_N) = e^{-\frac{N}{2}\log(2\pi\sigma^2)-\frac{1}{2\sigma^2}\sum_{k=1}^{N}(x_k-m)^2}. \tag{4.10}$$

Maximizing the term in the exponential with respect to m and σ^2, one finds:

$$\sum_{i=1}^{N}(x_k-m) = 0 \quad \text{and} \quad \frac{N}{\sigma^2} - \frac{1}{\sigma^4}\sum_{i=1}^{N}(x_k-m)^2 = 0, \tag{4.11}$$

which leads to the announced result.
- $P_\lambda(x)$ to be a power-law distribution:

$$P_\lambda(x) = \frac{\mu x_0^\mu}{x^{1+\mu}} \qquad x > x_0, \tag{4.12}$$

(with x_0 known), one has:

$$P_\lambda(x_1)P_\lambda(x_2)\dots P_\lambda(x_N) = e^{N\log\mu+N\mu\log x_0-(1+\mu)\sum_{k=1}^{N}\log x_k}. \tag{4.13}$$

The equation fixing μ^* is thus, in this case:

$$\frac{N}{\mu^*} + N\log x_0 - \sum_{k=1}^{N}\log x_k = 0 \Rightarrow \mu^* = \frac{N}{\sum_{k=1}^{N}\log(x_k/x_0)}. \tag{4.14}$$

In the above example, if x_0 is unknown, its most likely value is simply given by: $x_0 = \min\{x_1, x_2, \dots, x_N\}$. A related topic is the so-called **Hill estimator** for power-law *tails*. If one assumes that the random variable that one observes has power-law tails beyond a certain (unknown) value x_0, one can use the maximum likelihood estimator of $\mu = \mu^*$ given above using only the points exceeding the value x_0. This means that one only uses the rank ordered points x_k with $k \leq k^*$ such that $x_{k^*+1} > x_0$, and obtains:

$$\mu^*(k^*) = \frac{k^*}{\sum_{k=1}^{k^*}\log(x_k/x_{k^*})}. \tag{4.15}$$

This quantity, plotted as a function of k^*, should reveal a plateau region for $k^* \gg 1$, corresponding to the power-law regime of the distribution. We show in Figure 4.1 the quantity $\mu^*(k^*)$ for the Student distribution with $\mu = 3$, using 1000 points. This estimator indeed wanders (with rather large oscillations) around the value $\mu = 3$ before going astray when one reaches the core of the Student distribution, where the asymptotic power-law ceases to hold. A nicer looking estimate is obtained by averaging the previous estimator over a certain range of k^*.

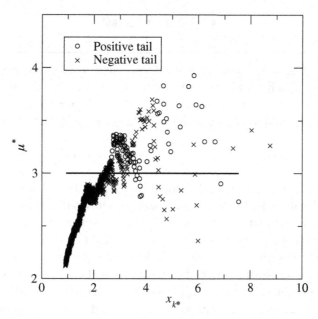

Fig. 4.1. Hill estimator $\mu^*(k^*)$ as a function of x_{k*} for a 1000 point sample of a $\mu = 3$ Student distribution. The two curves corresponds to the left and right tail. Note that for small x one leaves the tail region. The plain horizontal line is the expected value $\mu = 3$.

4.1.4 Relative likelihood

When the most likely value of the parameter is determined, one can assess the likelihood of the proposed distribution $P_{\lambda^*}(x)$, equal to $\prod_{i=1}^{N}(P_{\lambda^*}(x_i)dx)$, or more interestingly the log-likelihood per point \mathcal{L} (up to an additive constant involving dx):

$$\mathcal{L} = \frac{1}{N}\sum_{k=1}^{N} \log P_{\lambda^*}(x_k). \qquad (4.16)$$

This quantity is, for large N, equal to:

$$\mathcal{L} = \int P(x)\log P_{\lambda^*}(x)dx, \qquad (4.17)$$

where $P(x)$ is the 'true' probability. If $P_{\lambda^*}(x)$ is indeed a good approximation to $P(x)$, then the log likelihood per point is precisely minus the entropy of $P_{\lambda^*}(x)$. Note that the log-likelihood per point of the optimal Gaussian distribution is always equal to $\mathcal{L}_G = -1.419$ (for $\sigma = 1$), independently of P – see Eq. (2.28).

However, the log-likelihood per point has no intrinsic meaning, due to the additive constant that we swept under the rug above. In particular, a mere change of variables $y = f(x)$ changes the value of \mathcal{L}. On the other hand, the difference of log-likelihood for two competing distributions for the same data set *is* meaningful and invariant under a change of variable. If these two distributions are *a priori* equally likely, then the probability that the data is drawn from the distribution P_1^* rather than from P_2^* is equal to $\exp N(\mathcal{L}_1 - \mathcal{L}_2)$.

4.1.5 A general caveat

Although of genuine interest, statistical tests, such as the Kolmogorov-Smirnov test or the maximum likelihood method discussed above, usually provide a single number as a measure of the goodness of fit. This is in general dangerous since it does not give much information about *where* the fit is good, and where it is not so good.

As mentioned above, a theoretical distribution can pass the Kolmogorov-Smirnov test but be very bad in the tails. This is why the quality of a fit should always be assessed using a joint graphical representation of the data and of the theoretical fit, using different ways of plotting the data (log-log, linear-log, etc.) to emphasize different parts of the distribution. This procedure, although not systematic, allows one to spot obvious errors and to track systematic effects, and is often of great pragmatic value. For example, one can always fit the distribution of financial returns using Lévy stable distributions. The maximum likelihood method will provide a set of optimal parameters, and indeed Lévy stable distributions usually provide a good fit in the core region of empirical distributions. However, graphing the data in log-log coordinates soon reveals that the essence of Lévy stable distributions, namely the power-law tails with an exponent $\mu < 2$, accounts very badly for the empirical tails. It is rather surprising that the use of graphical representation is so rarely used in econometric literature, where tables of numbers, giving the results of statistical tests, are preferred. This state of matters strongly contrasts with what happens in physics, where a graphical assessment of a theory is often considered to be most relevant, and helps in discussing the limitations of a given model.

4.2 Empirical moments: estimation and error

4.2.1 Empirical mean

Consider the empirical mean of the series of iid random variables $\{x_1, x_2, \ldots, x_N\}$, given by $m_e = \sum_k x_k / N$. The **standard error** on m_e, noted Δm_e, defined by the standard deviation of m_e, is easy to compute. One finds:

$$\Delta m_e = \frac{\sigma}{\sqrt{N}}, \tag{4.18}$$

where σ is the theoretical variance of the random variable X. As is well known, the error on the mean goes to zero as the square root of the system size. This assumes that the variables are independent. Correlations slow down the convergence to the true mean, sometimes by changing N into a reduced number of effectively independent random variables, sometimes by even changing the power $1/2$ into a smaller power.

4.2.2 Empirical variance and MAD

Similarly, the error on the empirical variance $\Delta(\sigma_e^2)$ can be computed as its standard deviation across different samples. One finds, for large N:

$$\left[\Delta\left(\sigma_e^2\right)\right]^2 = \frac{\langle x^4\rangle - \langle x^2\rangle^2}{N} = \frac{\sigma^2}{N}(2+\kappa), \tag{4.19}$$

where κ is the kurtosis of $P(x)$. Therefore, provided κ is finite, the relative error on the variance (and hence on the standard deviation) also decays as $1/\sqrt{N}$. When the kurtosis is infinite, the above formula is meaningless: another measure of the error, such as the width of the variance distribution, must be used. Note that from the above formula, the error on the standard deviation is given by:

$$\frac{\Delta\sigma_e}{\sigma} = \frac{1}{2\sqrt{N}}\sqrt{2+\kappa} \approx \frac{1}{\sqrt{2N}}\left(1+\frac{1}{4}\kappa\right), \tag{4.20}$$

where the last expression holds for small κ. It is interesting to compare the above result on the variance error to that on the mean absolute deviation $E_{\rm abs}$. The mean square error on this quantity reads:

$$[\Delta E_{\rm abs,e}]^2 = \frac{1}{N}\left(\sigma^2 - E_{\rm abs}^2\right). \tag{4.21}$$

In order to make sense of this formula and compare it to the error on the standard deviation computed just above, we will use the small kurtosis expansion explained in Section 2.3.3, according to which $E_{\rm abs} = \langle |x| \rangle = \sqrt{2/\pi}(1 - \kappa/24)\sigma$. Plugging in this expression, we find that the relative error on the MAD reads:

$$\frac{\Delta E_{\rm abs,e}}{E_{\rm abs}} = \sqrt{\frac{\pi-2}{2N}}\left(1+\frac{\pi}{24(\pi-2)}\kappa\right). \tag{4.22}$$

One therefore finds that the relative error on the MAD in the Gaussian case ($\kappa = 0$) is 7% larger than that on the standard deviation, but the latter grows faster with the kurtosis than the MAD, which is less sensitive to tail events and therefore a more robust estimator in the presence of tail events.

4.2.3 Empirical kurtosis

One can similarly estimate the error on the kurtosis. Not surprisingly, this error can be expressed in terms of the eighth moment of the distribution (!). Even in the best case, when X is a Gaussian variable, the error on the empirical kurtosis is as large as $\sqrt{96/N}$.

4.2.4 Error on the volatility

Suppose that one observes the time series of a random variable X obtained by summing iid random increments δX. The total number of increments is N. The variable X is observed on a certain coarse time resolution M (say one day for financial applications), so that the statistics

of the coarse grained increments $x_{(k+1)M} - x_{kM}$ can be studied using N/M different values. The volatility of the process on scale M is the standard deviation of these increments. Therefore, using the above results, the error on the volatility on scale M using N/M data points, is $\sqrt{2 + \kappa_M}/2\sqrt{N/M}$, where κ_M is the kurtosis on scale M. This formula suggests at first sight that the volatility would be more precisely determined if N/M was increased, i.e. if the process was observed at higher frequencies. However, since the increments are iid, the kurtosis κ_M on scale M is κ_1/M. Therefore, the error on the volatility as a function of the observation scale M finally reads:

$$\frac{\Delta \sigma_e}{\sigma_e} = \frac{1}{2\sqrt{N}}\sqrt{2M + \kappa_1}. \tag{4.23}$$

Hence, the error on the volatility can be indeed made very small for a Gaussian process for which $\kappa_1 = 0$, by going to higher and higher frequencies ($M \to 1$). In the continuous time limit, one can in principle determine the volatility of the process with arbitrary precision. This is yet another aspect of the Ito formula. However, in the case of a kurtic process, the precision saturates and high frequency data does not add much as soon as $M \leq \kappa_1$.

4.3 Correlograms and variograms

4.3.1 Variogram

Consider a certain stationary process $\{x_1, x_2, \ldots, x_N\}$. A natural question to ask is: how 'far' does the process typically escape from itself between times k and $k + \ell$? This propensity to move can be quantified by the variance of the difference $x_{k+\ell} - x_k$:

$$\mathcal{V}(\ell) \equiv \langle (x_{k+\ell} - x_k)^2 \rangle, \tag{4.24}$$

which is called in this context the **variogram** of the process. Empirically, this quantity is simply obtained by computing the variance of the difference of all data points separated by a time lag ℓ. If the process is a random walk, the variogram $\mathcal{V}(\ell)$ grows linearly with ℓ, with a coefficient which is the square volatility of the increments. If the series $\{x_1, x_2, \ldots, x_N\}$ is made of iid random variables, the variogram is independent of ℓ (as soon as $\ell \neq 0$), and equal to twice the variance of these variables. An interesting intermediate situation is that of mean reverting processes, such that $x_{k+1} = \alpha x_k + \xi_k$, where $\alpha \leq 1$ and ξ_k are iid random variables of variance σ^2. For $\alpha = 1$, this is a random walk, whereas for $\alpha = 0$, the x's are themselves iid random variables, identical to the ξ's. Suppose that $x_0 = 0$; the explicit solution of the above recursion relation is $x_k = \sum_{k'=0}^{k-1} \alpha^{k-1-k'} \xi_{k'}$, from which the variogram is easily computed. In the limit $k \to \infty$ where the process becomes stationary, one finds:

$$\mathcal{V}(\ell) = 2\sigma^2 \frac{1 - \alpha^\ell}{1 - \alpha^2}. \tag{4.25}$$

The limit $\alpha = 1$ must be taken carefully to recover the random walk result $\sigma^2 \ell$. Setting $\alpha = 1 - \epsilon$ in the above expression, one finds, for $\epsilon \ll 1$:

$$\mathcal{V}(\ell) = \frac{\sigma^2}{\epsilon}(1 - e^{-\epsilon \ell}). \tag{4.26}$$

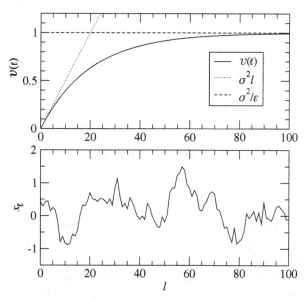

Fig. 4.2. Variogram of the Ornstein-Uhlenbeck process (top). The dotted line shows the short time linear behaviour, where the process is similar to a random walk with independent increments. The dashed line indicates the long term asymptotic value corresponding to independent random numbers. The bottom graph shows a realization of this process with the same correlation time (20 steps).

This is the variogram of a so-called **Ornstein-Uhlenbeck** process. When ℓ is much smaller than the correlation time $1/\epsilon$, one finds, as expected, the random walk result $\mathcal{V}(\ell) = \sigma^2 \ell$, whereas for $\ell \gg 1/\epsilon$, the variogram saturates at σ^2/ϵ (Fig. 4.2). On very large time scales, the variogram is therefore flat.

4.3.2 Correlogram

A related, well known quantity is the **correlation function** of the variables x_k. If the average value $\langle x \rangle$ of x_k is well defined, the correlation function (or **correlogram**) is defined as:

$$C(\ell) = \langle x_{k+\ell} x_k \rangle - \langle x \rangle^2. \tag{4.27}$$

Simple manipulations then show that:

$$\mathcal{V}(\ell) = 2\left(C(0) - C(\ell)\right). \tag{4.28}$$

However, in many cases, the average value of x is undefined (for example in the case of a random walk), or difficult to measure because when the correlation time is large, it is hard to know whether the process diffuses away to infinity or remains bounded (this is the case, for example, of the time series of volatility of financial assets). In these cases, it is much better to use the variogram to characterize the fluctuations of the process, since this quantity does not require the knowledge of $\langle x \rangle$. Another caveat concerning the empirical correlation function when the correlation time is somewhat large is the following. From the data, one

computes $C(\ell)$ as:

$$C_e(\ell) = \frac{1}{N - \ell} \sum_{k=1}^{N-\ell} \tilde{x}_k \tilde{x}_{k+\ell} \tag{4.29}$$

with

$$\tilde{x}_k = x_k - \frac{1}{N} \sum_{k=1}^{N} \tilde{x}_k. \tag{4.30}$$

From this definition, one finds the following exact sum rule:

$$\sum_{\ell=0}^{N} \left(1 - \frac{\ell}{N}\right) C_e(\ell) = 0 \tag{4.31}$$

Assume now that the true correlation time ℓ_0 is large compared to one and not very small compared to the total sample size N. The theoretical value of the left hand side should be of the order of $\sigma^2 \ell_0$. The above sum rule can only be satisfied in this case by having an empirical correlation function slightly negative when ℓ is a fraction of N, whereas the true correlation function should still be positive if ℓ_0 is large. Therefore, spurious anti-correlations are generated in this procedure.

4.3.3 Hurst exponent

Another interesting way to characterize the temporal development of the fluctuations is to study a quantity of direct relevance in financial applications, that measures the average span of the fluctuations. Following Hurst, one can define the average distance between the 'high' and the 'low' in a window of size $t = \ell\tau$:[†]

$$\mathcal{H}(\ell) = \langle \max(x_k)_{k=n+1,n+\ell} - \min(x_k)_{k=n+1,n+\ell} \rangle_n. \tag{4.32}$$

The **Hurst exponent** H is defined from $\mathcal{H}(\ell) \propto \ell^H$. In the case where the increments δX are Gaussian, one finds that $H \equiv \frac{1}{2}$ (for large ℓ). In the case of a TLD distribution of increments of index $1 < \mu < 2$, one finds (see the discussion in Section 2.3.6):

$$\mathcal{H}(\ell) \propto A\ell^{\frac{1}{\mu}} \quad (\ell \ll N^*); \qquad \mathcal{H}(\ell) \propto \sqrt{D\tau\ell} \quad (\ell \gg N^*). \tag{4.33}$$

The Hurst exponent therefore evolves from an anomalously high value $H = 1/\mu$ to the 'normal' value $H = \frac{1}{2}$ as ℓ increases.

Finally, for a fractional Brownian motion (defined in Section 2.5), one finds that $H = 1 - \nu/2$. In this case, the variogram also grows as $\mathcal{V}(\ell) \propto \ell^{1-\nu/2}$.

4.3.4 Correlations across different time zones

An effect that can be very important for risk management is the existence of non triv-ial correlations between different time zones. For example, there is a significant positive

[†] Note however that $\mathcal{H}(\ell)$ is different form the 'R/S statistics' considered by Hurst. The difference lies in an ℓ-dependent normalization, which in some cases can change the scaling with ℓ.

Table 4.1. Same day variance and covariance (in $\%^2$ per day). Data from July 1, 1999 to June 30, 2002.

	S&P 500	FTSE 100	Nikkei 225
S&P 500	1.62	0.64	0.23
FTSE 100		1.40	0.33
Nikkei 225			2.31

Table 4.2. Covariance between t and $t + 1$ day (in $\%^2$ per day). Data from July 1, 1999 to June 30, 2002. Numbers in boldface are statistically significant.

	S&P 500	FTSE 100	Nikkei 225
S&P 500 $(t + 1)$	0.01	0.02	−0.06
FTSE 100 $(t + 1)$	**0.42**	−0.01	−0.10
Nikkei 225 $(t + 1)$	**0.67**	**0.54**	−0.06

correlation between the U.S. market on day t and the European and Japanese markets on day $t + 1$, as shown in Table 4.2. This correlation can be strong: in the case of the Nikkei and the S&P 500, it is stronger between consecutive days than for the same (calendar) day. After a market closes, other world markets continue trading, taking into account (and generating) new information. For the closed market, this new information is only reflected in the opening price of the next day, generating a correlation between the returns of "late" markets and next day returns of "early" markets. On the other hand, the serial correlations of a (single time-zone) index with itself is very small. In the case of the S&P 500, one finds $\rho(t, t + 1) = 0.006$ for the period considered, with a standard error on this number of 0.03.

Due to these correlations across time zones, multi-time zone indices such as the MSCI world index can have significant serial correlations. On the same period, these correlations are found to be as large as 0.18 between two consecutive days.

These correlations are also important when one estimates the risk of a portfolio. The relevant quantities are not so much the daily variances and covariances but rather the variances and covariances of longer time scale (month, year). These can be estimated using daily data:

$$\Delta x \equiv \sum_{k=1}^{N} \delta x_k \qquad \Delta y \equiv \sum_{k=1}^{N} \delta y_k \tag{4.34}$$

$$
\begin{aligned}
\langle \Delta x \Delta y \rangle &= \sum_{k,\ell=1}^{N} \langle \delta x_k \delta y_\ell \rangle \\
&= N \langle \delta x_0 \delta y_0 \rangle + (N - 1)(\langle \delta x_0 \delta y_1 \rangle + \langle \delta x_1 \delta y_0 \rangle) + \cdots \\
&\approx N (\langle \delta x_0 \delta y_0 \rangle + \langle \delta x_0 \delta y_1 \rangle + \langle \delta x_1 \delta y_0 \rangle).
\end{aligned}
$$

One sees that the correlations between assets is increased by the presence of this time zone effect. To correctly account for correlations between assets in different time-zones, one should add to the daily covariance ($\langle \delta x_0 \delta y_0 \rangle$) the sum of the one-day-lagged covariances ($\langle \delta x_0 \delta y_1 \rangle + \langle \delta x_1 \delta y_0 \rangle$). This modified covariance can then be used to compute properly annualized variances. The same formula should also be used to compute the variance of a multi-time zone index or portfolio. In this case the two cross terms are equal and one should add $2 \langle \delta x_0 \delta x_1 \rangle$ to the daily variance ($\langle \delta x_0 \delta x_0 \rangle$). Failing to use formula (4.34) can greatly underestimate the annualized standard deviation of a world index or portfolio. In the case of the MSCI world index, the above formula leads to a annualized volatility of 17.6%, instead of 15.1% when neglecting the 'retarded' term.

4.4 Data with heterogeneous volatilities

Let us end this chapter on empirical data analysis by commenting on the problem of data sets with different volatilities, that very frequently occurs in financial analysis. Suppose for example that one studies the time series of different stocks, each one with a different volatility σ_i, with the expectation that the returns of these different stocks have the *same* distribution, up to a rescaling factor. In other words, one assumes that the return $\eta_i(k)$ of stock i at time k can be written as:

$$\eta_i(k) = \sigma_i \xi_i(k), \tag{4.35}$$

where σ_i is the volatility of stock i and $\xi_i(k)$ are random variables of unit variance, with a common distribution for all i's that one would like to determine. A similar situation would be that of a single asset when its volatility changes over time.

The naive way to proceed would be to rescale the empirical return $\eta_i(k)$ by the empirical volatility of stock i, i.e. to define scaled returns as:

$$\hat{\xi}_i(k) = \frac{\sqrt{N} \eta_i(k)}{\sqrt{\sum_{k'=1}^{N} (\eta_i(k'))^2}}, \tag{4.36}$$

where N is the common size of the time series and we have assumed that the empirical mean can be neglected. This definition however leads to a systematic underestimation of the tails of the distribution of the ξ's, which is obvious from the above definition, since even when one of the $\eta_i(k)$ is very large, $\hat{\xi}_i(k)$ is bounded by \sqrt{N}, because the same large $\eta_i(k)$ also contributes to the empirical variance.

A trick to get rid of this effect is to remove the contribution of $\eta_i(k)$ itself form the variance, and to define new rescaled variables as:[†]

$$\tilde{\xi}_i(k) = \frac{\sqrt{N-1} \eta_i(k)}{\sqrt{\sum_{k'=1, k' \neq k}^{N} (\eta_i(k'))^2}}. \tag{4.37}$$

[†] This trick is due to Martin Meyer, and was used in V. Plerou et al. (1999) to determine the distribution of returns of a large pool of stocks.

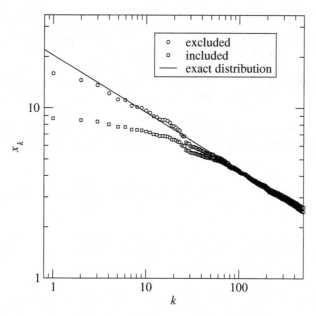

Fig. 4.3. Rank histogram of $\mu = 3$ Student variables, rescaled naively as in (4.36), or following the trick (4.37). The naive rescaling severely truncates the true tails, which are correctly captured using the $\tilde{\xi}$ variables.

This trick is illustrated in Figure 4.3, where $M = 400$ samples of $N = 100$ random variables are drawn according to a $\mu = 3$ Student distribution. The rank histogram of the naively scaled variables $\hat{\xi}_i(k)$ bends down for high ranks, and saturates around $\sqrt{N} = 10$, whereas the rank histogram of the $\tilde{\xi}_i(k)$ does indeed perfectly follow the expected behaviour of the $\mu = 3$ tail. Note however that the variance of $\tilde{\xi}_i(k)$ is positively biased by an amount proportional to $1/N$.

4.5 Summary

- This chapter offers guidelines useful for analyzing financial data: how to fit a probability distribution, determine an empirical variance, or a correlation function.
- The relative error on the mean, MAD, RMS, skewness and kurtosis of a distribution decreases like $1/\sqrt{N}$ (whenever higher moments of the distribution exist), but the prefactor can be quite large for high empirical moments (variance, fourth moment), especially when the true distribution is non-Gaussian.
- An important point is made in Section 4.1.5: although of genuine interest, statistical tests usually provide only a single measure of the goodness of fit. This is generally dangerous, as it does not give much information about *where* the fit is good, and where it is not so good. Representing the data graphically allows one to spot obvious errors and to track systematic effects, and is often of great pragmatic value.

- FURTHER READING

W. T. Vetterling, S. A. Teukolsky, W. H. Press, B. P. Flannery, *Numerical recipes in C: the art of scientific computing*, Cambridge University Press, 1993.

- SPECIALIZED PAPERS

B. Hill, *A simple general approach to inference about the tail of a distribution*, Ann. Stat. **3**, 1163 (1975).

V. Plerou, P. Gopikrishnan, L. A. Amaral, M. Meyer, H. E. Stanley, *Scaling of the distribution of price fluctuations of individual companies*, Phys. Rev., **E 60**, 6519 (1999).

5
Financial products and financial markets

> We should not be content to have the salesman stand between us – the salesman who
> knows nothing of what he is selling save that he is charging a great deal too much for it.
>
> (Oscar Wilde, *House Decoration.*)

5.1 Introduction

This chapter takes on finance. The first part offers an overview of the major products in modern finance. It describes some of the possible complex characteristics of each of these products, and suggests tricks on how to take these particularities into account in data analysis. The second part of the chapter turns to the practical functioning of financial markets: who are the players who buy and sell, and how are the markets organized. This chapter is essentially a source of definitions and references, covering the basic knowledge necessary to understand our book as a whole, knowledge familiar to practitioners in finance, but perhaps new to people outside the field. More details on the all the subjects alluded to in this chapter can be found in standard textbooks (see further reading at the end of the chapter).

5.2 Financial products

5.2.1 Cash (Interbank market)

Putting cash in the bank is the simplest form of investment. Banks pay interest rates on deposits as well as charge interest rates on loans. Interest is paid at the end of a given period, quoted in annual rate, although actual deposits and loans can have any lifetime. **Rate conventions** determine how the annual interest rate relates to the actual payment for the given time period. The most common rate convention is the **actual/360**. According to this convention, the amount of interest to be paid is equal to $P \times rT/360$ where P is the 'principal', r the annual rate, and T the number of calendar days (including week-ends and holidays) of the deposit/loan. Other conventions differ according to how the calendar is defined: 30/360, for example, differs from actual/360 in that it assumes all months to have thirty days, whereas the latter takes into account the exact number of investment days.

Table 5.1. Libor rates for selected currencies and maturities on
October 3, 2002.

Maturity	EUR	USD	JPY	GBP
Overnight	3.30%	1.77%	0.05%	3.97%
1 month	3.30%	1.80%	0.05%	3.95%
12 months	3.05%	1.70%	0.10%	3.97%

Reference services exist to inform customers of the daily standard interest rate. The most popular source is **Libor** (the London interbank offered rate) published by the BBA (British banker's association) every day at 11am London time. Libor is the standard interest rate for different currencies and different maturities (the varying investment lifetimes), which the BBA calculates based on up to 16 banks average after removing top and bottom quartile; see Table 5.1. In this calculation, the BBA uses the actual/360 convention rate, or, in the case of GBP, the actual/365. An alternative reference source, specific to the Euro currency, is the **Euribor** by EBF (European banking federation), which calculates its standard by surveying over 50 banks, also using the actual/360. Note that interest rates are quoted in percents. When comparing interest rates, comparisons are made in fractions of a percent, the difference is calculated in **basis points**, or bp. One basis point is equal to 0.01 % (10^{-4}); for example a .07 % difference in interest rate is a 7 basis point difference.

Rates are either given as yield rates, that apply to a loan starting now until a given expiry date, as in Table 5.1, but can also be described in terms of **forward rates**. A forward rate is the rate for a loan starting in the future, for a given period of time (typically 3 months). For example, the 1 year forward rate is the rate for a loan starting in a year from now, and that will expire in 1 year and 3 months. More details on forward rates and on their relation with the yield rates are given in Chapter 19.

There are seasonality effects in interest rates. The most striking one is that overnight interest rates can become very high over the end of the year. This then makes all forward rates higher if they include the year-end period. For example December Eurodollar and Euribor contracts are always lower (corresponding to higher rates) than the interpolation between the September and March contracts. Corporations and banks are reluctant to lend money over the year-end for balance sheet purpose.

Interest rates are linked with currencies but not with countries. It is possible to arbitrage a difference in interest rates in the same currency: borrow at the low rate and lend at the high rate. This operation carries a counterparty risk but no financial risk if the maturity of the loan and the deposit are the same. If the two rates are not in the same currency, then the above arbitrage is no longer risk free. If one borrows yen (low interest rates), converts the sum to British pound and makes a deposit (high rate), there is a risk that at the end of the period the yen has increased in value, creating a loss potentially greater than the gain from the interest rate difference.

Conversely, one has to take into account the difference in interest rates between the two traded currencies. For example a speculative bet that currency A will depreciate with respect

to currency B can be offset by the higher interest rates for loans in currency A respective to B.

5.2.2 Stocks

Companies with more than one owner are broken down into **stocks** (or shares) which represent a fractional ownership of a company. Generally speaking, the terms "stocks" and "shares" are subsumed under the term **equity**. In a public company, equity can be bought and sold freely in the financial markets, whereas in a private company one must contact the owners directly, and the transaction might be governed (or ruled impossible) by private ownership rules. As no data is available on private companies, this book deals exclusively with shares of public companies.

Sometimes more than one type of share is available for a given company. They might differ in voting rights or dividend. For example, **preferred stocks** differ from **common stocks** in that they are entitled to a fixed dividend that must be paid first if the company pays dividend. Preferred shares also have precedence over common shares in case the company is liquidated. On the other hand, common shares can receive an unlimited amount of dividends and carry voting rights.

A company that makes a profit can do one of two things, either reinvest the profit (build up infrastructure, buy other companies, etc.) or return the profit to its owners in the form of **dividend** payments. Companies that pay dividends do so regularly, the standard frequency varying from country to country (e.g., quarterly in the US, annually in Italy). Since share-holders of a publicly traded company change constantly, a mechanism exists to determine who will receive the next dividend, the seller or the buyer. Hence we have the **ex-dividend date**: on this date, the stock is traded without the dividend right. In other words, on or after this date, the new buyer has no claim to the upcoming dividend, and whoever owned the stock the evening prior to the ex-dividend date is entitled to the dividend.

Between the closing of the market prior to an ex-dividend date and the opening on the ex-date, a stock will lose a part of its value, namely the dividend amount. The previous owner of the stock does not lose anything on the ex-dividend date, because although the stock itself is lowered in value (now detached from its dividend), the owner still owns the right to a divident payment.

Note that the drop in price of a stock on the ex-date is hard to detect since by normal daily fluctuations, a price can go up and down much larger amounts (i.e., 2–5%) than a typical dividend (0.5%).

Still dividend payments add up, and must be taken into account when computing the long term returns on dividend paying stocks (see Section 5.2.3).

Dividends are associated with rather complex tax issues. In most countries dividend income is not taxed at the same rate as capital gains. Investors who are residents of the same country as the dividend paying company often receive a tax credit along with the dividend, so as to avoid the double taxation of profits. Take the example of a French corporation that has made a 1 EUR/share pre-tax profit, it will pay about 0.34 EUR in corporate tax and may then pay a 0.66 EUR dividend. A French investor would receive a

50% tax credit along with the dividend ($1.5 \times 0.66 \approx 1$) and pay income tax on the whole amount.

Other **corporate events** can change the price of a stock.[†] A **stock split** is when every share of a company gets multiplied to become n shares (e.g. 1.5, 2, 3 or 10). Reverse-split (rarer) is when the factor is smaller than one–the price then increases. Splits are used to bring the price per share in an acceptable range. In the US, it is customary to have share price between $10 and $100. Splits are less common in Europe, where prices can be as low as 0.40 EUR (Finmecanica Spa) or as high as 240 EUR (Electrabel). From the investors point of view, the payment of dividends in shares (**stock dividends**) is very similar to splits with a multiplier very close to 1, although at the firm's accounting level the two operations are quite different. A **spin-off** is when a company separates itself from one of its subsidiaries by giving shares of the newly independent company to its shareholders. Finally, when a company issues new shares, it often grants existing shareholders the right to buy new shares at a discounted price in proportion to their current holdings. This **right** has a monetary value and can often be bought and sold very much like a standard option (see below).

Just like dividends, splits, stock dividends, spin-offs and rights are valuable assets that detach themselves from the stock ownership on a specific date (the ex-date).

> On the ex-date, the price of the stock drops by an amount equal to the value of the newly received asset.

In practice this relationship is never exact because of investors preferences, taxation and, more importantly, the usual daily fluctuations of prices. Most data providers multiply historical prices by an adjustment factor so as to remove the price discontinuity due to these corporate events.

A new public company can appear when a private company decides to make its share available publicly through an **initial public offering** (IPO) or as the result of a spin-off (see above). Stocks can sometime disappear from databases. A **takeover** or acquisition is when a company is bought by another one (public or private), a **merger** is when the owners of two company exchange shares as to all become the owners of a single company encompassing the original two. Companies may also go bankrupt and be liquidated. Companies with financial difficulties might be suspended or even delisted from an exchange without being bankrupt. Since a bankruptcy or a takeover is usually a violent event in the life of a stock, the absence of data on stocks having undergone these events can lead to a risk underestimation. This is an example of selection bias (see below).

5.2.3 Stock indices

Stock indices are indices that reflect the price movements of equity markets. They are computed and published by stock exchanges (e.g. NYSE, Euronext), publishing companies

[†] In French, corporate events are called 'opérations sur titre', or OST.

(e.g. Dow Jones, McGraw Hill (S&P), Financial Times), or investment banks (e.g. Morgan Stanley). They serve as benchmarks for portfolio managers. Equity derivatives such as futures and options are linked to a given stock index, the most famous being the S&P 500 index futures. More complex derivatives, including those sold to the general public like Guaranteed Capital funds, are also often linked to the performance of stock indices.

Each index is meant to reflect the performance of a given category of stocks. This category is typically defined in terms of geography (world, continent, country, specific exchange), market capitalization (large cap, mid cap, small cap), economical sector (e.g. technology, transportation, banking). The number of stocks represented in a given index varies from a few tens (Bel-20, DJIA) to several thousands (MSCI World, NYSE Composite, Russell 3000).

Most stock indices are price averages weighted by market capitalization. The **market capitalization** represents the total market value of a company; it is given by its current share price multiplied by the number of outstanding shares (i.e. the number of shares required to own 100% of the company). Therefore the value of such an index is given by:

$$I(t) = C \sum_{i=1}^{N} n_i \, x_i(t), \tag{5.1}$$

where n_i and x_i are the number of shares and current price of stock i. The constant C is chosen such that the index equals a certain reference value (say 100) at a given date. When stocks are added to or removed from the index, the constant C is modified so that the value of the index computed with the new constituent list equals the old index value on the same day. Note that a number of indices are now basing their computation on the **float** market capitalization: they are excluding from the number of outstanding shares, the shares that cannot be bought in the open market, such as shares held by the state, by a parent company or other core investors.

A notable exception to the market capitalization method is the Dow Jones Industrial Average (DJIA). The DJIA is an index composed of 30 of the largest corporations in the US. It is computed from the sum of the prices of its constituents (normalized by a single constant). Table 5.2 shows the constituents of the DJIA on August 15, 2001. Note that General Electric has a weight of about a third of 3M while its market capitalization is ten times larger. This is due to the fact that the current price of 3M ($109.10) is about three times larger than that of GE ($41.85). The reason why DJIA, the best known stock index, is following such a strange methodology is purely historical. At the time of its creation, in 1896, all computations were done by hand, summing the prices of 30 stocks (actually 12 at the time) was definitely easier than anything else. The fact that this methodology is still used today shows the extreme weight of history in financial development.

Traditional stock indices are not good indicators of the total returns on a stock investment for they do not include dividend payments made by stocks. In a standard **price index**, historical prices are adjusted for splits, spin-offs and exceptional dividends but not for regular dividends. Some more recent indices do include dividend payments, they are called **total return indices**. For example, The Dow Jones Stoxx indices exist in both methodology, their Europe Broad Index (total return) had an annualized return of 9.3% from January 1992

Table 5.2. Weights of the constituents of the Dow Jones Industrial Average Index on August 15, 2001.

Alcoa Inc	2.45	Intel Corp	2.02
American Express Company	2.62	International Paper Co	2.70
AT&T Corp	1.32	Intl Business Machines Corp	7.06
Boeing Co	3.71	Johnson & Johnson	3.79
Caterpillar Inc	3.58	JP Morgan Chase & Co	2.81
Citigroup Inc	3.24	Mcdonald's Corporation	1.86
Coca–Cola Company	3.06	Merck & Co., Inc.	4.65
Du Pont De Nemours	2.78	Microsoft corp	4.30
Eastman Kodak Co	2.91	Minnesota Mining & Mfg Co	7.25
Exxon Mobil Corp	2.73	Philip Morris Companies Inc	2.95
General Electric Co.	2.78	Procter & Gamble Co	4.82
General Motors Corp	4.22	SBC Communications Inc	2.92
Hewlett-Packard Co.	1.65	United Technologies Corp	4.83
Home Depot Inc	3.28	Wal–Mart Stores Inc	3.48
Honeywell International Inc	2.44	The Walt Disney Co.	1.80

to September 2002, while its *price* counterpart only had a return of 6.4% over the same period.

An issue related to the construction of stock indices is the problem of **selection bias**. When doing historical analysis on a large pool of assets such as stocks, one analyzes data that is available now. But the presence or absence of a stock's history in a database is often decided based on criteria that are applied now or in the near past. For instance it is very difficult to find historical data on companies that no longer exist (bankruptcy, mergers). The application of such ex-post criteria can change the statistical properties of the studied ensemble. As an illustration, we have computed the value of an index based on the average returns of the current members of the Dow Jones Industrial Index (Fig. 5.1). The average return of this index is about 50% higher than that of the historical Dow Jones. The reason for the discrepancy is that the composition of the index has changed over time, always including the largest industrial American companies of the time, while the synthetic index includes in 1986 the companies that will be among the largest in 2002 even though they were not necessarily the largest then. In other words, we have selected future winners. Selection bias becomes very important when dealing with returns of funds. Not only is the attrition rate high (especially for hedge funds,[†] where it is found to be 20% per year – see e.g. Fung & Hsieh (1997)), there is also a reporting bias, funds typically start reporting to databases after one year of good performance. A fund might also suspend reporting when in difficulty to start again if the results are good, or disappear if difficulties persist.

[†] **Hedge funds** are funds that use both long (buy) and short (sell) positions, derivative instruments and/or leverage. These funds are typically very active in their trading. By opposition, traditional funds only buy and hold securities (stocks and bonds).

Fig. 5.1. Example of selection bias. The full line shows the value over time of the Down Jones Industrial Average (DJIA) index. The dotted line is an index obtained by averaging the historical returns of the current 30 members of the DJIA.

5.2.4 Bonds

Bonds are a form of loans contracted by corporations and governments from financial markets. Unlike a bank loan which ties the bank to the corporation that contracted it, bonds are securities – which means that they can be bought and sold freely in the financial markets. One distinguishes the **primary market**, where the issuers sells the bonds to investors, usually through an investment bank, from the **secondary market** where investors buy and sell existing bonds.

The simplest form of a bond is the **zero-coupon** (or **pure discount**) bond. It is a certificate that entitles the bond owner (the bearer) to receive a nominal amount P from the issuer at a fixed date T in the future (the bond **maturity**). Since bonds can be bought and sold, the bearer at maturity need not be the original purchaser of the bond. The price of bond depends on the current level of interest rate, one defines the **yield** y of a zero-coupon bond via the formula:

$$B(t) = \frac{P}{(1 + y(t))^{T-t}} \approx P e^{-y(t)(T-t)}, \tag{5.2}$$

where $B(t)$ is the current price of the bond. The yield is thus the compounded rate of interest that one would effectively receive by buying the bond now and holding it to maturity. The yield is not a fundamental quantity – it is inferred from the prices at which bonds trade – but it follows very closely the level of interest rates. Obviously, $B(t) < P$, the ratio of the two is called the discount factor $d = (1 + y)^{-(T-t)}$.

> One should remember that the discount factor, and hence the bond price goes down when the yield goes up (and vice versa).

Most bonds also make regular payments (**coupons**) before the final payment at maturity (**principal**). For example, a five year 4% semi-annual coupon bond with a $1000 face value would pay a total of ten $20 coupons every six months and a final payment of $1000 along with the last coupon. The (effective) yield y on a coupon bearing bond can be computed by inverting a formula where the rate is assumed to be maturity independent:

$$B(t) = \sum_{i=1}^{N} \frac{P_i}{(1+y)^{T_i-t}} \approx \sum_{i=1}^{N} P_i e^{-y(T_i-t)}, \tag{5.3}$$

where the set of $\{P_i\}$ are the payments made by the bond at the times $\{T_i\}$. In the example above P_i are all equal to $20 except the last one which is $1020. For a coupon bearing bond, the price can be greater than the principal. This is the case when the coupon rate is higher than the current level of interest rates. One then says the that bond is trading at a premium. It is important to understand that the coupon *rate* of a bond stays fixed in time, while its **yield** fluctuates with supply and demand.

An important quantity for risk management is the so-called **duration** of a bond, which measures the sensitivity of the bond price to a change of the yield y: one therefore tries to assess the change of the bond price if the interest rate curve was shifted parallel to itself by an amount Δy. One would then have:

$$\frac{\Delta B}{B} \approx \frac{\partial \log B}{\partial y} \Delta y \equiv -\mathcal{D}\Delta y, \tag{5.4}$$

where \mathcal{D} is the duration of the bond, defined as:

$$\mathcal{D} = \frac{1}{B} \sum_{i=1}^{N} P_i(T_i - t)e^{-y(T_i-t)}. \tag{5.5}$$

Note that the duration is an effective expiry time, i.e. the average time of the future cash-flows weighted by their contribution to the bond price. Note the minus sign in Eq. (5.4): as noted above, bond prices go down when rates go up.

Bonds often have built-in options. In a **callable bond**, the issuers keeps the right to buy back the bond (usually at parity) after a certain date (first call date). In a **convertible bond**, the owner can exchange the bond for company stock under certain condition. Owning a callable bond can therefore be modelled as owing a bond and being short a bond-option, whereas a convertible bond is like owing a bond plus a stock-option.

Bonds come with **credit risk**. A company (or more rarely a government) might not be able to make a coupon or principal payment to its bond holders. The market prices this risk into the bond price. Bonds issued by companies that are more likely to default are cheaper (higher yield) than those that are financially trustworthy. Independant financial companies, called **rating agency**, give a grade to corporate and government bonds. Standard and Poor's and Moody's are the better known ones. S&P's rating systems goes from AAA for very strong capacity to repay to C (highly vulnerable to non-payment) or D (in payment default); Moody's follows a similar system of letters from Aaa to C.

5.2.5 Commodities

Commodities are goods, usually raw materials, that are available in standard quality and format from many different producers and needed by a large number of users. Like financial products, commodities are bought and sold in organized markets. In addition, the commodities market led to the development of derivatives (i.e., forwards and options), which were later adopted by the financial markets (see below). The most common types of commodities are energy (crude oil, heating oil, gasoline, electricity, natural gas), precious and industrial metals (gold, silver, platinum, copper, aluminum, lead, zinc), grains (soy, wheat, corn, oat, rapeseed), livestock and meat (cattle, hogs, pork bellies), other foods (sugar, cocoa, coffee, orange juice, milk) and various other raw materials (rubber, cotton, lumber, ammonia).

5.2.6 Derivatives

We give in this section a few definitions of some **derivative** contracts, also called **contingent claims** because their value depend on the future dynamics of one or several underlying assets. We will expand on the theory of these contracts in the second part of this book (see Chapters 13 to 18).

A **forward** contract amounts to buying or selling today a particular asset (the **underlying**) with a certain delivery date in the future. This enables the buyer (or seller) of the contract to secure today the price of a good that he will have to buy (or sell) in the future. Historically, forward contracts were invented to hedge the risk of large price variations on agricultural products (e.g. wheat) and enable the producers to sell their crop at a fixed price before harvesting.

In practice **futures** contracts are now much more common than forwards. The difference is that forwards are over-the-counter contracts, whereas futures are traded on organized markets. This means that one can pull out of a future contract at any time by selling back the contract on the market, rather than having to negotiate with a single counterparty. Also, forward contracts (like all over-the-counter contracts) carry a counterparty (or default) risk, which is absent on futures markets. For forwards there are typically no payments from either side before the expiry date whereas futures are marked-to-market and compensated every day, meaning that payments are made daily by one side or the other to bring the value of the contract back to zero: this is the **margin call** mechanism, see Table 5.3.

A **swap** contract is an agreement to exchange an uncertain stream of payments against a predefined stream of payments. One of the most common swap contracts is the interest

Table 5.3. Example of how margin calls work for futures contract. The contract is bought on Day 1 at $290. At the end of this day, the price is $300, and the $10 gain is transfered to the buyer. Similarly, the daily gain/loss of the subsequent days are immediately materialized. On the last day, the buyer sells back his contract at $300, while the contract finishes the day at $307. The total gain is $300 − $290 = $10, which is also the sum of the margin calls $(+10 + 10 − 15 + 5)$.

Days	Day 1	Day 2	Day 3	Day 4
Action	Buy 1 at $ 290			Sell 1 at $ 300
Close price	300	310	295	307
Δ margin	+10	+10	−15	+5

rate swap. For example, two parties A and B agree on the following: A will receive every three months (90 days) during a certain time T an amount $P \times r90/360$, where r is a fixed interest rate decided when the swap is contracted and P a certain nominal value. Party B, on the other hand, will receive at the end of the ith three months interval an amount $P \times r_i90/360$, where r_i is the value of the three month rate at the beginning of the ith three month interval. The swap in this case is an exchange between a non risky income and a risky income. Of course, the value of the swap depends on the value of the fixed interest rate r that is being swapped. Usually, this fixed interest rate (or the constant 'leg' of the swap) is fixed such that the initial value of the swap is nil.

A buy **option** (or **call** option) is the right, but not the obligation, to buy an asset (the underlying) at a certain date in the future at a certain maximum price, called the **strike price**, or **exercize price**. A call option is therefore a kind of insurance policy, protecting the owner against the potential increase of the price of a given asset, which he will need to buy in the future. The call option provides to its owner the certainty of not paying the asset more than the strike price. Symmetrically, a **put** option protects against losses, by insuring to the owner a minimum value for his asset.

Plain vanilla options (or European options) have a well defined maturity or expiry date. These options can only be exercized on a given date; for call options:

- either the price of the underlying on that date is above the strike and the option is exercized: its owner receives the underlying asset that he has to pay the strike price. He can then choose to sell back the asset immediately and pocket the difference;
- or the price is below the strike and the option is worthless.

There are many other types of options, the most common ones being **American options** which can be exercized any time between now and their maturity. European and American options are traded on organized markets and can be quite liquid, such as currency options or index options. But some derivative contracts can have more complex, special purpose clauses. These so-called 'exotic' derivatives are usually not traded on organized markets,

but over the counter (OTC). For example, **barrier options** that get inactivated whenever the price exceeds some predefined threshold (the 'barrier') are considered to be exotic options.

There is of course no end to complexity: one can trade an option on an interest swap (called a **swaption**), which is characterized by the maturity of the option, the period and the condition of the swap. One can even consider buying options on options!

5.3 Financial markets

5.3.1 Market participants

In principle, financial markets allow two complementary populations to meet: **entrepreneurs** that have some industrial projects but need funding, and therefore want some **investors** that have money to invest and share, on the long run, both the future profits and the risk of these industrial projects. Note that 'entrepreneurs' can be private individuals, companies or states. The investment instruments are usually stocks or bonds.

Financial markets can also play the role of insurance companies. As explained above, the availability of forward (or futures) contracts, swaps and options offer the possibility of hedging, i.e. securing against losses. This is useful for many market participants, from farmers to oil or metal producers, and to financial institutions themselves that can use options, for example, to protect themselves against market risks, or interest swaps to hedge against interest rate variations. This population of market participants can be called **hedgers**. There are two further broad categories of market participants. One is made of the **broker-dealers**, or **market makers** that provide liquidity to the markets by offering to buy or to sell some stocks, at any given instant of time. The dealers act as intermediaries between other market participants, and compensate any momentary lack of counterparties that would make trading less fluid – and therefore more risky, since frequent trading ensures, to some extent, the continuity (in time) of the price itself. The last category is that of **speculators**, that take short to medium term positions on markets (at variance with *investors* that take longer term positions). Speculators essentially make bets, that are based on extremely different types of considerations. For example, speculators can consider that the recent price variations are excessive and bet on a short term reversal: this is called a **contrarian** strategy. Conversely, **trend following** strategies posit that recent trends will persist in the future. These decisions can be based on intuition or on more or less sophisticated statistical approaches. Other speculators make their mind based on the interpretation of economical or political news. The role of speculators in the above 'ecological' system is both to provide liquidity and, in principle, to enforce that assets are correctly priced, since large mispricing should give rise to arbitrage opportunities (at least in a statistical sense). However, that speculators indeed help the markets to set a fair price is far from obvious, since this population is most sensitive to rumors, herding effects and panic. Arguably, bubbles and crashes are the result of the inevitable presence of speculators in financial markets. Finally, let us note that with the appearance of fully electronic markets, the distinction between dealers and speculators becomes more fuzzy. As will be discussed below, the possibility of placing limit orders allow any market participant to effectively play the role of a market maker.

5.3.2 Market mechanisms

As explained above, some contracts, such as forward contracts or exotic options, are sold **over-the-counter**, that is between two well defined parties, such as a private investor and a financial institution, or two financial institutions. The transaction data (price, volume) is usually not made public. By definition, the corresponding contracts are not very liquid, but they can be tailor made for the customer. On the other hand, organized markets provide liquidity to the traded contracts, which nonetheless have to be standardized in terms of their strike, maturity, etc. The way open markets fix the price of any traded asset very much depends on the market. The two main market mechanisms are **continuous trading** and **auctions**.

Auctions

Every market has its own set of rules for how auctions are conducted, but the general principle is that during the auction, no trade takes place but buyers and sellers notify the market of their intentions (price and volume of the trades). After a certain time, the auction is closed and all the transactions are made at the auction price, determined such that the number of transactions is maximized. Auctions are the norm in the primary markets, when new bonds or stocks are issued. Auctions also take place in the secondary markets for some less liquid securities or at special times (open, close) along with continuous trading. A typical opening/closing auction procedure is that during the auctions, orders to buy or sell can be entered or cancelled, and are recorded in an order book. The auction terminates at a random time during a set interval (say 9:00 to 9:05) and a price is determined as to maximize the number of trades. All buy orders above or equal to the auction price are executed against all the sell order at or below it. During the auction the order book is usually kept hidden with only an indicative price and volume being shown. The indicative price is the auction price were it to close now.

Continuous trading

In continuous trading markets potential buyers notify the other participants of the maximum price at which they are willing to buy (called the **bid** price), for sellers the minimum selling price is called the offer, or **ask** price. The market keeps track of the highest bid (best bid, or simply the bid) and lowest ask (best ask or the ask). At a given instant of time, two prices are quoted: the bid price and the ask price, which are called the quotes. The collection of all orders to buy and to sell is called the **order book** (see below), which is the list of the total volume available for a trade at a given price. New buy orders below the bid price or sell orders above the ask price add to the queue at the corresponding price.

If someone comes in with an unconditional order to buy (market order) or with a limit price above the current ask price, he will then make a trade at the ask price with the person with the lowest ask.[†] The corresponding price is printed as the transaction price. If the buy

[†] When there are multiple sell orders at the same price, priority is normally given to the oldest order, however some markets such as the LIFFE fill orders on a pro rata basis.

volume is larger than the volume at the ask, a succession of trades at increasingly higher prices is triggered, until all the volume is executed. The activity of a market is therefore a succession of **quotes** (quoted bid and ask) and **trades** (transaction prices). In most markets, any participant can place either limit or market orders, but in **quote driven** markets, final clients can only place market orders, and thus must trade on the bids and asks of market markers (see below). Continuous trading can be done by open outcry, where traders gather together in the exchange or by an electronic trading system. In Europe and Japan, essentially all trading is now done electronically while in the US, the transition to electronic markets is slowly taking place.

5.3.3 Discreteness

The smallest interval between two prices is fixed by the market, and is called the **tick**. For example, on EuroNext (Stock exchange for France, Belgium and the Netherlands), the tick size is 0.01 Euros for stocks worth less than 50 Euros, 0.05 Euros for stocks between 50 and 100 Euros, and 0.10 Euros for stocks beyond 100 Euros. For British stocks, the tick size is typically 0.25 pence for stocks worth 300 pence. The order of magnitude of the tick size is therefore 10^{-3} relative to the price of the stock, or 10 basis points. On OTC markets there is no fixed tick size, so that orders can be launched at a price with an arbitrary precision.

When analyzing prices on a tick-by-tick level, one finds distributions of returns which are very far from continuous with most returns being equal to zero or plus or minus one tick. Even on longer time scales: minutes, hours and even days, the discreteness of prices shows up in empirical analysis of returns. For example, prior to decimalization, when US stocks were still quoted in 1/8th or 1/16th of a dollar, the probability that the closing price be equal to the opening price was 20% for a typical stock. In continuous price models, this probability is zero.

5.3.4 The order book

Let us describe in more details the structure of the order book, and a few basic empirical observations. As mentioned above, the order book is the list of all buy and sell 'limit' orders, with their corresponding price and volume, at a given instant of time. A snapshot of an order book (or in fact of the part of the order book close to the bid and ask prices) is given in Figure 5.2.

When a new order appears (say a buy order), it either adds to the book if it is below the ask price, or generates a trade at the ask if it is above (or equal to) the ask price. The dynamics of the quotes and trades is therefore the result of the interplay between the flow of limit orders and market orders. An important quantity is the **bid-ask spread**: the difference between the best sell price and the best offer price. This gives an order of magnitude of the transaction costs, since the 'fair' price at a given instant of time is the midpoint between the bid and the ask. Therefore, the cost of a market order is one half of the bid-ask spread. The bid-ask spread is itself a dynamical quantity: market orders tend to deplete the order book and increase the spread, whereas limit orders tend to fill the gap and decrease the spread. The role of market makers (or dealers) is to place buy and sell limit orders in order to provide

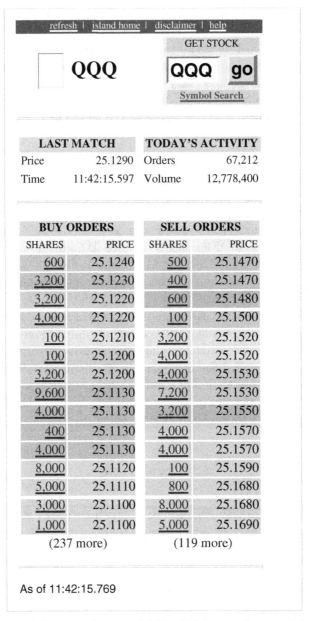

Fig. 5.2. Restricted part of the order book on 'QQQ', which is an exchange traded fund that tracks the Nasdaq 100 index. Only the 15 best offers are shown. The first column gives the number of offered shares, the second column the corresponding price. Note that the tick size here is 0.001. Source: www.island.com/bookviewer.

Table 5.4. Some data for the two liquid French stocks in February 2001. The transaction volume is in number of shares.

Quantity	France-Telecom	Total
Initial/final price (Euros)	90-65	157-157
Tick size (Euros)	0.05	0.1
Total # orders	270,000	94,350
# trades	176,000	60,000
Transaction volume	$75,6 \times 10^6$	$23,4 \times 10^6$
Average bid-ask (ticks)	2.0	1.4

some liquidity and reduce the average spread, and therefore reduce the transaction costs. Typically, the average spread on stocks ranges from one tick (for the most liquid stocks) to ten ticks for less liquid stocks (see Table 5.4). We also show in Table 5.4 the total number of orders (limit and market). In fact, although most of these orders are either market orders, or placed in the immediate vicinity of the last trade, one can observe that some limit orders are placed very far from the current price. The full distribution of the distance between the current midpoint and the price of a limit order reveals an interesting slow *power-law* tail for large distances, as if some some patient investors are willing to wait until the price goes down quite significantly before they buy. Note finally that limit orders can be cancelled at any time if they have not been executed.

The way limit orders are placed, meet market orders, or are cancelled lead to very interesting dynamics for the order book. One of the simplest question to ask is: what is the (average) 'shape' of the order book, i.e., the average number of shares in the queue at a certain distance from the best bid (or best ask). Perhaps surprisingly, this shape is empirically found to be non trivial, see Fig. 5.3. The size of the queue reaches a maximum away from the best offered price. This results from the competition between two effects: (a) the order flow is maximum around the current price, but (b) an order very near to the current price has a larger probability to be executed and disappear from the book. Theoretical models that explain this shape are currently being actively studied: see references below.

5.3.5 The bid-ask spread

As mentioned above, in order driven markets, the bid-ask spread appears as a consequence of the way limit orders are stored and executed by market orders. In quote driven or **specialists** markets, normal participants (clients) can only send market orders while the bid-ask spread is fixed by the 'market makers' (or 'broker-dealers') that offer simultaneously a buy price and a sell price and make a profit on the difference. Foreign exchange markets work this way and so did the NASDAQ before the appearance of electronic trading platforms (ECNs). In this case, the bid-ask spread is fixed by the competition between different broker-dealers,

Fig. 5.3. Average size of the queue in the order book as a function of the distance from the best price, in a log-linear plot for three liquid French stocks. The 'buy' and 'sell' sides of the distribution are found to be identical (up to statistical fluctuations). Both axis have been rescaled such as to superimpose the data. Interestingly, the shape is found to be very similar for all three stocks studied. Inset: same in log-log coordinates, showing a power-law behaviour.

and also by the market conditions: larger uncertainties lead to larger bid-ask spreads. Note that market makers, because they have to quote prices first, take the risk of being arbitraged by market participants that have access to a piece of information not available to market makers. This is a general phenomenon: liquidity providers (market makers or traders using limit orders) try to earn the bid-ask spread but give away information, whereas liquidity takers typically pay the spread but can take advantage of news and information.

5.3.6 Transaction costs

Trading costs money. When buying and selling securities and derivatives, one has to pay all the agents and intermediaries involved in the transaction (in finance, no one works for free...). Commissions for buying and selling stocks can be as high as 1 or 2% for individuals buying over the phone and as low as 1 or 2 basis points for institutions with very high turnover and fully electronic processing. Commissions for futures contracts typically vary between \$2 and \$20 per contract. For index futures (\sim \$50 000) the lower figure corresponds to 0.4 bp, and 0.2 bp for bond and currency futures (\sim \$100 000).

As we mentioned above, the bid-ask spread is a form of transaction cost. The half-spread can be as low as 0.5–2 bp for liquid futures or 1–5 bp for liquid stocks, but it can reach 50 bp for less liquid stock or even 10 to 20% of the price for equity options. Finally, when

one buys (sells) very large quantities, the action of buying can increase the price leading to a total cost per share higher than the current ask price, one then speaks of **price impact**. As one buys, one trades first with all the immediate sellers, the price has then to increase to convince more people to sell. Another mechanism is that as participants see that there is a large buyer in the market, they pull back their offers out of fear (that the buyer knows something they don't) or greed (trying to extract more money from the large buyer).

> As a rule of thumb, commissions are the most important cost for individuals, bid-ask spreads for moderate size traders and price impact dominates for very large funds.

5.3.7 Time zones, overnight, seasonalities

As compared to many simplified models of price changes, real markets display many idiosyncrasies and 'imperfections'. One is the fact that markets are not open all day long, but only have a limited period of activity (market hours). Therefore, news that becomes known when the market is closed cannot affect the price immediately, but rather accumulate until the next morning and influence the 'open' price and lead to an overnight 'gap', i.e. a difference between the close of the previous day and the new open. This effect is significant: in terms of volatility, the overnight corresponds to roughly two hours of trading on stock markets. Even during opening hours, the trading activity is not uniform: it peaks both at the open and at the close, with a minimum around lunch time. Similarly, the activity is not uniform across days of the week, or months of the year. Finally, not all markets are simultaneously open, and different time zones behave differently. A well-known effect is for example an increased activity at 2:30 p.m. European time, due to the fact that major U.S. economic figures are announced at 8:30 a.m. east-coast time.

5.4 Summary

- Among the large variety of financial products and financial markets, some are quite simple (like stocks), whereas others, like options on swaps, can be exceedingly complex. Both financial markets and financial products display many idiosyncrasies. These details may be unimportant for a rough understanding, but are often crucial when it comes to the final implementation of any quantitative theory or model.

• FURTHER READING

J. C. Hull, *Futures, Options and Other Derivatives*, Prentice Hall, Upper Saddle River, NJ, 1997.
W. F. Sharpe, G. J. Alexander, J. V. Bailey, *Investments (6th Edition)*, Prentice Hall, 1998.

- SPECIALIZED PAPERS: SURVIVAL BIAS

S. J. Brown, W. N. Goetzmann, J. Park, *Conditions for survival: Changing risk and the performance of Hedge fund managers and CTAs*, Review of Financial Studies, **5**, 553 (1997).

W. Fung, D. A. Hsieh, *The information content of performance track records*, Journal of Portfolio Management, **24**, 30 (1997).

- SPECIALIZED PAPERS: STATISTICAL ANALYSIS OF ORDER BOOKS

B. Biais, P. Hilton, C. Spatt, *An empirical analysis of the limit order book and the order flow in the Paris Bourse*, Journal of Finance, **50**, 1655 (1995).

J.-P. Bouchaud, M. Mézard, M. Potters, *Statistical properties of stock order books: empirical results and models*, Quantitative Finance, **2**, 251 (2002).

D. Challet, R. Stinchcombe, *Analyzing and modelling 1+1d markets*, Physica A, **300**, 285 (2001).

D. Challet, R. Stinchcombe, *Exclusion particle models of limit order financial markets*, Physica A, **324**, 141 (2003).

M. G. Daniels, J. D. Farmer, G. Iori, E. Smith, *Quantitative model of price diffusion and market friction based on trading as a mechanistic random process*, Phys. Rev. Lett., **90**, 108102 (2003).

H. Luckock, *A steady-state model of the continuous double auction.* Quantitative Finance, **3**, 385 (2003).

S. Maslov, *Simple model of a limit order-driven market*, Physica A, **278**, 571 (2000).

S. Maslov, M. Millis, *Price fluctuations from the order book perspective – empirical facts and a simple model*, Physica A, **299**, 234 (2001).

M. Potters, J. P. Bouchaud, *More statistical properties of order books and price impact*, Physica A, **324**, 133 (2003).

F. Slanina, *Mean-field approximation for a limit order driven market model*, Phys. Rev., **E 64**, 056136 (2001).

E. Smith, J. D. Farmer, L. Gillemot, and S. Krishnamurthy. *Statistical theory of the continuous double auction.* Quantitative Finance, **3**, 481 (2003).

R. D. Willmann, G. M. Schuetz, D. Challet, *Exact Hurst exponent and crossover behavior in a limit order market model*, Physica A, **316**, 430 (2002).

I. Zovko, J. D. Farmer, *The power of patience: A behavioral regularity in limit order placement*, Quantitative Finance, **2**, 387 (2002).

- SOME RECENT READS

J. P. Bouchaud, Y. Gefen, M. Potters, M. Wyart, *Fluctuations and response in financial markets: The subtle nature of 'random' price changes*, Quantitative Finance, **4**, 176 (2004).

F. Lillo, J. D. Farmer, *The long memory of efficient markets*, Studies in Nonlinear Dynamics and Econometrics, **8**, 1 (2004).

J. P. Bouchaud, J. Kockelkoren, M. Potters, *Random walks, liquidity molasses and critical response in financial markets*, Quantitative Finance, **6**, 115 (2006).

J. Farmer, A. Gerig, F. Lillo, and S. Mike, *Market efficiency and the long-memory of supply and demand: Is price impact variable and permanent or fixed and temporary?* Quantitative Finance, **6**, 107 (2006).

6

Statistics of real prices: basic results

Le marché, à son insu, obéit à une loi qui le domine: la loi de la probabilité.[†]
(Louis Bachelier, *Théorie de la spéculation.*)

6.1 Aim of the chapter

The easy access to enormous financial databases, containing thousands of asset time series, sampled at a frequency of minutes or sometimes seconds, allows one to investigate in detail the statistical features of the time evolution of financial assets. The description of any kind of data, be it of physical, biological or financial origin requires, however, an interpretation framework, needed to order and give a meaning to the observations. To describe necessarily means to simplify, and even sometimes betray: the aim of any empirical science is to approach reality progressively, through successive improved approximations.

The goal of the present and subsequent chapters is to present in this spirit the statistical properties of financial time series. We shall propose some plausible mathematical modelling, as faithful as possible (though imperfect) to the observed properties of these time series. The models we discuss are however not the only possible models; the available data is often not sufficiently accurate to distinguish, say, between a truncated Lévy distribution and a Student distribution. The choice between the two is then guided by mathematical convenience. In this respect, it is interesting to note that the word 'modelling' has two rather different meanings within the scientific community. The first one, often used in applied mathematics, engineering sciences and financial mathematics, means that one represents reality using appropriate mathematical formulae. This is the scope of the present and next two chapters.

The second meaning, more common in the physical sciences, is perhaps more ambitious: it aims at finding a set of plausible causes sufficient to explain the observed phenomena, and therefore, ultimately, to justify the chosen mathematical description. We will defer to the last chapter of this book the discussion of several 'microscopic' mechanisms of price formation and evolution (e.g. adaptive traders' strategies, herding behaviour between traders, feedback of price variations onto themselves, etc.) which are certainly at the origin of the interesting statistics that we shall report below. This line of research is still in its infancy and evolves very rapidly.

[†] The market, without knowing it, obeys a law which overwhelms it: the law of probability.

We shall describe several types of market:

- Very liquid, 'mature' markets of which we take three examples: a US stock index (S&P 500), an exchange rate (British pound against U.S. dollar), and a long-term interest rate index (the German Bund);
- Single liquid stocks, which we choose to be representative U.S. stocks;
- Very volatile markets, such as emerging markets like the Mexican peso;
- Volatility markets: through option markets, the volatility of an asset (which is empirically found to be time dependent) can be seen as a price which is quoted on markets (see Chapter 13);
- Interest rate markets, which give fluctuating prices to loans of different maturities, between which special types of correlations must however exist. This will be discussed in Chapter 19.

We choose to limit our study to fluctuations taking place on rather short time scales (typically from minutes to months). For longer time scales, the available data-set is in general too small to be meaningful. From a fundamental point of view, the influence of the average return is negligible for short time scales, but becomes crucial on long time scales. Typically, a stock varies by several per cent within a day, but its average return is, say, 10% per year, or 0.04% per day. Now, the 'average return' of a financial asset appears to be unstable in time: the past return of a stock is seldom a good indicator of future returns. Financial time series are intrinsically non-stationary: new financial products appear and influence the markets, trading techniques evolve with time, as does the number of participants and their access to the markets, etc. This means that taking very long historical data-set to describe the long-term statistics of markets is *a priori* not justified. We will thus avoid this difficult (albeit important) subject of long time scales.

The simplified model that we will present in this chapter, and that will be the starting point of the theory of portfolios and options discussed in later chapters, can be summarized as follows. The variation of price of the asset X between time $t = 0$ and $t = T$ can be decomposed as:

$$\log \left(\frac{x(T)}{x_0} \right) = \sum_{k=0}^{N-1} \eta_k \qquad N = \frac{T}{\tau}, \tag{6.1}$$

where:

- *In a first approximation*, the relative price increments η_k are random variables which are (i) independent as soon as τ is larger than a few tens of minutes (on liquid markets) and (ii) identically distributed, according to a rather broad distribution, which can be chosen to be a truncated Lévy distribution, Eq. (1.26) or a Student distribution, Eq. (1.38): both fits are found to be of comparable quality. The results of Chapter 1 concerning sums of random variables, and the convergence towards the Gaussian distribution, allows one

to understand qualitatively the observed 'narrowing' of the tails as the time interval T increases.

- A more refined analysis, presented in Chapter 7, however reveals important systematic deviations from this simple model. In particular, the kurtosis of the distribution of $\log(x(T)/x_0)$ decreases more slowly than $1/N$, as would be found if the increments η_k were iid random variables. This suggests a certain form of temporal dependence. The volatility (or the variance) of the price returns η is actually itself time dependent: this is the so-called 'heteroskedasticity' phenomenon. As we shall see in Chapter 7, periods of high volatility tend to persist over time, thereby creating long-range higher-order correlations in the price increments.
- Finally, on short time scales, one observes a systematic skewness effect that suggests, as discussed in details in Chapter 8, that in this limit a better model is to consider price themselves, rather than log-prices, as additive random variables:

$$x(T) = x_0 + \sum_{k=0}^{N-1} \delta x_k \qquad N = \frac{T}{\tau}, \tag{6.2}$$

where the price increments δx_k (rather than the returns η_k) have a fixed variance. We will actually show in Chapter 8 that reality must be described by an intermediate model, which interpolates between a purely additive model, Eq. (6.2), and a purely multiplicative model, Eq. (6.1).

Studied assets

The chosen stock index is the Standard and Poor's 500 (S&P 500) US stock index. During the time period chosen (from September 1991 to August 2001), the index rose from 392 to 1134 points, with a high at 1527 in March 2000 (Fig. 6.1 (top)). Qualitatively, all the conclusions reached on this period of time are more generally valid, although the value of some parameters (such as the volatility) can change significantly from one period to the next.

The exchange rate is the US dollar ($) against the British pound (GBP). During the analysed period, the pound varied between 1.37 and 2.00 dollars (Fig. 6.1 (middle)). Since the interbank settlement prices are not available, we have defined the price as the average between the bid and the ask prices, see Section 5.3.5.

Finally, the chosen interest rate index is the futures contract on long-term German bonds (Bund), first quoted on the London International Financial Futures and Options Exchange (LIFFE) and later on the EUREX German futures market. It is typically varying between 70 and 120 points (Fig. 6.1 (bottom)).

For intra-day analysis of the S&P 500 and the GBP we have used data from the futures market traded on the Chicago Mercantile Exchange (CME). The fluctuations of futures prices follow in general those of the underlying contract and it is reasonable to identify the statistical properties of these two objects. Futures contracts exist with several fixed maturity dates. We have always chosen the most liquid maturity and suppressed the artificial difference of prices when one changes from one maturity to the next (roll).

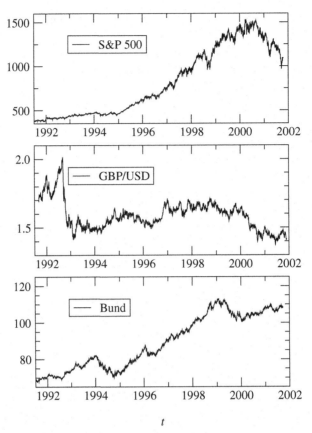

Fig. 6.1. Charts of the studied assets between September 1991 and August 2001. The top chart is the S&P 500, the middle one is the GBP/$ and the bottom one is the long-term German interest rate (Bund).

We have also neglected the weak dependence of the futures contracts on the short time interest rate (see Section 13.2.1): this trend is completely masked by the fluctuations of the underlying contract itself.

6.2 Second-order statistics

6.2.1 Price increments vs. returns

In all that follows, the notation δx represents an absolute price increment between two instants separated by a certain time interval τ:

$$\delta x_k = x_{k+1} - x_k = x(t + \tau) - x(t) \qquad (t \equiv k\tau); \tag{6.3}$$

whereas η denotes the relative price increment (or return):

$$\eta_k = \frac{\delta x_k}{x_k} \approx \log x_{k+1} - \log x_k, \tag{6.4}$$

where the last equality holds for τ sufficiently small so that relative price increments over that time scale are small. For example, the typical change of prices for a stock over a day is a few percent, so that the difference between $\delta x_k/x_k$ and $\log x_{k+1} - \log x_k$ is of the order of 10^{-3}, and totally insignificant for the purpose of this book. Even when $\delta x_k/x_k = 0.10$, $\log x_{k+1} - \log x_k = 0.0953$, which is still quite close.

In the whole of modern financial literature, it is postulated that the relevant variable to study is not the increment δx itself, but rather the *return* η. It is certainly correct that different stocks can have completely different prices, and therefore unrelated absolute daily price changes, but rather similar daily returns. Similarly, on long time scales, price changes tend to be proportional to prices themselves: when the S&P 500 is multiplied by a factor ten, absolute price changes are also multiplied roughly by the same factor (see Fig. 8.2).

However, on shorter time scales, the choice of returns rather than price increments is much less justified, as will be argued at length in Chapter 8. The difference leads to small, but systematic skewness effects that must be discussed carefully. For the purpose of the present chapter, these small effects can nevertheless be neglected, and we will follow the tradition of studying price returns, or more precisely difference of log prices.

6.2.2 Autocorrelation and power spectrum

The simplest quantity, commonly used to measure the correlations between price increments, is the temporal two-point correlation function $C^\tau(k, \ell)$, defined as:[†]

$$C^\tau(\ell) = \frac{1}{\sigma_1^2} \langle \eta_k \eta_{k+\ell} \rangle_e; \qquad \sigma_1^2 \equiv \sigma^2 \tau = \langle \eta_k^2 \rangle_e, \qquad (6.5)$$

where the notation $\langle ... \rangle_e$ refers to an empirical average over the whole time series. Figure 6.2 shows this correlation function for the three chosen assets, and for $\tau = 5$ min. For uncorrelated increments, the correlation function $C^\tau(\ell)$ should be equal to zero for $k \neq l$, with an empirical RMS equal to $\sigma_e = 1/\sqrt{N}$, where N is the number of independent points used in the computation. Figure 6.2 also shows the $3\sigma_e$ error bars. We conclude that beyond 30 min, the two-point correlation function cannot be distinguished from zero. On less liquid markets, however, this correlation time is longer. On the US stock market, for example, this correlation time has significantly decreased between the 1960s (where it was equal to a few hours) and the 1990s.

On very short time scales, however, weak but significant correlations do exist. These correlations are however too small to allow profit making: the potential return is smaller than the transaction costs involved for such a high-frequency trading strategy, even for the operators having direct access to the markets (cf. Section 13.1.3). Conversely, if the transaction costs are high, one may expect significant correlations to exist on longer time scales.

[†] In principle, one should subtract the average value $\langle \eta \rangle = \tilde{m}\tau \equiv \tilde{m}_1$ from η. However, if τ is small (for example equal to a day), \tilde{m}_1 (of the order of $10^{-4} - 10^{-3}$ for stocks) is usually very small compared to σ_1 (of the order of a few percents).

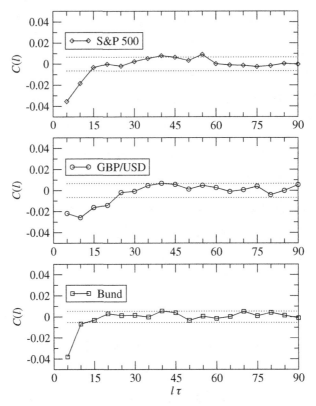

Fig. 6.2. Normalized correlation function $C^\tau(\ell)$ for the three chosen assets, as a function of the time difference $\ell\tau$, and for $\tau = 5$ min. Up to 30 min, some weak but significant negative correlations do exist (of amplitude ~ 0.05). Beyond 30 min, however, the two-point correlations are not statistically significant. (Note the horizontal dotted lines, corresponding to $\pm 3\sigma_e$.)

Note actually that the high frequency correlations are *negative* before reverting back to zero. There are therefore significant intra day anticorrelations in price fluctuations. This is not related to the so-called 'bid-ask' bounce,[†] since this effect is found on the high frequency dynamics of the midpoint (defined as the mean of the bid price and the ask price) for individual stocks. This effect can be simply understood from the persistence of the order book: orders that have accumulated above and below the current price have a *confining* effect on short scales, since they oppose a kind of 'barrier' to the progression of the price.

One can perform the same analysis for the daily increments of the three chosen assets ($\tau = 1$ day). The correlation function (not shown) is always within $3\sigma_e$ of zero, indicating that the daily increments are not significantly correlated, or more precisely that such correlations, if they exist, can only be measured using much longer time series. Indeed, using data over several decades, one can detect significant correlations on the scale of a few days: individual stocks tend to be positively correlated, whereas stock indices tend to be negatively correlated.

[†] Suppose the true price (midpoint) is fixed but one observes a sequence of trades alternating between the bid and the ask. The resulting time series shows strong anticorrelations.

The order of magnitude of these correlations is however small: one finds that daily returns have a ~ 0.05 correlation from one day to the next.

Price variogram

Another way to observe the near absence of linear correlations is to study the variogram of (log-)price changes, defined as:

$$\mathcal{V}(\ell) = \langle (\log x_k - \log x_{k+\ell})^2 \rangle_e, \tag{6.6}$$

which measures how much, on average, the logarithm of price differs between two instant of times. If the returns η_k have zero mean and are uncorrelated, it is easy to show that the variogram \mathcal{V} grows linearly with the lag, with a prefactor equal to the volatility (see Section 4.3).

It is interesting to study this quantity for large time lags, using very long time series. However, since on long time scales the level of – say – the Dow Jones grows on average exponentially with time, one has to subtract this average trend from $\log x_k$. In fact, the average annual return of the U.S. market has itself increased with time. A reasonable fit of $\log x_k$ over the twentieth century is given by:

$$\log x_k = a_1 k + a_2 k^2 + \log \tilde{x}_k, \tag{6.7}$$

where k is in years, $a_1 = 9.21\ 10^{-3}$ per year and $a_2 = 3.45\ 10^{-4}$ per squared year (we have rescaled the Dow-Jones index by an arbitrary factor such as to remove the constant term a_0). The quantity $\log \tilde{x}_k$ is the fluctuation away from this mean behaviour. The average return given by this fit is $a_1 + 2a_2 k$, found to be around 1% annual for $k = 0$ (corresponding to year 1900) and around 8% for year 1999. The variogram of $\log \tilde{x}_k$ computed between 1950 and 1999 is shown in Fig. 6.3, for ℓ ranging from a day to four market years (1000 days). A linear behaviour of this quantity is equivalent to the absence of autocorrelations in the fluctuating part of the returns. Fig. 6.3 shows, in a log-log plot, \mathcal{V} versus ℓ, together with a linear fit of slope one. The data corresponding to the beginning of the century (1900–1949) is also linear for small enough ℓ but shows stronger signs of saturation for ℓ of the order of a year. (This saturation might also be present in Fig. 6.3.)

Power spectrum

*Let us briefly mention another equivalent way of presenting the same results, using the so-called **power spectrum**, defined as:*

$$\mathcal{S}(\omega) = \frac{1}{N} \sum_{k,\ell} \langle \eta_k \eta_{k+\ell} \rangle e^{i\omega\ell}. \tag{6.8}$$

This quantity is simple to calculate in the case where the correlation function decays as an exponential: $C(\ell) = \alpha\sigma^2 \exp -\alpha|\ell|$. In this case, one finds:

$$\mathcal{S}(\omega) = \frac{\sigma^2 \alpha^2}{\omega^2 + \alpha^2}. \tag{6.9}$$

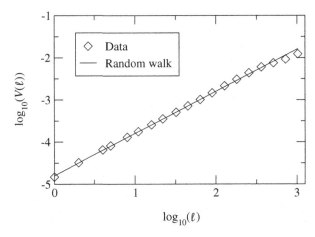

Fig. 6.3. Variogram of the detrended logarithm of the Dow Jones index, during the period 1950–2000. ℓ is in days. Except perhaps for the last few points, the variogram shows no sign of saturation.

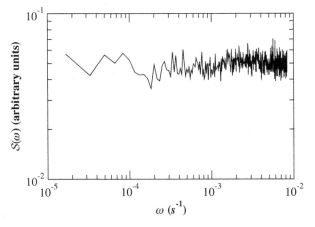

Fig. 6.4. Power spectrum $\mathcal{S}(\omega)$ of the time series DEM/$, as a function of the frequency ω. The spectrum is flat: for this reason one often speaks of white noise, where all the frequencies are represented with equal weights. This corresponds to uncorrelated increments.

The case of uncorrelated increments corresponds to the limit $\alpha \to \infty$, which leads to a flat power spectrum, $\mathcal{S}(\omega) = \sigma^2$. Figure 6.4 shows the power spectrum of the DEM/$ time series, where no significant structure appears.

6.3 Distribution of returns over different time scales

The results of the previous section are compatible with the simplest scenario where the price returns η_k are, beyond a certain correlation time of a few tens of minutes on liquid markets, independent random variables. A much finer test of this assumption consists in studying directly the probability distributions of the log-price differences $\log(x_N/x_0) = \sum_{k=0}^{N-1} \eta_k$ on

Table 6.1. Value of the average return for the 30 minutes price changes and the daily price changes for the three assets studied. The results are in % for the S&P 500 and GBP/$, and correspond to absolute price changes for the Bund. Note that unlike the daily returns, the 30 minute returns do not include the overnight price changes.

Asset	Average return	
	30 minutes	Daily
S&P 500	0.00115	0.0432
GBP/$	0.00211	−0.00242
Bund	0.000318	0.0117

different time scales $N = T/\tau$. If the increments were independent identically distributed random variables, the distributions on different time scales would be deduced from the one pertaining to the elementary time scale τ (chosen to be larger than the correlation time) using a simple convolution. More precisely (see Section 2.2.1), one would have in that case $P(\log x/x_0, N) = [P_1]^{\star N}$, where P_1 is the distribution of elementary returns. This is what we test in Section 6.3.3 below.

6.3.1 Presentation of the data

For the futures contracts, we have considered 5 minute data sets starting September 1st, 1991 until September 1st, 2001 (or slightly longer daily data sets (from 1990 to November 2001) for the S&P 500). The 5 minute data sets does *not* include the overnight return (the return form the close to the next day's open). One should remember that the overnight contribution to the total close to close variations is quite important. The variance of the overnight return is equal to 1/4 of the full day without overnight (i.e. it corresponds to about 8/4 = 2 hours of daily trading). For the GBP/$, the overnight corresponds to 80% of the daily variance, and for the Bund, it is 16%. In all the subsequent analysis, the returns are actually 'detrended' returns, for which the empirical average return has been subtracted. The value of this average return is given, both for 30 minute increments and for daily increments, in Table 6.1. (These two quantities are not simply related because the overnight is absent from the former but contributes to the latter.)

As far as individual stocks are concerned, we used daily price changes of the 320 most liquid U.S. stocks. This leads to a data set containing 800 000 points. The most negative event is -58σ. It was Enron in November 2001. In order to be comparable, the average return of each stock was subtracted and RMS of the log-returns of each stock was normalized to one using the trick explained in Section 4.4. From this data set, we also considered monthly returns, using log returns over non overlapping 22 day periods. This generates 36 100 points, which are treated exactly as for daily data. The worst events were again Enron (-18σ) and Providian (-12σ). Those two events happened at the end of our data set (autumn 2001);

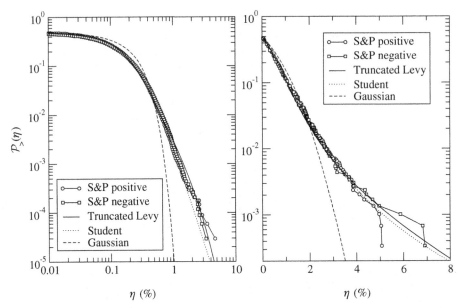

Fig. 6.5. Elementary cumulative distribution $\mathcal{P}_{1>}(\eta)$ (for $\eta > 0$) and $\mathcal{P}_{1<}(\eta)$ (for $\eta < 0$), for the S&P 500 returns, with $\tau = 30$ min (left, log-log scale) and 1 day (right, semi-log scale). The thick line corresponds to the best fit using a symmetric TLD $L^{(t)}_{\mu}$, of index $\mu = \frac{3}{2}$. We have also shown the best Student distribution and the Gaussian of same RMS.

this could be a selection bias effect, in the sense that the data set contains only stocks which survived until 2001, that had by definition no violent accident before.

6.3.2 The distribution of returns

High frequency distribution

The 'elementary' high frequency (30 minutes) cumulative probability distribution $\mathcal{P}_{1>}(\delta x)$ is represented in Figures 6.5, 6.6 and 6.7. One should notice that the tail of the distribution is broad, in any case much broader than a Gaussian. A fit using a truncated Lévy distribution (TLD) with $\mu = \frac{3}{2}$ as given by Eq. (1.26), or alternatively a Student distribution are both quite satisfying (see Fig. 1.5).[†] The corresponding fitting parameters are reported in Tables 6.2 and 6.3. In both cases, one has a priori two free parameters (since we fixed the value of μ for the TLD). We imposed that these two parameters are such that the variance of the theoretical distribution exactly reproduces the empirical variance, and determined the only remaining parameter using a maximum likelihood criterion. From the results reported in Tables 6.2 and 6.3, one sees that both fits are of comparable quality, with a slight advantage for the Student fit.

For the TLD, some useful relations are, in the present case ($\mu = \frac{3}{2}$):

[†] A more refined study of the tails actually reveals the existence of a small asymmetry, which we neglect here. Therefore, the skewness ς is taken to be zero – but see Chapter 8.

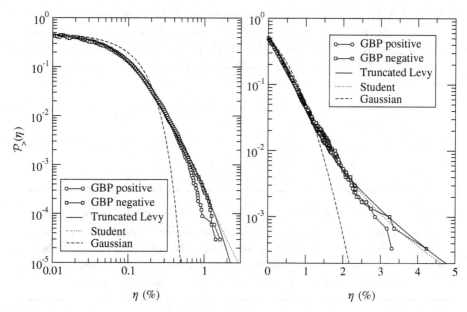

Fig. 6.6. Elementary cumulative distribution for the GBP/$ returns, for $\tau = 30$ min (left, log-log scale) and 1 day (right, semi-log scale), and best fit using a symmetric TLD $L_\mu^{(t)}$, of index $\mu = \frac{3}{2}$. We again show the best Student distribution and the Gaussian of same RMS.

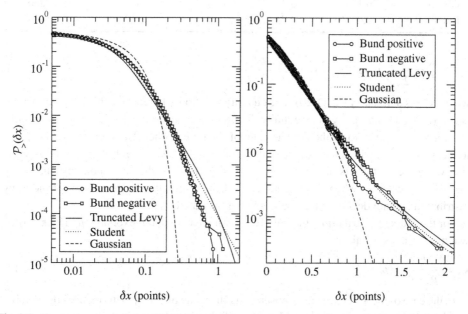

Fig. 6.7. Elementary cumulative distribution for the price increments (i.e. not the returns) of the Bund, for $\tau = 30$ min (left) and 1 day (right), and best fit using a symmetric TLD $L_\mu^{(t)}$, of index $\mu = \frac{3}{2}$, Student and Gaussian. In this case, the TLD and Student are not very good fits. A better fit is obtained using a TLD with a smaller μ.

Table 6.2. Value of the parameters obtained by fitting the 30 minutes data with a symmetric TLD $L_\mu^{(t)}$, of index $\mu = \frac{3}{2}$. The kurtosis is either measured directly or computed using the parameters of the best TLD. The data is returns in percent for the S&P 500 and GBP/\$, and absolute price changes for the Bund.

Asset	Variance σ_1^2 Measured	$a^{2/3}\alpha$ Fitted	Kurtosis κ_1 Measured	Computed	Log likelihood (per point)
S&P 500	0.240	0.096	15.8	23.8	−1.2964
GBP/\$	0.115	0.091	11.3	25.8	−1.2886
Bund	0.0666	0.046	11.9	72.2	−1.1197

Table 6.3. Value of the tail parameter μ obtained by fitting the 30 minute data with a Student distribution. The kurtosis is either measured directly or computed using the parameters of the best Student. For the three assets, the value of μ is found to be less than four, hence leading to an infinite theoretical kurtosis. The data is returns in % for the S&P 500 and GBP/\$, and absolute price changes for the Bund.

Asset	Variance σ_1^2 Measured	μ Fitted	Kurtosis κ_1 Measured	Computed	Log likelihood (per point)
S&P 500	0.240	3.24	15.8	∞	−1.2927
GBP/\$	0.115	3.20	11.3	∞	−1.2860
Bund	0.0666	2.74	11.9	∞	−1.1176

$$c_2 = \sigma_1^2 = \frac{3}{2\sqrt{2}} a_{3/2}\alpha^{-1} \qquad \kappa_{TLD} = \frac{\sqrt{2}}{2} \frac{1}{a_{3/2}\alpha^{3/2}}, \qquad (6.10)$$

where $a_{3/2}$ is defined by Eq. (1.20). The quantity we choose to report in the following Tables is $(a_{3/2})^{2/3}\alpha$. This quantity is essentially the fitted kurtosis to the power minus two thirds, and represents the ratio of the typical scale of the core of the TLD – given by $(a_{3/2})^{2/3}$ – to the scale α^{-1} at which the exponential cut-off takes place. When this parameter is small (i.e. when the kurtosis is large), the TLD is 'very close' to the corresponding pure Lévy distribution, whereas when this parameter is of order one, the power-law tail of the Lévy distribution cannot develop much before being truncated.

For the Student distribution, the only free parameter is the tail exponent μ. The kurtosis, when it exists, is equal to:

$$\kappa_S = \frac{6}{\mu - 4} \quad (\mu > 4). \qquad (6.11)$$

In the case of the TLD, we have chosen to fix the value of μ to $\frac{3}{2}$. This reduces the number of adjustable parameters, and is guided by the following observations:

- A large number of empirical studies that use pure Lévy distributions to fit the financial market fluctuations report values of μ in the range 1.6–1.8, obtained by fitting the whole

Table 6.4. Value of the parameters obtained by fitting the daily data with a symmetric TLD $L_\mu^{(t)}$, of index $\mu = \frac{3}{2}$. The kurtosis is either measured directly or computed using the parameters of the best TLD. The data is returns in % for the S&P 500 and GBP/\$, and absolute price changes for the Bund.

Asset	Variance σ_1^2 Measured	$a^{2/3}\alpha$ Fitted	Kurtosis κ_1 Measured	Kurtosis κ_1 Computed	Log likelihood (per point)
S&P 500	0.990	0.159	4.39	11.15	−1.3487
GBP/\$	0.612	0.185	3.40	8.9	−1.3617
Bund	0.3353	0.253	2.55	5.55	−1.3782

Table 6.5. Value of the tail parameter μ obtained by fitting the daily data with a Student distribution. The kurtosis is either measured directly or computed using the parameters of the best Student. The data is returns in % for the S&P 500 and GBP/\$, and absolute price changes for the Bund.

Asset	Variance σ_1^2 Measured	μ Fitted	Kurtosis κ_1 Measured	Kurtosis κ_1 Computed	Log likelihood (per point)
S&P 500	0.990	3.84	4.39	∞	−1.3461
GBP/\$	0.612	4.06	3.40	100	−1.3592
Bund	0.3353	4.78	2.55	7.7	−1.3770

distribution using maximum likelihood estimates, which focuses primarily on the 'core' of the distribution. However, in the absence of truncation (i.e. with $\alpha = 0$), the fit strongly overestimates the tails of the distribution. Choosing a higher value of μ partly corrects for this effect, since it leads to a thinner tail.

• If the exponent μ is left as a free parameter of the TLD, it is in many cases found to be in the range 1.4–1.6, although sometimes smaller, as in the case of the Bund ($\mu \approx 1.2$).

• The particular value $\mu = \frac{3}{2}$ has a simple (although by no means compelling) theoretical interpretation, which we shall discuss in Chapter 20.

Daily distribution of returns

The shape of the daily distribution of returns is also shown in Figures 6.5, 6.6 and 6.7. Again, the distributions are much broader than Gaussian, and can be fitted using a TLD or a Student distribution. The parameters of the fits are given in Tables 6.4 and 6.5, respectively. As expected on general grounds, the non-Gaussian character of these daily distributions is already weaker than at high frequency, as can be seen, for example, from the value of the Student index μ. Note that the daily distribution has a noticeable asymmetry in the case of the S&P 500, not captured by the fitted distributions. This asymmetry will be further discussed in Chapter 8.

Table 6.6. Value of the parameters corresponding to a TLD or Student (S) fit
to the daily and monthly returns of a pool of U.S. stocks. The variance of
each stock is normalized to one using the method explained in Section 4.4.

Time lag	$a^{2/3}\alpha$	μ	Kurtosis κ_1			Log likelihood per point	
			Measured	TLD	S	TLD	S
Daily	0.174	4.16	38.3	9.7	37.5	-1.3383	-1.3369
Monthly	0.332	5.68	5.38	3.7	3.6	-1.3818	-1.3818

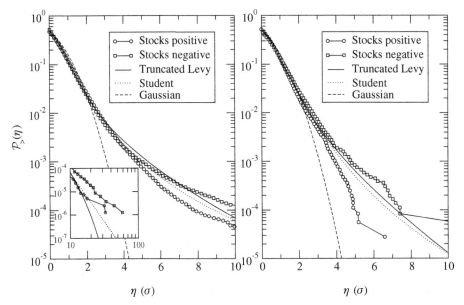

Fig. 6.8. Daily (left) and monthly (right) distribution returns of a pool of U.S. stocks. The RMS of
the log-returns of each stock was normalized to one using the method explained in Section 4.4.
Daily extreme events are shown in left inset. Note power law with $\mu = 3$ for the negative tails (open
squares).

Individual stocks

The analysis of individual stocks returns lead to comparable conclusions. We show in
Figure 6.8 the distribution of daily and monthly returns of a pool of U.S. stocks, together
with a fit, again using TLD or Student distributions. We show in the left inset the extreme
tail of the distribution of daily returns, in log-log coordinates. The power law nature of this
tail, with an exponent $\mu \approx 3$, is rather convincing. Note that the distribution of monthly
returns (22 days) is still strongly non-Gaussian. The parameters corresponding to the two
fits are reported in Table 6.6.

6.3.3 Convolutions

The parameterization of $P_1(\eta)$ as a TLD or a Student distribution, and the assumption that the increments are iid random variables, allows one to reconstruct the distribution of (log-)price increments for all time intervals $T = N\tau$. As discussed in Chapter 2, one then has $P(\log x/x_0, N) = [P_1]^{\star N}$. Figure 6.9 shows the cumulative distribution for the S&P 500 for $T = 1$ hour, 1 day and 5 days, reconstructed from the one at 15 min, according to the simple iid hypothesis. The symbols show empirical data corresponding to the same time intervals. The agreement is acceptable; one notices in particular the slow progressive deformation of $P(\log x/x_0, N)$ towards a Gaussian for large N. However, a closer look at Figure 6.9 reveals systematic deviations: for example the tails at 5 days are distinctively fatter than they should be. The evolution of the empirical kurtosis as a function of N confirms this finding: κ_N decreases much more slowly than the $1/N$ behaviour predicted by an iid hypothesis. This effect will be discussed in much more details in Chapter 7, see in particular Fig. 7.3.

Fig. 6.9. Evolution of the distribution of the price increments of the S&P 500, $P(\log x/x_0, N)$ (symbols), compared with the result obtained by a simple convolution of the elementary distribution P_1 (dark lines). The width and the general shape of the curves are rather well reproduced within this simple convolution rule. However, systematic deviations can be observed, in particular in the tails. This is also reflected by the fact that the empirical kurtosis κ_N decreases more slowly than κ_1/N.

6.4 Tails, what tails?

The asymptotic tails of TLD are exponential, and, as demonstrated by the above figures, satisfactorily fit the tails of the empirical distributions $P(\eta, N)$. However, as mentioned in Section 1.9 and in the above paragraph, the distribution of price changes can also be fitted using Student distributions, which have asymptotically much thicker, power-law tails. In many cases, for example for the distribution of losses of the S&P 500 (Fig. 6.5), one sees an upward bend in the plot of $\mathcal{P}_>(\eta)$ versus η in a linear-log plot. This indeed suggests that the decay could be slower than exponential. Recently, several authors have quite convincingly established that the tails of *stock* returns (and indices) are well described by a power-law with an exponent μ in the range 3–5. For example, the most likely value of μ using a Student distribution to fit the whole distribution of daily variations of the S&P in the period 1991–95 is $\mu \sim 4$ (see Table 6.5). The estimate of μ also somewhat depends on the time lag: tails tend to get thinner (i.e. μ increases) as one goes from intra-day time scales to week or months. Nevertheless, there is now good evidence that on short time scales, and using long time series, the tail exponent for stocks is around $\mu = 3$ on several different markets (U.S., Japan, Germany). We show in Figure 6.10 the Hill estimator of both the positive and negative tails of U.S. stocks (using the same data as in Fig. 6.8), where one indeed sees that the positive tail exponent is around $\mu = 4$ whereas the negative tail exponent is around $\mu = 3$.

Even if it is rather hard to distinguish empirically between an exponential and a high power-law, the precise characterisation of the tails is very important from a theoretical point of view. In particular, the existence of a finite kurtosis requires μ to be larger than 4. We have insisted on the kurtosis as an interesting measure of the deviation from Gaussianity. This idea will furthermore be quantitatively used in Section 13.3.4 in the context of option pricing. If $\mu < 4$ however, as seems to be the case for high frequency data, this quantity is

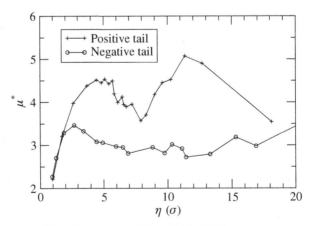

Fig. 6.10. Hill estimator of the tail exponent of the individual U.S. stock returns, normalized by their volatility using the trick explained in Eq. (4.37). The negative tail exponent is $\mu \approx 3$, whereas the positive tail exponent is $\mu \approx 4$.

at best ill-defined and one should worry about its interpretation.[†] This important topic will be further discussed in Section 7.1.5.

> *On the other hand, as far as applications to risk control are concerned, the difference between – say – a TLD with an exponential tail and a Student distribution with a high power-law tail is significant but not crucial. For example, suppose that one generates $N_1 = 1000$ positive random variables x_k with a $\mu = 4$ power-law distribution (representing – say – 8 years of daily losses), but fits the tail rank histogram as if the distribution tail was exponential. What would be the extrapolated value of the most probable 'worst loss' for a sample of $N_2 = 10\,000$ days (80 years) using this procedure?*
>
> *The rank ordering procedure leads to $x[n] = x_0(N_1/n)^{1/4}$, whereas an exponential tail would predict $x[n] = a_0 + a_1 \log(N_1/n)$. Fitting the values of a_0 and a_1 in the tail region $n \sim 10$ by forcing a logarithmic dependence leads to:*

$$a_1 = \frac{x_0}{4}\left(\frac{N_1}{10}\right)^{1/4} \tag{6.12}$$

$$a_0 = \frac{x_0}{4}\left(\frac{N_1}{10}\right)^{1/4}\left(1 - \frac{1}{4}\log\frac{N_1}{10}\right). \tag{6.13}$$

> *Using these values, the extrapolated worst loss for a series of size N_2 using an exponential tail is given by $a_0 + a_1 \log(N_2)$, whereas the true value should be $x_0 N_2^{1/4}$. The ratio φ of these two estimates, as a function of the ratio N_2/N_1, reads:*

$$\varphi = \left(\frac{N_1}{10N_2}\right)^{1/4}\left(1 + \frac{1}{4}\log\frac{10N_2}{N_1}\right), \tag{6.14}$$

> *which is equal to 0.68 for $N_2 = 10N_1$. Therefore, the exponential approximation underestimates the true risk by 30% for this case.[†] Note that as expected this ratio diverges when $N_2/N_1 \to \infty$: the local exponential approximation becomes worse and worse for deeper extrapolations. For a more general value of μ, the factor $1/4$ is replaced by $1/\mu$ and one can see that for large μ this ratio becomes exactly one. We again find that a power-law with a large exponent is close to an exponential.*

6.5 Extreme markets

We have considered, up to now, very liquid markets, where extreme price fluctuations are rather rare. On less liquid/less mature markets, the probability of extreme moves is much larger. The case of short-term interest rates is also interesting, since the evolution of, say, the 3-month rate is directly affected by the decision of central banks to increase or to decrease the day to day rate. As discussed in Section 3.1.3, these jumps lead to a rather high kurtosis, related to the fact that the short rate often does not change at all, but sometimes changes a lot. The distribution in this case is not well fitted, neither by a TLD nor by a Student distribution (Fig. 6.11 left). The kurtosis extracted from a TLD fit is 88.5, whereas the most probable Student exponent μ is found to be $\mu = 2.64$, corresponding to infinite kurtosis.

[†] Note that for a finite size sample of length N, the empirical kurtosis is finite but grows as $N^{4/\mu-1}$ for $\mu < 4$.
[†] This number would raise to 45% if $\mu = 3$.

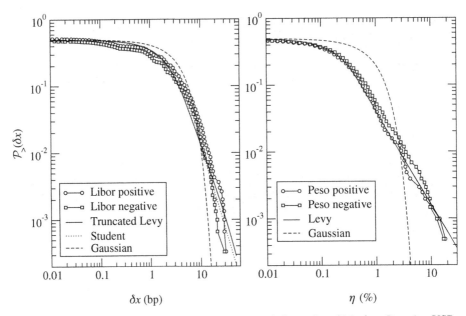

Fig. 6.11. Eurodollar front-month (ie. USD Libor futures, left panel) and Mexican Peso (vs. USD, right panel) daily returns cumulative distributions. Note that neither the TLD nor the Student fit are very good for Libor. For the Mexican peso a pure Lévy distribution $\mu = 1.34$ works remarkably well.

Emerging markets (such as Latin America or Eastern Europe markets) are even wilder. The example of the Mexican peso (MXN) is interesting, because the cumulative distribution of the daily changes of the rate MXN/\$ is extremely well fitted by a *pure Lévy distribution*, with no obvious truncation, with an exponent $\mu = 1.34$ (Fig. 6.11 right), and a daily MAD (mean absolute deviation) equal to 0.495%.

6.6 Discussion

The above statistical analysis already reveals very important differences between the simple model usually adopted to describe price fluctuations, namely the geometric (continuous-time) Brownian motion and the rather involved statistics of real price changes. The geometric Brownian motion description is at the heart of most theoretical work in mathematical finance, and can be summarized as follows:

- One assumes that the relative returns are independent identically distributed random variables.
- One assumes that the elementary time scale τ *tends to zero*; in other words that the price process is a continuous-time process. It is clear that in this limit, the number of independent price changes in an interval of time T is $N = T/\tau \to \infty$. One is thus in the limit where the CLT applies whatever the time scale T.

If the variance of the returns is finite on the shortest time increment τ, then according to the CLT, the only possibility is that prices obey a log-normal distribution. Therefore, the logarithm of the price performs a Brownian random walk. The process is *scale invariant*, that is that its statistical properties do not depend on the chosen time scale (up to a multiplicative factor – see Section 2.2.3).

The main difference between this basic model and real data is not only that the tails of the distributions are very poorly described by a Gaussian law (they seem to be power laws), but also that several important time scales appear in the analysis of price changes (see the discussion in the following chapters):

- A 'microscopic' time scale τ below which price changes are correlated. This time scale is of the order of several minutes even on very liquid markets.
- Several time scales associated to the volatility fluctuations, of the order of 10 days to a month or perhaps even longer (see Chapter 7).
- A time scale corresponding to the negative return-volatility correlations, which is of the order of a week for stock indices (see Chapter 8).
- A time scale governing the crossover from an additive model, where absolute price changes are the relevant random variables for short time scales, to a multiplicative model, where relative returns become relevant. This time scale is on the order of months (see Chapter 8).

It is clear that the existence of all these effects is extremely important to take into account in a faithful representation of price changes, and play a crucial role both for the pricing of derivative products, and for risk control. Different assets differ in the value of their higher cumulants (skewness, kurtosis), and in the value of the different time scales. For this reason, a description where the volatility is the only parameter (as is the case for Gaussian models) are bound to miss a great deal of the reality.

6.7 Summary

- Price returns are to a good approximation uncorrelated beyond a time scale of a few tens of minutes in liquid markets. On shorter time scales, strong correlation effects are observed. For time scales on the order of days, weak residual correlations can be detected in long time series of stocks and index returns.
- Price returns have strongly non-Gaussian distributions, which can be fitted by truncated Lévy or Student distributions. The tails are much fatter than those of a Gaussian distribution, and can be fitted by a Pareto (power-law) tail with an exponent μ in the range 3 to 5 in liquid markets. This exponent can be much smaller in less mature markets, such as emerging country markets, where pure Lévy distributions are observed.
- The distribution of returns corresponding to a time difference equal to $N\tau$ is given to a first approximation, by the Nth convolution of the elementary distribution at scale τ. However, systematic deviations from this simple rule can be observed, and are related to correlated volatility fluctuations, discussed in the following chapters.

• FURTHER READING

L. Bachelier, *Théorie de la spéculation*, Gabay, Paris, 1995.

J. Y. Campbell, A. W. Lo, A. C. McKinley, *The Econometrics of Financial Markets*, Princeton University Press, Priceton, 1997.

R. Cont, *Empirical properties of asset returns: stylized facts and statistical issues*, Quantitative Finance, **1**, 223, (2001).

P. H. Cootner, *The Random Character of Stock Market Prices*, MIT Press, Cambridge, 1964.

M. M. Dacorogna, R. Gençay, U. A. Muller, R. B. Olsen, O. V. Pictet, *An Introduction to High-Frequency Finance*, Academic Press, San Diego, 2001.

J. Lévy-Vehel, Ch. Walter, *Les marchés fractals*, P.U.F., Paris, 2002.

B. B. Mandelbrot, *Fractals and Scaling in Finance*, Springer, New York, 1997.

R. Mantegna, H. E. Stanley, *An Introduction to Econophysics*, Cambridge University Press, Cambridge, 1999.

S. Rachev, S. Mittnik, *Stable Paretian Models in Finance*, Wiley, 2000.

• SPECIALIZED PAPERS

N. H. Bingham, R. Kiesel, *Semi-parametric modelling in finance: theoretical foundations*, Quantitative Finance, **2**, 241 (2002).

P. Carr, H. Geman, D. Madan, M. Yor, *The fine structure of asset returns: an empirical investigation*, Journal of Business **75**, 305 (2002).

R. Cont, M. Potters, J.-P. Bouchaud, *Scaling in stock market data: stable laws and beyond*, in *Scale invariance and beyond*, B. Dubrulle, F. Graner, D. Sornette (Edts.), EDP Sciences, 1997.

M. M. Dacorogna, U. A. Muller, O. V. Pictet, C. G. de Vries, *The distribution of extremal exchange rate returns in extremely large data sets*, Olsen and Associate working paper (1995), available at http://www.olsen.ch.

P. Gopikrishnan, M. Meyer, L. A. Amaral, H. E. Stanley, *Inverse cubic law for the distribution of stock price variations*, European Journal of Physics, **B 3**, 139 (1998).

P. Gopikrishnan, V. Plerou, L. A. Amaral, M. Meyer, H. E. Stanley, *Scaling of the distribution of fluctuations of financial market indices*, Phys. Rev., **E 60**, 5305 (1999).

F. Longin, *The asymptotic distribution of extreme stock market returns*, Journal of Business, **69**, 383 (1996).

T. Lux, *The stable Paretian hypothesis and the frequency of large returns: an examination of major German stocks*, Applied Financial Economics, **6**, 463, (1996).

B. B. Mandelbrot, *The variation of certain speculative prices*, Journal of Business, **36**, 394 (1963).

B. B. Mandelbrot, *The variation of some other speculative prices*, Journal of Business, **40**, 394 (1967).

V. Plerou, P. Gopikrishnan, L. A. Amaral, M. Meyer, H. E. Stanley, *Scaling of the distribution of price fluctuations of individual companies*, Phys. Rev., **E 60**, 6519 (1999).

Lei-Han Tang and Zhi-Feng Huang, *Modelling High-frequency Economic Time Series*, Physica A, **28**, 444 (2000).

7

Non-linear correlations and volatility fluctuations

For in a minute there are many days.

(William Shakespeare, *Romeo and Juliet*.)

7.1 Non-linear correlations and dependence

7.1.1 Non identical variables

The previous chapter presented a statistical analysis of financial data that implicitly assumes *homogenous* time series. For example, when we constructed the empirical distributions of price returns η_k, we pooled together all the data, thereby assuming that $P_1(\eta_1)$, $P_2(\eta_2), \ldots, P_N(\eta_N)$ are all identical.

It turns out that the distributions of the elementary random variables $P_1(\eta_1)$, $P_2(\eta_2), \ldots, P_N(\eta_N)$ are often not all identical. This is the case, for example, when the variance of the random process itself depends upon time – in financial markets, it is a well-known fact that the daily volatility is time dependent, taking rather high levels in periods of uncertainty, and reverting back to lower values in calmer periods. This phenomena is called **heteroskedasticity**. For example, the volatility of the bond market has been very high during 1994, and decreased in later years. Similarly, the stock markets since the beginning of the century have witnessed periods of high volatility separated by rather quiet periods (Fig. 7.1). Another interesting analogy is turbulent flows which are well known to be *intermittent*: periods of relative tranquility (laminar flow), where the velocity field is smooth, are interrupted by intense 'bursts' of activity.[†]

If the distribution P_k varies sufficiently 'slowly', one can in principle measure some of its moments (for example its mean and variance) over a time scale which is long enough to allow for a precise determination of these moments, but short compared to the time scale over which P_k is expected to vary. The situation is less clear if P_k varies 'rapidly'. Suppose for example that $P_k(\eta_k)$ is a Gaussian distribution of variance σ_k^2, which is itself a random variable. We shall denote as $\overline{(\cdots)}$ the average over the random variable σ_k, to distinguish it from the notation $\langle \cdots \rangle_k$ which we have used to describe the average over the

[†] On this point, see Frisch (1997).

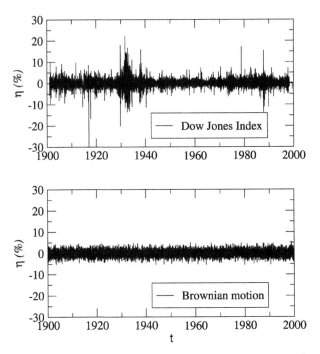

Fig. 7.1. Top panel: daily returns of the Dow Jones index since 1900. One very clearly sees the intermittent nature of volatility fluctuations. Note in particular the sharp feature in the 1930s. Lower panel: daily returns for a geometric Brownian random walk, showing a featureless pattern.

probability distribution P_k. If σ_k varies rapidly, it is impossible to separate the two sources of uncertainty. Thus, the empirical histogram constructed from the series $\{\eta_1, \eta_2, \dots, \eta_N\}$ leads to an 'apparent' distribution \overline{P} which is non-Gaussian even if each individual P_k is Gaussian. Indeed, from:

$$\overline{P}(\eta) \equiv \int P(\sigma) \frac{1}{\sqrt{2\pi\sigma^2}} \exp\left(-\frac{\eta^2}{2\sigma^2}\right) d\sigma, \tag{7.1}$$

one can calculate the kurtosis of \overline{P} as:

$$\overline{\kappa} = \frac{\overline{\langle \eta^4 \rangle}}{(\overline{\langle \eta^2 \rangle})^2} - 3 \equiv 3\left(\frac{\overline{\sigma^4}}{(\overline{\sigma^2})^2} - 1\right). \tag{7.2}$$

Since for any random variable one has $\overline{\sigma^4} \geq (\overline{\sigma^2})^2$ (the equality being reached only if σ^2 does not fluctuate at all), one finds that $\overline{\kappa}$ is always positive. Volatility fluctuations thus lead to 'fat tails'.

More precisely, let us assume that the probability distribution of the RMS, $P(\sigma)$, decays itself for large σ as $\exp(-\sigma^c)$, $c > 0$. Assuming P_k to be Gaussian, it is easy to obtain,

using a saddle-point method (cf. Eq. (2.47)), that for large η one has:

$$\log[\overline{P}(\eta)] \propto -\eta^{\frac{2c}{2+c}}. \tag{7.3}$$

Since $c < 2 + c$, this asymptotic decay is always much slower than in the Gaussian case, which corresponds to $c \to \infty$. The case where the volatility itself has a Gaussian tail ($c = 2$) leads to an exponential decay of $\overline{P}(\eta)$.

Another interesting case is when σ^2 is distributed as a completely asymmetric Lévy distribution ($\beta = 1$) of exponent $\mu < 1$. Using the properties of Lévy distributions, one can then show that \overline{P} is itself a symmetric Lévy distribution ($\beta = 0$), of exponent equal to 2μ. When σ^2 is distributed according to an inverse gamma distribution, \overline{P} is a Student distribution.

7.1.2 A stochastic volatility model

If the fluctuations of σ_k are themselves correlated, one observes an interesting case of *dependence*. For example, if σ_k is large, σ_{k+1} will probably also be large. The fluctuation η_k thus has a large probability to be large (but of arbitrary sign) twice in a row.

We shall often refer, in the following, to a rather general **stochastic volatility model** where the random part of the return can be written as a product:

$$\frac{\delta x_k}{x_k} \equiv \eta_k = m + \epsilon_k \sigma_k, \tag{7.4}$$

where ϵ_k are iid random variables of zero mean and unit variance (but not necessarily Gaussian), and σ_k corresponds to the local 'scale' of the fluctuations, which can be correlated in time and also, possibly, with the random variables ϵ_j with $j < k$ (see Chapter 8).

Assuming for the moment that $m = 0$ and that the random variables ϵ and σ are independent from each other, the correlation function of the returns η_k is given by:

$$\overline{\langle \eta_i \eta_j \rangle} = \overline{\sigma_i \sigma_j} \langle \epsilon_i \epsilon_j \rangle = \delta_{i,j} \overline{\sigma^2}. \tag{7.5}$$

Hence the returns are uncorrelated random variables, but they are *not* independent since higher-order, non-linear, correlation functions reveal a richer structure. Let us for example consider the correlation of the square returns:

$$C_2(i, j) \equiv \overline{\langle \eta_i^2 \eta_j^2 \rangle} - \overline{\langle \eta_i^2 \rangle \langle \eta_j^2 \rangle} = \overline{\sigma_i^2 \sigma_j^2} - \overline{\sigma_i^2} \, \overline{\sigma_j^2} \qquad (i \neq j), \tag{7.6}$$

which indeed has an interesting temporal behaviour in financial time series, as we shall discuss below. Notice that for $i \neq j$ this correlation function can be zero either because σ is identically equal to a certain value σ_0, or because the fluctuations of σ are completely uncorrelated from one time to the next.

7.1.3 GARCH(1,1)

As an illustration, let us discuss a model often used in finance, called GARCH(1,1), where the volatility evolves through a simple feedback mechanism.[†] More precisely, one writes the following set of equations:

$$\eta_i = \sigma_i \epsilon_i \tag{7.7}$$

$$\sigma_i^2 = \sigma_0^2 + \alpha \left(\sigma_{i-1}^2 - \sigma_0^2 \right) + g \left(\eta_{i-1}^2 - \sigma_{i-1}^2 \right). \tag{7.8}$$

where ϵ_i are Gaussian iid random numbers of zero mean and unit variance. The three parameters of the model have the following interpretation: σ_0^2 is the unconditional average variance, α measures the strength with which the volatility reverts to its average value and g is a coupling parameter describing how the difference between the realized past square return η_{i-1}^2 and its expected value σ_{i-1}^2 feeds back on the present value of the volatility. The logic is that large squared returns are a source for higher future volatility. Note that unlike the stochastic volatility model discussed above, this model only has a single noise source ϵ_i which determines both the price and the volatility processes. For this reason, in this section we will use a single type of averaging denoted by $\langle \ldots \rangle$.

In this model returns have zero mean ($\langle \eta_i \rangle = 0$) and are uncorrelated ($\langle \eta_i \eta_j \rangle = 0$, for $i \neq j$), and the unconditional variance is given by

$$\langle \eta_i^2 \rangle = \langle \sigma_i^2 \rangle = \sigma_0^2. \tag{7.9}$$

To compute the auto-correlation function of σ_i^2 and that of η_i^2, one should realize that after squaring Eq. (7.7), the model is a linear coupled system with noise. One can then find a change of variables to obtain two linear equations coupled only through the noise term. More precisely, setting $\delta_i = \eta_i^2 - \sigma_i^2$ and $\psi_i = g\eta_i^2 + (\alpha - g)\sigma_i^2 + \alpha\sigma_0^2$ transforms the system into

$$\delta_i = \left(\psi_{i-1} + \sigma_0^2 \right) \tilde{\xi}_i \tag{7.10}$$

$$\psi_i = \alpha\psi_{i-1} + g \left(\psi_{i-1} + \sigma_0^2 \right) \tilde{\xi}_i, \tag{7.11}$$

where $\tilde{\xi}_i = \epsilon_i^2 - 1$ a zero mean (non-Gaussian) noise. From then on, it is relatively straightforward (but tedious) to compute the squared auto-correlation functions of this model. For the volatilities, one finds:

$$\langle \sigma_i^2 \sigma_j^2 \rangle - \langle \sigma_i^2 \rangle \langle \sigma_j^2 \rangle = \frac{2g^2\sigma_0^4}{1 - \alpha^2 - 2g^2} \alpha^{|i-j|}. \tag{7.12}$$

This result holds provided $\alpha < 1$ and $2g^2 < 1 - \alpha^2$, which are necessary conditions to obtain a stationary model. For g too large (large feedback), the volatility process blows up. In the regular case, the above correlation function decays exponentially with a characteristic time equal to $1/|\log\alpha|$. In fact, the volatility process follows a (non-Gaussian)

[†] GARCH stands for Generalized Autoregressive Conditional Heteroskedasticity model, as the name suggests GARCH is a generalization of the simpler ARCH model, where $\sigma_i^2 = \sigma_0^2 + g\eta_{i-1}^2$.

Ornstein-Uhlenbeck process (compare with Section 4.3.1). Similarly, one finds that the squared return correlation function is given by:

$$C_2(i, j) \equiv \langle \eta_i^2 \eta_j^2 \rangle - \langle \eta_i^2 \rangle \langle \eta_j^2 \rangle = \begin{cases} \dfrac{2\sigma_0^4(1 - \alpha^2 + g^2)}{1 - \alpha^2 - 2g^2} & i = j \\[3mm] \dfrac{2g\sigma_0^4(1 - \alpha^2 + g\alpha)}{1 - \alpha^2 - 2g^2} \alpha^{|i-j|-1} & i \neq j \end{cases},$$ (7.13)

which also decays exponentially with the same correlation time. Using the $i = j$ term, we can compute the unconditional kurtosis

$$\kappa = \frac{6g^2}{1 - \alpha^2 - 2g^2},$$ (7.14)

which is zero in the absence of feedback ($g = 0$).

7.1.4 Anomalous kurtosis

Suppose that the correlation function $\overline{\sigma_i^2 \sigma_j^2} - \overline{\sigma^2}^2$ decreases (even very slowly) with $|i - j|$. One can then show that the sum of the η_k, given by $\sum_{k=1}^{N} \epsilon_k \sigma_k$ is still governed by the CLT, and converges for large N towards a Gaussian variable. A way to see this is to compute the average unconditional kurtosis of the sum, κ_N. As detailed below, one finds the following result:

$$\kappa_N = \frac{1}{N} \left[\kappa_0 + (3 + \kappa_0)g(0) + 6 \sum_{\ell=1}^{N} \left(1 - \frac{\ell}{N} \right) g(\ell) \right],$$ (7.15)

where κ_0 is the kurtosis of the variable ϵ, and $g(\ell)$ the correlation function of the variance, defined as:

$$\overline{\sigma_i^2 \sigma_j^2} - \overline{\sigma^2}^2 = \overline{\sigma^2}^2 g(|i - j|).$$ (7.16)

It is interesting to see that for $N = 1$, the above formula gives:

$$\kappa_1 = \kappa_0 + (3 + \kappa_0)g(0) > \kappa_0,$$ (7.17)

which means that even if $\kappa_0 = 0$, a fluctuating volatility is enough to produce some kurtosis, as we already mentioned: for $\kappa_0 = 0$ the previous formula for κ_1 reduces to Eq. (7.2) above.

More importantly, one sees that as soon as the variance correlation function $g(\ell)$ decays with ℓ, the kurtosis κ_N tends to zero with N, thus showing that the sum indeed converges towards a Gaussian variable. For example, if $g(\ell)$ decays as a power-law $\ell^{-\nu}$ for large ℓ, one finds that for large N:

$$\kappa_N \propto \frac{1}{N} \quad \text{for} \quad \nu > 1; \qquad \kappa_N \propto \frac{1}{N^\nu} \quad \text{for} \quad \nu < 1.$$ (7.18)

The difference comes form the fact that the sum over ℓ that appears in the expression of κ_N is convergent for $\nu > 1$ but diverges as $N^{1-\nu}$ for $\nu < 1$.[†] Hence, long-range correlations in the variance considerably slow down the convergence towards the Gaussian: the kurtosis and all the 'hypercumulants' defined in Eq. (2.43) decay very slowly to zero. This remark will be of importance in the following, since financial time series often reveal long-ranged volatility fluctuations, with ν rather small. Therefore, the non Gaussian corrections do remain anomalously high even for relatively long time scales (months or even years).

An important remark must be made here. The quantity κ_N defined above is an unconditional kurtosis, where we averaged over both sources of uncertainty: ϵ and σ. However, since the correlation function of the volatility is assumed to be non trivial, σ must have some degree of predictability. The *conditional* kurtosis κ_N^c of $\log(x_N/x_0)$, computed using all the information available at time 0, is in general smaller than the unconditional kurtosis computed above. Let us assume for example that all the future σ_i are in fact perfectly known. The conditional kurtosis in this case reads:

$$\kappa_N^c = \kappa_0 \frac{\sum_{i=1}^{N} \sigma_i^4}{\left(\sum_{i=1}^{N} \sigma_i^2\right)^2}. \tag{7.19}$$

Note that if the ϵ are Gaussian variables, $\kappa_0 = 0$ and hence $\kappa_N^c \equiv 0$, whereas the unconditional kurtosis κ_N is always positive. The kurtosis can be seen as a measure of our ignorance about the future value of the volatility.

Let us show how Eq. (7.15) can be obtained. We calculate the unconditional average kurtosis of the sum $\sum_{i=1}^{N} \eta_i$, assuming that the η_i can be written as $\sigma_i \epsilon_i$. We define the rescaled correlation function g as in Eq. (7.16).

Let us first compute $\langle (\sum_{i=1}^{N} \eta_i)^4 \rangle$, where $\langle \cdots \rangle$ means an average over the ϵ_i's and the overline means an average over the σ_i's. If $\langle \epsilon_i \rangle = 0$, and $\langle \epsilon_i \epsilon_j \rangle = 0$ for $i \neq j$, one finds:

$$\overline{\left\langle \left(\sum_{i,j,k,l=1}^{N} \eta_i \eta_j \eta_k \eta_l \right) \right\rangle} = \sum_{i=1}^{N} \overline{\langle \eta_i^4 \rangle} + 3 \sum_{i \neq j=1}^{N} \overline{\langle \eta_i^2 \rangle \langle \eta_j^2 \rangle}$$

$$= (3 + \kappa_0) \sum_{i=1}^{N} \overline{\langle \eta_i^2 \rangle^2} + 3 \sum_{i \neq j=1}^{N} \overline{\langle \eta_i^2 \rangle \langle \eta_j^2 \rangle}, \tag{7.20}$$

where we have used the definition of κ_0 (the kurtosis of ϵ). On the other hand, one must estimate $\overline{\left\langle \left(\sum_{i=1}^{N} \eta_i \right)^2 \right\rangle}^2$. One finds:

$$\overline{\left\langle \left(\sum_{i=1}^{N} \eta_i \right)^2 \right\rangle}^2 = \sum_{i,j=1}^{N} \overline{\langle \eta_i^2 \rangle} \, \overline{\langle \eta_j^2 \rangle}. \tag{7.21}$$

[†] This sum diverges as $\log N$ for $\nu = 1$.

Gathering the different terms and using the definition Eq. (7.16), one finally establishes the following general relation:

$$\kappa_N = \frac{1}{N^2 \overline{\sigma^2}^2} \left[N \overline{\sigma^2}^2 (3 + \kappa_0)(1 + g(0)) - 3N \overline{\sigma^2}^2 + 3 \overline{\sigma^2}^2 \sum_{i \neq j = 1}^{N} g(|i - j|) \right], \qquad (7.22)$$

which readily gives Eq. (7.15).

7.1.5 The case of infinite kurtosis*

What survives from the above analysis if the kurtosis of the process on elementary time scales is infinite? An illustrative case is that of a Student process with $\mu = 3$, as already considered in Section 2.3.5. Suppose that the return η_k at time k is distributed according to:

$$P_k(\eta_k) = \frac{2a_k^3}{\pi \left(\eta_k^2 + a_k^2 \right)}. \qquad (7.23)$$

The local volatility of this process is $\sigma_k \equiv a_k$, but its local kurtosis is infinite. The characteristic function of this distribution is given by Eq. (2.56). From this result, one obtains the characteristic function for the sum of N returns as:

$$\hat{P}(z, N) = \exp \sum_{k=1}^{N} [\log(1 + |z|a_k) - |z|a_k]. \qquad (7.24)$$

Anticipating that for large N the distribution becomes Gaussian, we set $z = \hat{z}/\sqrt{N}$ and expand for large N. One finds:

$$\hat{P}(z, N) \simeq \exp \left[-\frac{\hat{z}^2}{2N} \sum_{k=1}^{N} a_k^2 + \frac{|\hat{z}|^3}{3N^{3/2}} \sum_{k=1}^{N} a_k^3 - \frac{\hat{z}^4}{4N^2} \sum_{k=1}^{N} a_k^4 + \cdots \right]. \qquad (7.25)$$

Let us find the unconditional distribution by averaging the above result over the volatility fluctuations. Much as in the case above where the kurtosis is finite, the fluctuations of a_k^2 give rise to a term proportional to \hat{z}^4 that reads $\sum_{ij} C_2(i, j)\hat{z}^4/4N^2$. If the exponent ν describing the decay of the variance fluctuations is less than unity, then, as discussed above, this term is of order $N^{-\nu}$ and dominates the term $\sum_{k=1}^{N} a_k^4 \hat{z}^4/4N^2 \simeq \overline{a^4} \hat{z}^4/4N$, which is of order $1/N$ provided $\overline{a^4}$ is finite. Now, one has to compare $N^{-\nu}$ with the term $\sum_{k=1}^{N} a_k^3 |\hat{z}|^3/3N^{3/2} \simeq \overline{a^3} |\hat{z}|^3/3N^{1/2}$, which is of order $1/N^{1/2}$ and, as discussed in Section 2.3.5, dominates the convergence towards the Gaussian if volatility fluctuations are absent.

One therefore finds that if volatility fluctuations are such that $\nu > 1/2$, the dominant factor slowing down the convergence to the Gaussian are the inverse cubic tails of the elementary distribution, whereas if $\nu < 1/2$, the dominant factor becomes the volatility correlations. In this case, the fact that the kurtosis is infinite on short time scales does not prevent the

convergence to the Gaussian to be dominated, for large N by an effective kurtosis κ_N given by:

$$\kappa_N = \frac{6}{N} \sum_{\ell=1}^{N} \left(1 - \frac{\ell}{N}\right) g(\ell). \tag{7.26}$$

This remark is very important in practice since for many assets $\mu = 3$ whereas the volatility fluctuation exponent ν is indeed often found to be smaller than $1/2$. The long term deviations form Gaussianity, relevant for the shape of the option smiles for long maturities, can therefore be estimated using an effective kurtosis, even if the kurtosis is infinite on short time scales.

Note finally that if the tails of the elementary distribution decay with a power-law μ in the range $]2, 4]$, one can argue similarly that whenever $\nu < \mu/2 - 1$, long range volatility fluctuations are dominant for large N. (The above case corresponds to $\mu = 3$.)

7.2 Non-linear correlations in financial markets: empirical results

7.2.1 Anomalous decay of the cumulants

As mentioned in Chapter 6, one sees from Figure 6.9 that $P(\eta, N)$ systematically deviates from $[P_1(\eta_1)]^{*N}$. In particular, the tails of $P(\eta, N)$ are anomalously 'fat' compared to what should be expected if the price increments were iid. A way to quantify this effect is to study the kurtosis κ_N of $P(\eta, N)$, which is found to be larger than κ_1/N. This can be seen from Figures 7.2 and 7.3, where κ_N is plotted as a function of N, for the assets considered

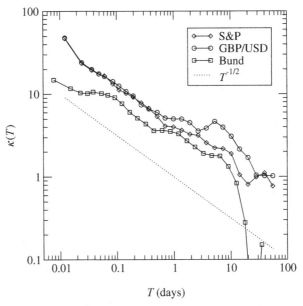

Fig. 7.2. Kurtosis κ_N at scale $N = \frac{T}{\tau}$ as a function of N, for the S&P 500, the GBP/USD and the Bund. If the iid assumption were true, one should find $\kappa_N = \kappa_1/N$. A straight line of slope of $-1/2$, is shown for comparison. The decay of the kurtosis κ_N is much slower than $1/N$.

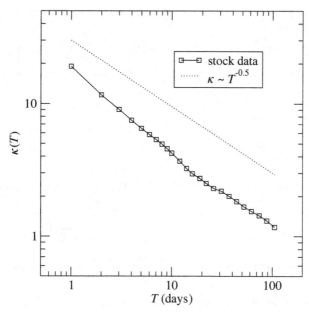

Fig. 7.3. Average kurtosis as a function of time for individual stocks. Time is expressed in trading days. The kurtosis was computed individually for each stock and then averaged over all stocks. The dataset consisted of 10 to 12 years of daily prices from a pool of 288 U.S. large cap stocks. The dotted line shows a $1/\sqrt{T}$ scaling for comparison.

in Chapter 6: the S&P 500, the GBP/USD and the Bund on the one hand, and for a pool of U.S. stocks on the other hand. However, as we have seen in Section 6.4, the tails of $P(\eta, N)$ are so fat that the kurtosis might actually be ill defined on short time scales. A more robust estimation of the convergence towards the Gaussian is provided by lower moments. More precisely, consider the time dependent 'hypercumulants' Λ_q defined in Section 2.3.3, which read:

$$\Lambda_q(N) = \frac{\sqrt{\pi}}{2^{q/2}\Gamma\left(\frac{1+q}{2}\right)} \frac{\langle |\log x_N/x_0|^q \rangle}{\langle |\log x_N/x_0|^2 \rangle^{q/2}} - 1. \tag{7.27}$$

Figure 7.4 shows Λ_q as a function of $T = N\tau$ for $q = 1, 3, 4$. Although the time dependence of Λ_q depends on q for small time lags T, one sees clearly that for longer times, all these hypercumulants behave in the same way. This can be understood following Section 7.1.5 above: for short times, high moments are affected by the power-law tail of high frequency returns and contain an information not revealed by lower moments, whereas at long times, the dominant effect is the slow decay of the volatility fluctuations, which affects all the moments in a similar fashion: see Eq. (2.43). It is interesting to test directly the small kurtosis expansion formula (2.43) by plotting $\Lambda_q(N)/\Lambda_{q^*}(N)$ as a function of q for different

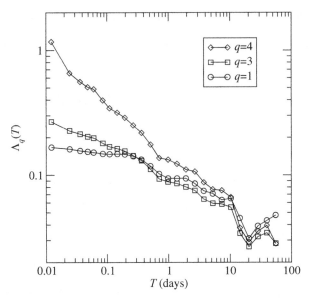

Fig. 7.4. 'Hypercumulants' Λ_1, $\Lambda_3/3$ and $\Lambda_4/8$ for the S&P 500 as a function of the time scale $T = N\tau$ (τ is equal to 5 minutes). The rescaling factors 3 and 8 above were chosen according to formula (2.43), which suggest that at long times these quantities should be equal. One sees from this graph this approximately holds for $T > 1$ day.

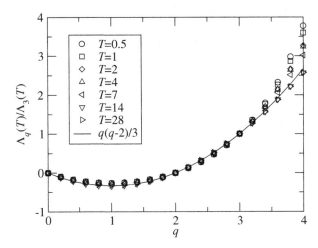

Fig. 7.5. Rescaled hypercumulants Λ_q/Λ_3 (for the S&P 500), as a function of q for different time scales T from half a day to 30 days (symbols). These curves should collapse at long times on top of each other and coincide with the parabola $q(q-2)/3$ (plain line). Note that even for $q \sim 4$, the convergence towards the theoretical prediction is observed for long times.

N (Fig. 7.5). Here q^* is a fixed number that we choose to be $q^* = 3$. The prediction of Eq. (2.43) is:

$$\frac{\Lambda_q(N)}{\Lambda_3(N)} = \frac{1}{3}q(q-2), \tag{7.28}$$

independently of N. From Figure 7.5, we see that the rescaling of the curves correspond-ing to different N onto the theoretical prediction (plain line) is rather good, although the convergence is slower for $q \sim 4$, as expected.

We conclude that the convergence to Gaussian is indeed, for medium to long time lags (say beyond a few days) well represented by a single number, the effective kurtosis, which captures the behaviour of all the even moments. (For odd moments, related to small skewness effects, another mechanism comes into play – see Chapter 8.)

7.2.2 Volatility correlations and variogram

The fact that the η_i's are *not* iid, as the study of the cumulants illustrates, also appears when one looks at non-linear correlations functions, such as that of the *squares* of η_i, which reveal a non-trivial structure. These correlations are associated to a very interesting dynamics of the volatility that we discuss in the present section.

A volatility proxy

Although of great practical interest for many purposes, such as risk control or option pricing, the true volatility process is not directly observable – one can at best study approximate estimates of the volatility. For example, it is interesting to define a proxy of the volatility using the daily average of 5-min price increments, as:

$$\sigma_{hf} = \frac{1}{N_d} \sum_{k=1}^{N_d} |\eta_k(5')|, \tag{7.29}$$

where σ_{hf} is our notation for a high frequency volatility proxy, $\eta_k(5')$ is the 5-min increment, and N_d is the number of 5-min intervals within a day. It is reasonable to expect that the true volatility σ is proportional to σ_{hf}, plus a noise term, which can be thought of as a measurement error. Indeed, even for a Brownian random walk with a constant volatility, the quantity σ_{hf} depends, for N_d finite, on the particular realization of the Gaussian process. Other proxies for the daily volatility can be constructed, for example the absolute daily return $|\eta_i|$, or the difference between the 'high' and the 'low', divided by the 'open' of that given day. This last proxy will be noted σ_{HL}.

The quantity σ_{hf} as a function of time for the S&P 500 in the period 1991–2001 is plotted in Figure 7.6, and reveals obvious correlations: the periods of strong volatility persist far beyond the day time scale. It is interesting to compare this graph with that shown in Figure 7.1 for the Dow Jones over a much longer time span (100 years): the same alternation of highly volatile and rather quiet periods can be observed on all time scales.

Probability distribution of the empirical volatility

The probability density of σ_{hf} for the S&P 500 is shown in Figure 7.7, together with two very good fits: a log-normal distribution and an inverse gamma distribution:

$$P_\Gamma(\sigma) = \frac{\sigma_0^\mu}{\Gamma(\mu)\sigma^{1+\mu}} \exp\left(-\frac{\sigma_0}{\sigma}\right). \tag{7.30}$$

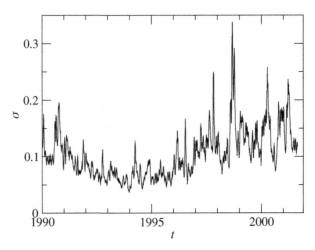

Fig. 7.6. Evolution of the 'volatility' σ_{hf} as a function of time, for the S&P 500 in the period 1990–2001. One clearly sees periods of large volatility, which persist in time. Here the signal σ_{hf} has been filtered over two weeks.

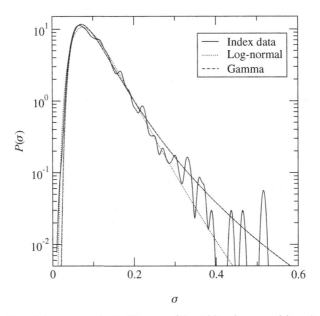

Fig. 7.7. Distribution of the measured volatility σ_{hf} of the S&P using a semi-log plot, and fits using a log-normal distribution, or an inverse gamma distribution. Note that the log-normal distribution underestimates the right tail.

This distribution is motivated by simple models of volatility feedback, such as the one discussed in Chapter 20. The two fits were performed using maximum likelihood on $\log \sigma_{hf}$, and are of comparable quality. The total number of points is 3000. The log-likelihood per point for the log-normal fit is found to be -1.419 (see Section 4.1.4), whereas the log-likelihood

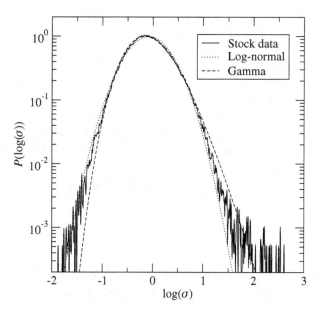

Fig. 7.8. Distribution of the measured volatility σ_{HL} of individual stocks, in a log-log scale, and fits using a log-normal distribution, or an inverse gamma distribution. Again, the log-normal distribution underestimates the right tail, and is not very good in the centre.

per point for the inverse gamma distribution is slightly better, equal to -1.4177, with an exponent μ equal to 5.4.[†] The log-normal tends to underestimate the large volatility tail, whereas the inverse gamma distribution tends to overestimate the same tail. Identical conclusions are drawn from the study of the empirical volatility σ_{HL} of individual stocks, as shown in Figure 7.8. Again, the inverse gamma distribution (with $\mu = 6.43$) is found to be superior to the log-normal fit (the relative likelihood is now 10^{-60}, with 150 000 points), and the same systematics are observed.

Correlations and variograms

A more quantitative characterisation of these long-ranged correlations can be obtained using the so-called volatility variogram, which measures how 'far' the volatility at time $i + \ell$ has escaped away from the value it had at time i. More precisely, we will study the quantity:

$$V_\sigma(\ell) = \left\langle \left(\sigma_{e,i}^2 - \sigma_{e,i+\ell}^2\right)^2 \right\rangle_e, \tag{7.31}$$

where $\langle \ldots \rangle_e$ refers to an empirical average over the initial time i, and σ_e to an empirical volatility proxy. Let us first discuss how this quantity is related to the correlation of the square of price increments that we introduced above in relation with the kurtosis (see Eq. (7.6)).

[†] Although the difference per point is small, the overall relative likelihood of the log-normal compared to the inverse gamma distribution is $\exp -(3000 \times 0.0013) = 0.02$. Note that the volatility estimators used here, in particular σ_{HL}, tend to overestimate the probability of small volatilities.

Consider the quantity:

$$V_2(\ell) = \overline{\langle (\eta_i^2 - \eta_{i+\ell}^2)^2 \rangle}. \tag{7.32}$$

As can be seen by expanding the square, the above quantity is simply equal to $2[C_2(i, i) - C_2(i, i + \ell)]$. Therefore the variogram of the square increments is closely related to their correlation function. However, the empirical determination of the variogram does not require the knowledge of $\langle \eta_i^2 \rangle$, needed to determine the correlation function. This is interesting since the latter quantity is often noisy and does not allow a precise determination of the baseline of the correlations. Using the fact that $\eta_i = \sigma_i \epsilon_i$ where the ϵ are iid, we obtain:

$$V_2(\ell) = \overline{(\sigma_i^2 - \sigma_{i+\ell}^2)^2} + v_0(1 - \delta_{\ell,0}), \tag{7.33}$$

where the last term comes from the contribution of the ϵ. Now, since σ and σ_e are proportional up to a measurement noise term which is presumably uncorrelated from one day to the next, one also obtains:

$$V_\sigma(\ell) \propto \overline{(\sigma_i^2 - \sigma_{i+\ell}^2)^2} + v_0'(1 - \delta_{\ell,0}), \tag{7.34}$$

where v_0' comes from the measurement noise term. Therefore, the non trivial part of the σ_e variogram (i.e. the $\ell \neq 0$ part) is proportional to the variogram of η^2, up to an additive constant, and in turn contains all the information on $C_2(i, i + \ell)$.

Figures 7.9 and 7.10 show these variograms (normalized by $\overline{\sigma^2}^2$) for the S&P 500 and for individual stocks (averaged over stocks), together with a fit that correspond to power-law decay of the correlation function C_2:

$$V_\sigma(\ell) = \left[v_\infty - \frac{v_1}{\ell^\nu} \right] \qquad (\ell \neq 0). \tag{7.35}$$

The fit is rather good, and leads to $\nu \approx 0.3$ for the S&P 500, and $\nu \approx 0.22$ for individual stocks. The ratio $\phi = v_1/v_\infty$ is found to be $\phi \approx 0.77$ for the S&P 500 and $\phi \approx 0.35$ for stocks.

On these graphs, we have also shown the variogram of the log-volatility, that turns out to be of interest in the context of the multifractal model discussed in the following section. Within this model, the log-volatility variogram is expected to behave as:

$$\left\langle \left(\log \frac{\sigma_{i+\ell}}{\sigma_i} \right)^2 \right\rangle \approx 2K_0^2 \log(\alpha\ell), \tag{7.36}$$

also in reasonable agreement with the data.

Fig. 7.9. Temporal variogram of the high frequency volatility proxy of the S&P 500 (lower curve, right scale). The value $\ell = 1$ corresponds to an interval of 1 day. For comparison, we have shown a power-law fit using Eq. (7.35). We also show the variogram of the log-volatility (left scale), together with the multifractal prediction, Eq. (7.39), with $K_0^2 = 0.0145$. We also show another fit of comparable quality, motivated by a simple model explained in Chapter 20.

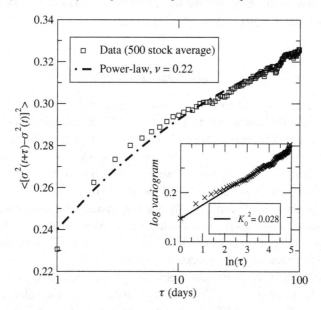

Fig. 7.10. Temporal variogram of the volatility proxy for individual stocks. The value $\tau = 1$ corresponds to an interval of 1 day. We have shown a power-law fit using Eq. (7.35). We also show the variogram of the log-volatility, together with the multifractal prediction, Eq. (7.39), with $K_0^2 = 0.028$.

Two important remarks are in order:

(i) A fit of V_σ using a power law works quite nicely. However, a fit of similar quality is obtained using other functional forms. For example, a constant v_∞ minus the sum of two exponentials $\exp(-\ell/\ell_{1,2})$ with adjustable amplitudes is also acceptable. One then finds two rather different time scales: ℓ_1 is short (a day or less), and a long correlation time ℓ_2 of the order of months.

> The important point here is that the dynamics of the volatility cannot be accounted for using a single time scale: the volatility fluctuations is a multi-timescale phenomenon.

(ii) The short time behaviour of the volatility correlation function is to a certain extent a matter of choice. One can either insist that the variables ϵ are Gaussian but allow the true volatility process to have a high frequency random component unrelated to the long term process, or equivalently consider that the ϵ have a non zero kurtosis and that the volatility process only has low frequency components. We have indeed shown in Section 7.1 that a positive kurtosis can be thought of as a random volatility effect. A relative measure of the long time volatility fluctuations, which has some degree of predictability, as compared to the short time component, is given by $V_\sigma(\infty)/V_\sigma(\ell = 1) - 1 = 1/(1 - \phi)$, which is found to be roughly equal to two.

Back to the kurtosis

From the empirical determination of $V_\sigma(\ell)$, one can determine, up to a multiplicative constant, the correlation function $g(\ell)$ defined by Eq. (7.16) and use this information to fit the anomalous behaviour of the empirical kurtosis κ_N using Eq. (7.15) that relates κ_N and $g(\ell)$. The result is shown in Figure 7.11 for U.S. individual stocks. Since the variogram $V_\sigma(\ell)$ does not tell us anything about $g(0)$, we have treated the quantity $\kappa_0 + (3 + \kappa_0)g(0)$ as an adjustable parameter along with the overall scale of $g(\ell)$ for $\ell > 1$.

The agreement shows quite convincingly that long range volatility correlations and the persistence of fat tails are two sides of the same coin. As we shall see in Section 13.3.4, these long ranged volatility correlations have direct consequences for the dynamics of the volatility smile observed on option markets.

Although $g(0)$ cannot be estimated directly because of measurement noise, it is reasonable, from the value of $\phi = v_1/v_\infty$ found from the fits of the volatility variograms, to assume that $g(0) \sim 2g_1 \sim 1.5$. From the fitted value of $\kappa_0 + (3 + \kappa_0)g(0) = 23.0$, we estimate that the 'intrinsic' value of the daily kurtosis is $\kappa_0 \sim 7.5$ for individual stocks.

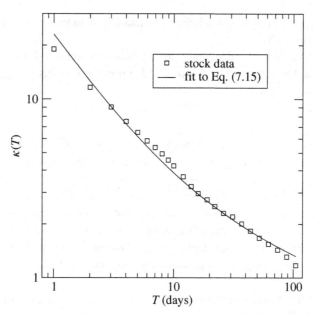

Fig. 7.11. Average kurtosis as a function of time for individual stocks and two parameter fit of Eq. (7.15) using $g(\ell) = g_1 \ell^{-0.22}$ for $\ell \geq 1$. The fit gives $\kappa_0 + (3 + \kappa_0)g(0) = 23.0$ and $g_1 = 0.73$. Time is expressed in trading days and the stock data is the same as in Fig. 7.3.

7.3 Models and mechanisms

7.3.1 Multifractality and multifractal models (*)

As we have emphasized several times, the distribution of price increments on scale N strongly depends on N, at variance with what would be observed for a Brownian random walk where this distribution would be Gaussian for all N. Empirical distributions of price changes are strongly non-Gaussian for small time delays, and progressively (but rather slowly) become Gaussian as N increases. This means that the different unconditional moments of the price changes, defined as $\overline{\langle |\log(x_N/x_0)|^q \rangle}$ depend on a non trivial way on N, even when rescaled by $\overline{\langle (\log(x_N/x_0))^2 \rangle}^{q/2}$. (Again, for a Brownian random walk, these rescaled moments would be given by q dependent, but N independent constants.) A multifractal (or multiaffine) process corresponds to the case where:

$$\overline{\langle |\log(x_N/x_0)|^q \rangle} \simeq A_q N^{\zeta_q} \qquad \zeta_q \neq \frac{q}{2}\zeta_2. \tag{7.37}$$

In this case, rescaled moments behave as a (q-dependent) power-law of N. Such a behaviour for price (or log-price) changes has been reported by many authors. Let us first show that this 'multifractal' effect is another manifestation of the intermittent nature of the volatility, which generically leads to an apparent (or approximate) multifractal behaviour. We will

then discuss a particular situation where multifractality is indeed exact, and for which ζ_q can be computed for all q.

We showed in Section 7.1.4 that long-ranged volatility correlations can lead to an anomalous decay of the kurtosis, as $AN^{-\nu}$ with $\nu < 1$. From the very definition of the kurtosis, this implies that, for large N:

$$\overline{\langle |\log(x_N/x_0)|^4 \rangle} = \overline{\sigma^2}^2 N^2 (1 + AN^{-\nu}). \tag{7.38}$$

The fourth moment of the price difference therefore behaves as the *sum* of two power-laws, not as a unique power-law as for a multifractal process. However, when ν is small and in a restricted range of N, this sum of two power-laws is indistinguishable from a unique power-law with an effective exponent ζ_4 somewhere in-between 2 and $2 - \nu$; therefore $\zeta_4 < 2\zeta_2 = 2$. If σ is a Gaussian random variable, one can show explicitly that this scenario occurs for all even q: the qth moment appears as a sum of $q/2$ power-laws that can be fitted by an effective exponent ζ_q which systematically bends downwards, away from the line $q\zeta_2/2$, leading to an *apparent* multifractal behaviour (see Bouchaud et al. (1999)).

On the other hand, one can define a truly multifractal stochastic volatility model by choosing the volatility σ to be a log-normal random variable such that $\sigma_i = \sigma_0 \exp \xi_i$, with ξ Gaussian and:

$$\overline{\xi_i} = -K_0^2 \log \alpha = m_0; \qquad \overline{\xi_i \xi_j} - m_0^2 = -K_0^2 \log [\alpha(|i - j| + 1)], \tag{7.39}$$

for $|i - j| \ll \alpha^{-1}$, where $\alpha^{-1} \gg 1$ is a large cut-off time scale. This defines the Bacry-Muzy-Delour (BMD) model. One can then show, using the properties of multivariate Gaussian variables, that:

$$\overline{\sigma^2} = \sigma_0^2. \tag{7.40}$$

The rescaled correlation of the squared volatilities, defined as in Eq. (7.16), is found to be:

$$g(\ell) = [\alpha(1 + \ell)]^{-\nu} \qquad \nu \equiv 4K_0^2. \tag{7.41}$$

We can choose K_0 small enough such that $\nu < 1$ and be in the 'anomalous' kurtosis regime discussed above, where Eq. (7.38) holds. The interesting point here is that for small α, the constant A appearing in Eq. (7.38) is much larger than unity: $A = \alpha^{-\nu}/(2 - \nu)(1 - \nu)$. Therefore, the second term in Eq. (7.38) is dominant whenever $\alpha N \ll 1$. In the regime $1 \ll N \ll \alpha^{-1}$, one thus find a *unique* power-law for $\overline{\langle |\log(x_N/x_0)|^4 \rangle}$, with $\zeta_4 \equiv 2 - \nu$.

One can actually compute exactly all even moments $\overline{\langle |\log(x_N/x_0)|^q \rangle}$, assuming that the random 'sign' part ϵ is Gaussian. For $qK_0^2 < 1$, the corresponding moments are finite, and one finds, in the scaling region $1 \ll N \ll \alpha^{-1}$, a true multifractal behaviour with $\zeta_q = q/2[1 - K_0^2(q - 2)]$. Note that $\zeta_2 \equiv 1$ and $\zeta_4 = 2 - \nu$. For $qK_0^2 > 1$, the moments diverge, suggesting that the unconditional distribution of $\log(x_N/x_0)$ has power-law tails with an exponent $\mu = 1/K_0^2$. For $N \gg \alpha^{-1}$, one the other hand, the scaling becomes that of a standard random walk, for which $\zeta_q = q/2$. Any multifractal scaling can actually only hold over a finite time interval.

Let us end this section with several remarks.

- The BMD multifractal model has some similarities with the multiplicative cascade models introduced by Mandelbrot and collaborators. The cascade model assumes that the volatility can be constructed as a *product* of random variables associated with different time scales τ_k, these time scales being in a geometrical progression, i.e. $\tau_{k+1}/\tau_k = \omega$, where ω is a fixed number. Using the central limit for the log-volatility, this construction naturally leads to a log-normal distribution for the volatility, and to a logarithmic correlation function. This cascade model however violates causality. In this respect the BMD model is attractive since it can be given, using the construction explained in Section 2.5, a purely causal interpretation, where the volatility only depends on the past, and not on the future of the process.[†] On the other hand, this additive construction does not account naturally for the log-normal nature of the volatility, which has to be put 'by hand'.
- The study of the empirical distribution of the volatility is indeed compatible with the assumption of log-normality. This is illustrated in Figures 7.7 and 7.8, where the distribution of volatility proxies σ_e is compared with a log-normal distribution. Note however that an inverse gamma distribution also fits the data very well, and is actually motivated by simple (non cascade) models of volatility feedback (see Section 20.3.2).
- The direct study of the correlation function of the log-volatility can indeed be fitted by a logarithm of the time lag (Figs. 7.9 and 7.10), as assumed in Eq. (7.39), with K_0^2 in the range 0.01–0.10 for many different markets, and α^{-1} on the order of a few years.
- On the other hand, the empirical tail of the distribution of price increments is described by a power-law with an exponent μ in the range 3–5, much smaller than the value $1/K_0^2 \sim$ 10–100 predicted by the BMD model. This suggests that the sign part ϵ itself is non Gaussian and further fattens the tails. Stochastic volatility is an important cause for fat tails, but is alone insufficient to generate the observed high frequency tails. On the other hand, the low frequency kurtosis is indeed dominated by long range volatility fluctuations. A similar conclusion was already reached above, in Sections 7.2.1, 7.1.5.
- Finally, one can extend the above multifractal model to account for a skewed distribution of returns. A simple possibility is to use the above construction to obtain *retarded* returns (as defined in the next chapter) rather than instantaneous returns.

7.3.2 The microstructure of volatility

Markets can be volatile for two, rather different reasons. The first is related to the market activity itself: if the 'atomic' price increments were all equal to $\epsilon_k = \pm 1$ (or more realistically one 'tick'), with a random sign for each trade, the volatility of the process would be equal to the average frequency of trades per unit time. One would indeed have:

$$X_t - X_0 = \sum_{k=1}^{N(t)} \epsilon_k \longrightarrow \langle (X_t - X_0)^2 \rangle = \langle N(t) \rangle \tag{7.42}$$

[†] An alternative causal construction has been proposed in Calvet & Fisher (2001).

where $N(t)$ is the actual number of trades during the time interval t, the average of which is simply t/τ_0, where τ_0 is the average time between trades. Typically, a tick is 0.05% of the price, and τ_0 is a fraction of a minute on liquid stocks. Periods of high volatilities would be, in this model, periods when the number of transactions is particularly high, corresponding to a large activity of the market. Although it is true that very large volumes often lead to high local volatilities, this is not the only cause. It is well known that crashes, on the contrary, correspond to very illiquid markets when the bid-ask spread widens anomalously, with no, or nearly no volume of transaction.

Therefore, another constituent of volatility is the average 'jump' size J_k between two trades. Suppose now that $\delta x_k = J_k \epsilon_k$, where the average value of J_k^2 is equal to J_0^2. The volatility of the process is now J_0^2/τ_0. Both the trading frequency $1/\tau_0$ and the jump size J_0 can be time dependent, and the volatility can be high either because J_0 is large or because τ_0 is small. Using tick data, where all the trades are recorded, one can measure both the number trades in a certain time interval Δt, and the average jump size in the same time window. Interestingly, J_0 shows long tails that can account for the tails in the price (or volatility) distribution, but has only short time autocorrelations. Conversely, the trading frequency $1/\tau_0$ has rather thin tails but has long term power law correlations that are similar to those observed on the volatility itself (see Plerou et al. (2001)).

We show for example in Figure 7.12 the variogram of the daily number of trades on the S&P 500 futures contract, together with a fit similar to those used for the volatility in Figures 7.9 (and which will be justified in Chapter 20).

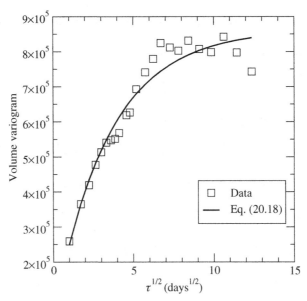

Fig. 7.12. Plot of the variogram of the volume of activity (number of trades) on the S&P500 futures contract during the years 1993–1999, as a function of the time lag $\sqrt{\tau}$. Also shown is the fit using Eq. (20.18), motivated by a microscopic model of trading agents.

Therefore, the explanation for large volatility values should be sought for in terms of large local jumps, where the liquidity of the market is severely reduced, whereas the explanation of the long-ranged nature of volatility correlations should be associated to long term correlations of the volume of activity. Again, we find a natural decomposition of the kurtosis in terms of a high frequency part (related to jumps) and a low frequency part (related to volume activity). As explained in Chapter 20, there might indeed be rather natural mechanisms that can account for such long-ranged volume correlations.

7.4 Summary

- In financial markets, the volatility is not constant in time, but has random, log-normal like, fluctuations. These fluctuations cannot be described by a single time scale. Rather, their temporal correlation function is well fitted by a power-law. This power-law reflects the multi-scale nature of these fluctuations: high volatility periods can persist from a few hours to several months.
- From a mathematical point of view, the sum of increments with a finite stochastic volatility still converge to a Gaussian variable, although the kurtosis decays much more slowly than in the case of iid variables when the volatility has long range correlations. Accordingly, the empirical moments of price variations approach the Gaussian limit only very slowly with the time horizon, and their long time value is dominated by volatility fluctuations.
- The slow decay of the volatility correlation function, together with the observed approximate log-normal distribution of the volatility, leads to a 'multifractal-like' behaviour of the price variations, much as what is observed in turbulent hydro-dynamical flows.

- FURTHER READING

C. Alexander, *Market Models: A Guide to Financial Data Analysis*, John Wiley, 2001.

J. Beran, *Statistics for long-memory processes*, Chapman & Hall, 1994.

T. Bollerslev, R. F. Engle, D. B. Nelson, *ARCH models*, Handbook of Econometrics, vol 4, ch. 49, R. F. Engle and D. I. McFadden Edts, North-Holland, Amsterdam, 1994.

J. Y. Campbell, A.W. Lo, A. C. McKinley, *The Econometrics of Financial Markets*, Princeton University Press, Priceton, 1997.

R. Cont, *Empirical properties of asset returns: stylized facts and statistical issues*, Quantitative Finance, **1**, 223 (2001).

M. M. Dacorogna, R. Gençay, U. A. Muller, R. B. Olsen, O. V. Pictet, *An Introduction to High-Frequency Finance*, Academic Press, San Diego, 2001.

U. Frisch, *Turbulence: The Legacy of A. Kolmogorov*, Cambridge University Press, 1997.

C. Gourieroux, A. Montfort, *Statistics and Econometric Models*, Cambridge University Press, Cambridge, 1996.

B. B. Mandelbrot, *Fractals and Scaling in Finance*, Springer, New York, 1997.

R. Mantegna, H. E. Stanley, *An Introduction to Econophysics*, Cambridge University Press, Cambridge, 1999.

- SPECIALIZED PAPERS: ARCH

T. Bollerslev, *Generalized ARCH*, Journal of Econometrics, **31**, 307 (1986).

R. F. Engle, *ARCH with estimates of the variance of the United Kingdom inflation*, Econometrica, **50**, 987 (1982).

R. F. Engle, A. J. Patton, *What good is a volatility model?*, Quantitative Finance, **1**, 237 (2001).

- SPECIALIZED PAPERS: VOLATILITY DISTRIBUTION AND CORRELATIONS

P. Cizeau, Y. Liu, M. Meyer, C.-K. Peng, H. E. Stanley, *Volatility distribution in the S&P500 Stock Index*, Physica A, **245**, 441 (1997).

Z. Ding, C. W. J. Granger, R. F. Engle, *A long memory property of stock market returns and a new model*, Journal of Empirical Finance, **1**, 83 (1993).

Z. Ding, C. W. J. Granger, *Modeling volatility persistence of speculative returns: a new approach*, Journal of Econometrics, **73**, 185 (1996).

B. LeBaron, *Stochastic volatility as a simple generator of apparent financial power laws and long memory*, Quantitative Finance, **1**, 621 (2001), and comments by B. Mandelbrot, T. Lux, V. Plerou and H. E. Stanley in the same issue.

Y. Liu, P. Cizeau, M. Meyer, C.-K. Peng, H. E. Stanley, *Correlations in economic time series*, Physica A, **245**, 437 (1997).

A. Lo, *Long term memory in stock market prices*, Econometrica, **59**, 1279 (1991).

S. Micciche, G. Bonanno, F. Lillo, R. N. Mantegna, *Volatility in financial markets: stochastic models and empirical results*, Physica A, **314**, 756 (2002).

- SPECIALIZED PAPERS: MULTISCALING AND MULTIFRACTAL MODELS

A. Arnéodo, J.-F. Muzy, D. Sornette, *Causal cascade in the stock market from the 'infrared' to the 'ultraviolet'*, European Journal of Physics, **B 2**, 277 (1998).

E. Bacry, J. Delour, J.-F. Muzy, *Modelling financial time series using multifractal random walks*, Physica A, **299**, 284 (2001).

J.-P. Bouchaud, M. Meyer, M. Potters, *Apparent multifractality in financial time series*, Eur. Phys. J. B **13**, 595 (1999).

M.-E. Brachet, E. Taflin, J. M. Tchéou, *Scaling transformation and probability distributions for financial time series*, e-print cond-mat/9905169.

L. Calvet, A. Fisher, *Forecasting multifractal volatility*, Journal of Econometrics, **105**, 27, (2001).

L. Calvet, A. Fisher, *Multifractality in asset returns: theory and evidence*, Review of Economics and Statistics, **84**, 381 (2002).

A. Fisher, L. Calvet, B. B. Mandelbrot, *Multifractality of DEM/$ rates*, Cowles Foundation Discussion Paper 1165.

Y. Fujiwara, H. Fujisaka, *Coarse-graining and Self-similarity of Price Fluctuations*, Physica A, **294**, 439 (2001).

S. Ghashghaie, W. Breymann, J. Peinke, P. Talkner, Y. Dodge, *Turbulent cascades in foreign exchange markets*, Nature, **381**, 767 (1996).

T. Lux, *Turbulence in financial markets: the surprising explanatory power of simple cascade models*, Quantitative Finance, **1**, 632 (2001).

U. A. Muller, M. M. Dacorogna, R. B. Olsen, O. V. Pictet, M. Schwartz, C. Morgenegg, *Statistical study of foreign exchange rates, empirical evidence of a price change scaling law, and intraday analysis*, Journal of Banking and Finance, **14**, 1189 (1990).

J.-F. Muzy, J. Delour, E. Bacry, *Modelling fluctuations of financial time series: from cascade process to stochastic volatility model*, Eur. Phys. J., B **17**, 537–548 (2000).

B. Pochart, J.-P. Bouchaud, *The skewed multifractal random walk with applications to option smiles*, Quantitative Finance, **2**, 303 (2002).

F. Schmitt, D. Schertzer, S. Lovejoy, *Multifractal analysis of Foreign exchange data*, Applied Stochastic Models and Data Analysis, **15**, 29 (1999).

G. Zumbach, P. Lynch, *Heterogeneous volatility cascade in financial markets*, Physica A, **298**, 521, (2001).

• SPECIALIZED PAPERS: TRADING VOLUME

G. Bonnano, F. Lillo, R. Mantegna, *Dynamics of the number of trades in financial securities*, Physica A, **280**, 136 (2000).

J.-P. Bouchaud, I. Giardina, M. Mézard, *On a universal mechanism for long ranged volatility correlations*, Quantitative Finance, **1**, 212 (2001).

V. Plerou, P. Gopikrishnan, L. A. Amaral, X. Gabaix, H. E. Stanley, *Price fluctuations, market activity and trading volume*, Quantitative Finance, **1**, 262 (2001).

• SOME RECENT READS

L. Borland, J. P. Bouchaud, *On a multi-timescale statistical feedback model for volatility fluctuations*, e-print cond-mat/0507073 (unpublished).

G. Zumbach, *Volatility processes and volatility forecast with long memory*, Quantitative Finance, **4**, 70 (2004).

8

Skewness and price-volatility correlations

It seemed to me that calculation was superfluous, and by no means possessed of the importance which certain other players attached to it, even though they sat with ruled papers in their hands, whereon they set down the coups, calculated the chances, reckoned, staked, and–lost exactly as we more simple mortals did who played without any reckoning at all.

(Fyodor Dostoevsky, *The Gambler*.)

8.1 Theoretical considerations

8.1.1 Anomalous skewness of sums of random variables

In the previous chapter, we have assumed that the 'sign' part of the fluctuations ϵ_i and the 'amplitude' part σ_i were independent. For financial applications, this might be somewhat restrictive. We will indeed discuss below that there are in financial time series interesting correlations between *past* price changes and *future* volatilities. These appear when one considers the following correlation function:[†]

$$\mathcal{L}(i, j) = \overline{\langle \eta_i \eta_j^2 \rangle} - \overline{\langle \eta_i \rangle}\,\overline{\langle \eta_j^2 \rangle} \equiv \overline{\sigma^2}^{3/2} g_L(|i - j|). \tag{8.1}$$

This correlation function is essentially zero when $i > j$, but significantly *negative* for $i < j$, indicating that price drops tend to be followed by higher volatilities (and vice versa).

> Much as volatility correlations induce anomalous kurtosis, these price-volatility correlations induce some anomalous skewness.

Taking into account the fact that in the stochastic volatility model defined by Eq. (7.4) the ϵ_i are iid, the third moment of the sum $\log(x_N/x_0) = \sum_{i=1}^{N} \eta_i$ reads:

$$\overline{\left\langle \left(\sum_{i,j,k=1}^{N} \eta_i \eta_j \eta_k \right) \right\rangle} = \sum_{i=1}^{N} \overline{\langle \eta_i^3 \rangle} + 3 \sum_{i=1}^{N} \sum_{j>i}^{N} \overline{\langle \eta_i \eta_j^2 \rangle}. \tag{8.2}$$

[†] The choice of the letter \mathcal{L} to denote this correlation refers to the 'leverage effect' to which it is usually, but somewhat incorrectly, associated – see the discussion below, Section 8.3.1.

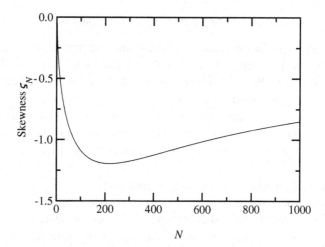

Fig. 8.1. Unconditional skewness ς_N at scale N as a function of N, when the rescaled price-volatility correlation function $g_L(\ell)$ is equal to $-\exp(-\alpha\ell)$, with $\alpha = 1/100$. Note that the skewness is non monotonous, and decays for $N \gg \alpha^{-1}$ as $N^{-1/2}$.

The unconditional skewness ς_N is then found to be:

$$\varsigma_N = \frac{\varsigma_0 g_L(0)}{\sqrt{N}} + \frac{3}{\sqrt{N}} \sum_{\ell=1}^{N} \left(1 - \frac{\ell}{N}\right) g_L(\ell), \tag{8.3}$$

where ς_0 is the skewness of ϵ.

The above formula is quite interesting, since it shows that even for an unskewed distribution of increments (i.e. for $\varsigma_0 = 0$), some skewness can appear on longer time scales. In this case, the skewness can actually be non monotonous if $g_L(\ell)$ decreases fast enough. We show this in Figure 8.1 for $g_L(\ell) = -\exp(-\alpha\ell)$ with $\alpha = 1/100$.

In the above calculation, we have assumed the price process to be multiplicative, and considered the skewness of the log-price distribution. Empirically, as discussed below, skewness effects are actually quite small, much smaller than the volatility fluctuation effects discussed in the previous chapter, that lead to a large kurtosis. This large kurtosis is seen both on price increments and on log-price increments: considering one or the other does not change the results much (if the time lag N is not too large). The skewness, on the other hand, is small and is therefore very sensitive to the quantity that one studies, absolute price increments δx or relative price increments η. Take for example η to be Gaussian with a volatility $\sigma \ll 1$. By definition, the skewness of η is zero, but the random variable $\exp\eta$ has a positive skewness equal to 3σ. We thus need at this stage to discuss in greater detail if one should choose absolute price changes, or relative price changes, or yet another quantity, as the 'fundamental' random variable.

8.1.2 Absolute vs. relative price changes

A standard hypothesis in theoretical finance is to assume that price changes are proportional to prices themselves, therefore leading to a *multiplicative* model $x_{k+1} = x_k(1 + \eta_k)$, where the random returns η_k are the fundamental objects. The logarithm of the price is thus a sum of random variables. Some justifications of this assumption have been given in Chapter 6; it is clear that the order of magnitude of relative fluctuations (a few percent per day for stocks) is much more stable in time and across stocks than absolute price changes. This fact is also, to some extent, related to important approximate symmetries in finance. For example, the price of stocks is somewhat arbitrary, and is controlled by the total number of stocks in circulation. One can for example *split* each share into ten, thereby dividing the price of each stock by ten. This split is often done to keep the share prices around $100. It is clear that the day just after the split, the typical absolute price variations will also be divided by a factor ten (dilation symmetry). Another symmetry exists in foreign exchange markets, between two similar currencies. In this case, there cannot be any distinction between expressing currency A in units of B and get a price X, or vice-versa, giving a price $1/X$. Relative returns, defined such that $\delta x/x = -\delta(1/x)/(1/x)$, indeed respect this symmetry.

However, these symmetries are not exactly enforced in financial markets. There are several effects that explicitly break the dilation symmetry. One is the fact that prices are discrete: price changes are quoted in ticks, which are fixed fraction of dollars (i.e. not in percent), which introduce a well defined *scale* for price changes. Another effect is round numbers: shares are usually bought and sold by lots of multiples of tens. The very mechanism leading to price changes is therefore not expected to vary continuously as prices evolve, but rather to adapt only progressively if prices are seen to rise (or decrease) significantly over a certain time window. This will be the motivation of the 'retarded' model developed below.

In foreign exchange markets, the symmetry between X and $1/X$ is obviously violated when the two currencies are very dissimilar. For example, the Mexican peso vs. U.S. dollar shows very sudden downward moves, not reflected on the other side of the distribution.

Finally, there are various financial instruments for which the dilation argument does not work. For example, the BUND contract is computed as the value of a ten year bond paying a fixed rate (currently 6% per annum) every three months on a nominal value of 100 000 Euros. The short term interest rates are modified by central banks, in general in steps of 0.25% or 0.50%, independently of the value of the rate itself. Another interesting example is the volatility itself, which is traded on option markets; there is not clear reason why volatility *changes* should be proportional to the volatility itself.

It is therefore interesting to study this issue empirically, by looking at the mean absolute deviation of δx, conditioned to a certain value of the price x itself, which we shall denote $\langle|\delta x|\rangle|_x$. If the return η were the natural random variable, one should observe that $\langle|\delta x|\rangle|_x = \sigma_1 x$, where σ_1 is constant (and equal to the MAD of η). Of course, since the volatility is

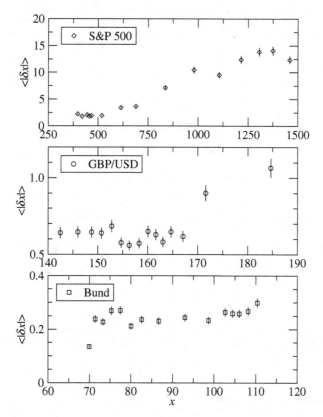

Fig. 8.2. MAD of the increments δx, conditioned to a certain value of the price x, as a function of x, for the three chosen assets. Although this quantity, overall, grows with the price x for the S&P 500 or the GBP/USD, there are plateau-like regions where the absolute price change is nearly independent of the price: see for example when the S&P 500 was around 400. In the case of the Bund, the absolute volatility is nearly flat.

itself random, there will be fluctuations; however, $\langle |\delta x| \rangle|_x$ and $\sqrt{\langle \delta x^2 \rangle|_x}$ should on average increase linearly with x.

Now, in many instances (Figs. 8.2 and 8.3), one rather finds that $\langle |\delta x| \rangle|_x$ is roughly independent of x for long periods of time. The case of the CAC 40 is particularly interesting, since during the period 1991–95, the index went from 1400 to 2100, leaving the absolute volatility nearly constant (if anything, it is seen to decrease with x).

On longer time scales, however, or when the price x rises substantially, the MAD of δx increases significantly, as to become indeed proportional to x (see Fig. 8.2 (a)). A way to model this crossover from an additive to a multiplicative behaviour is to postulate that the RMS of the increments progressively (over a time scale T_\times) adapt to the changes of price. A precise implementation of this idea is given in the next section. Schematically, for $T < T_\times$, the prices behave additively, whereas for $T > T_\times$, multiplicative effects start

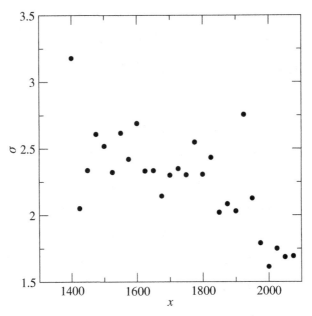

Fig. 8.3. RMS of the increments δx, conditioned to a certain value of the price x, as a function of x, for the CAC 40 index for the 1991–95 period; it is quite clear that during that time period $\langle \delta x^2 \rangle|_x$ was almost independent of x.

playing a significant role.[†] Mathematically, this reads:

$$\langle (x(T) - x_0)^2 \rangle = DT \quad (T \ll T_\times);$$

$$\left\langle \log^2 \left(\frac{x(T)}{x_0} \right) \right\rangle = \sigma^2 T \quad (T \gg T_\times). \tag{8.4}$$

On liquid stock markets, this time scale is on the order of months (see below).

8.1.3 The additive-multiplicative crossover and the q-transformation

A convenient way to parametrize the crossover between the two regimes is to introduce an *additive* random variable $\xi(T)$, and to represent the price $x(T)$ as:

$$x(T) = x_0 \left(1 + q(T)\xi(T) \right)^{1/q(T)} \qquad q(T) \in [0, 1]. \tag{8.5}$$

For $T \ll T_\times$, $q \to 1$, the price process is additive, with $x = x_0(1 + \xi)$, whereas for $T \gg T_\times$, $q \to 0$, $(1 + q\xi)^{1/q} \to \exp \xi$, which corresponds to the multiplicative limit. The index q can be taken as a quantitative measure of the 'additivity' of the process. This representation

[†] In the additive regime, where the variance of the increments can be taken as a constant, we shall write $\langle \delta x^2 \rangle = \sigma_1^2 x_0^2 = D\tau$.

can be inverted to define a 'fundamental' additive variable as:

$$\xi(T) = \frac{1}{q(T)} \left[\exp\left(q(T) \log\left[\frac{x(T)}{x_0} \right] \right) - 1 \right], \tag{8.6}$$

which we will refer to as the 'q-transformation'. It is interesting to compute the skewness $\varsigma(T)$ of the return $\log(x(T)/x_0)$ as a function of $q(T)$, assuming that $\xi(T)$ has zero skewness, and a volatility $\sigma(T) \ll 1$. One finds, to lowest order in σ:

$$\varsigma(T) = -3 q(T) \sigma(T). \tag{8.7}$$

If $q(T) = 0$, the skewness is zero as it should, since in this case $\log(x(T)/x_0) \equiv \xi(T)$. However, if $q(T) = 1$, the skewness of $\log(x(T)/x_0)$ is negative. Therefore, a negative skewness in the *log-price* can arise simply because as a consequence of taking the logarithm of an unskewed variable.

8.2 A retarded model

8.2.1 Definition and basic properties

Let us give some more quantitative flesh to the idea that price changes only progressively adapt if prices are seen to rise (or decrease) significantly over a certain time window. A way to describe this lagged response to price changes is generalize Eq. (7.4) and to postulate that price changes are proportional not to instantaneous prices but to a moving average x_i^R of past prices over a certain time window:

$$\delta x_i = x_i^R \sigma_i \epsilon_i, \qquad x_i^R = \sum_{\ell=0}^{\infty} \mathcal{K}(\ell) x_{i-\ell}, \tag{8.8}$$

where $\mathcal{K}(\ell)$ is a certain averaging kernel, normalized to one, $\sum_{\ell=0}^{\infty} \mathcal{K}(\tau) \equiv 1$, and decaying over a typical time T_\times. It will be more convenient to rewrite x_i^R as:

$$x_i^R = \sum_{\ell=0}^{\infty} \mathcal{K}(\ell) \left[x_i - \sum_{\ell'=1}^{\ell} \delta x_{i-\ell'} \right] \tag{8.9}$$

$$\equiv x_i - \sum_{\ell'=1}^{\infty} \widehat{\mathcal{K}}(\ell') \delta x_{i-\ell'}, \tag{8.10}$$

where $\widehat{\mathcal{K}}(\ell) = \sum_{\ell'=\ell}^{\infty} \mathcal{K}(\ell')$. Note that from the normalization of $\mathcal{K}(\ell)$, one has $\widehat{\mathcal{K}}(0) = 1$, independently of the specific shape of \mathcal{K}. (This will turn out to be important in the following.)

For the sake of simplicity, one can take for $\mathcal{K}(\ell)$ a simple exponential with a certain decay rate α, such that $T_\times = 1/\log \alpha$. In this case, $\widehat{\mathcal{K}}(\ell) = \alpha^\ell$. From Eq. (8.10), one sees that the limit $\alpha \to 1$ ($T_\times \to \infty$) corresponds to the case where x^R is a constant, equal to the initial price x_0. Therefore, in this limit, the retarded model corresponds to a purely *additive* model. The other limit $\alpha \to 0$ (infinitely small T_\times) leads to $x^R \equiv x$ and corresponds to a

purely *multiplicative* model. More generally, for time lags much smaller than T_\times, x^R does not vary much, and the model is additive, whereas for time lags much larger than T_\times, x^R roughly follows (in a retarded way) the value of x and the process is multiplicative.

> *The retarded model with an exponential kernel $\mathcal{K}(\ell)$ can indeed be reformulated in terms of a product of random 2×2 matrices. Define the vector \vec{V}_k as the set x_k^R, x_k. One then has:*
>
> $$\vec{V}_{k+1} = \mathcal{M}_k \vec{V}_k, \tag{8.11}$$
>
> *where \mathcal{M}_k is a random matrix whose elements are $\mathcal{M}_{11,k} = \alpha$, $\mathcal{M}_{12,k} = 1 - \alpha$, $\mathcal{M}_{21,k} = \sigma_k \epsilon_k$ and $\mathcal{M}_{22,k} = 0$. Using standard theorems on products of random matrices that generalize the classic results on products of random scalars, one directly finds that the large time statistics of both components of \vec{V} is log-normal.*[†]

In the following, we will assume that the relative difference between x and x^R is small. This is the case when $\sigma \sqrt{T_\times} \ll 1$, which means that the averaging takes place on a time scale such that the relative changes of price are not too large. Typically, for stocks one finds $T_\times = 50$ days, $\sigma = 2\%$ per square-root day, so that $\sigma \sqrt{T_\times} \sim 0.15$.

Let us calculate the 'leverage' correlation function defined by Eq. (8.1) above, to first order in $\sigma \sqrt{T_\times}$. One finds, using $\eta_i \equiv \delta x_i / x_i$ and Eq. (8.8), and assuming for a while that σ is constant:

$$\langle \eta_{i+\ell}^2 \, \eta_i \rangle = \sigma^2 \langle \epsilon_{i+\ell}^2 \rangle \left\langle \left(1 - 2 \sum_{\ell'=1}^{\infty} \widehat{\mathcal{K}}(\ell') \frac{\delta x_{i+\ell-\ell'}}{x_{i+\ell}} \right) \frac{\delta x_i}{x_i} \right\rangle. \tag{8.12}$$

To lowest order in the retardation effects, one can replace in the above expression $\delta x_{i+\ell-\ell'}/x_{i+\ell}$ by $\sigma \epsilon_{i+\ell-\ell'}$ and $\delta x_i / x_i$ by $\sigma \epsilon_i$, because $x_R / x = 1 + O(\sigma \sqrt{T_\times})$. Since the ϵ are assumed to be independent of zero mean and of unit variance, one has:

$$\langle \epsilon_{i+\ell-\ell'} \epsilon_i \rangle = \delta_{\ell,\ell'}. \tag{8.13}$$

Therefore, using the definition of g_L given in Eq. (8.1):

$$g_L(\ell) \approx -2\sigma \widehat{\mathcal{K}}(\ell). \tag{8.14}$$

Taking into account the volatility fluctuations would amount to replacing in the above result σ by $\overline{\sigma_i^2 \sigma_{i+\ell}^2} / \overline{\sigma_i^2}^{3/2}$.

8.2.2 Skewness in the retarded model

Using Eq. (8.14) in Eq. (8.3), we find for the skewness of the log-price distribution (assuming that ϵ has zero skewness):

$$\varsigma_N = -\frac{6\sigma}{\sqrt{N}} \sum_{\ell=1}^{N} \left(1 - \frac{\ell}{N} \right) \widehat{\mathcal{K}}(\ell). \tag{8.15}$$

[†] See, e.g., Crisanti et al. (1993).

It is interesting to compute the q-transform ξ_N (see Eq. (8.5)) of $\log x_N/x_0$, where x_N is obtained using the retarded model with a constant volatility and a Gaussian ϵ. In order to find an unskewed ξ_N, a natural choice for q, if σ is small, is given by Eq. (8.7). Using Eq. (8.15), one finds that q varies from 1 for small N to 0 at large N. Interestingly, the statistics of the resulting 'fundamental' variable ξ_N has a distribution which is found numerically to be very close to a Gaussian when the kernel is a pure exponential .

Therefore, for the Gaussian retarded model, the q-transformation with an appropriate value of q allows one to approximately account for the non Gaussian distribution of $\log(x_N/x_0)$.

8.3 Price-volatility correlations: empirical evidence

We now report some empirical study of this volatility-return correlation both for individual stocks and for stock indices. One indeed finds that correlations are between *future* volatilities and *past* price changes. For both stocks and stock indices, the volatility-return correlation is short ranged, with however a different decay time for stock indices (around 10 days) than for individual stocks (around 50 days). The amplitude of the correlation is also different, and is much stronger for indices than for individual stocks. We then argue that this effect for stocks can be interpreted within the retarded model defined above. This model does however not account properly for the data on stock indices, which seems to reflect a 'panic'-like effect, whereby large price drops of the market as a whole triggers a significant increase of activity.

Eq. (8.14) suggests a normalization for the 'leverage' correlation function different from the 'natural' one, as:

$$L(\ell) = \frac{\langle [\eta_{i+\ell}]^2 \eta_i \rangle_e}{\langle \eta_i^2 \rangle_e^2},$$ (8.16)

where $\langle \ldots \rangle_e$ refers to the empirical average. Within the retarded model with a constant volatility, one finds: $L \equiv -2\widehat{\mathcal{K}}$.

The analyzed set of data consists in 437 U.S. stocks, constituent of the S&P 500 index and 7 major international stock indices (S&P 500, NASDAQ, CAC 40, FTSE, DAX, Nikkei, Hang Seng). The data is daily price changes ranging from January 1990 to May 2000 for stocks and from January 1990 to October 2000 for indices. One can compute L both for individual stocks and stock indices. The raw results were rather noisy. If one assumes that all individual stocks behave similarly and average L over the 437 different stocks, one obtains a somewhat cleaner quantity noted L_S. Similarly, averaging over the 7 different indices, gives L_I. The results are given in Figure 8.4 and Figure 8.5 respectively. The insets of these figures show $L_S(\ell)$ and $L_I(\ell)$ for negative values of ℓ. For stocks the results are not significantly different from zero for $\ell < 0$. For indices, one finds some significant *positive* correlations for $\ell < 0$ up to $|\ell| = 4$ days, which might reflect the fact that large daily drops (that contribute significantly to $\langle \eta_{i+\ell}^2 \rangle_e$) are often followed by positive 'rebound' days.

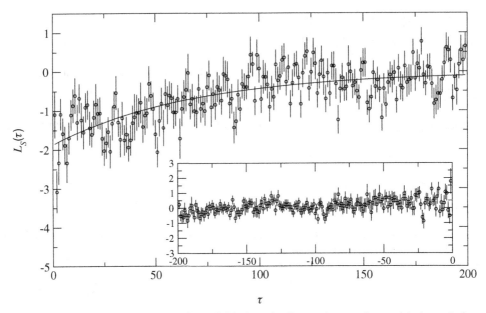

Fig. 8.4. Return-volatility correlation for individual stocks. Data points are the empirical correlation averaged over 437 U.S. stocks, the error bars are two sigma errors bars estimated from the inter-stock variability. The full line shows an exponential fit (Eq. (8.17)) with $A_S = 1.9$ and $T_{\times S} = 69$ days. Note that $L_S(0)$ is not far from the retarded model value -2. The inset shows the same function for negative times. τ is in days.

We now focus on positive values of ℓ. The correlation functions can rather well be fitted by single exponentials:

$$L(\ell) = -A \exp\left(-\frac{\ell}{T_\times}\right). \tag{8.17}$$

For U.S. stocks, one finds $A_S = 1.9$ and $T_{\times S} \approx 69$ days, whereas for indices the amplitude A_I is significantly larger, $A_I = 18$ and the decay time shorter: $T_{\times I} \approx 9$ days.

As can be seen from the figures, both L_S and L_I are significant and negative: price drops increase the subsequent volatility – this is the so-called "leverage" effect. The economic interpretation of this effect is still very much controversial.[†] Even the causality of the effect is sometimes debated: is the volatility increase induced by the price drop or conversely do prices tend to drop after a volatility increase? An early interpretation, due to Black, argues that a price drop increases the risk of a company to go bankrupt, and its stock therefore becomes more volatile. On the contrary, one can argue that an increase of volatility makes the stock less attractive, and therefore pushes its price down.

A more subtle interpretation still is the so-called 'volatility feedback' effect, where an anticipated increase of future volatility by market operators triggers sell orders, and therefore decreases the price. If the volatility indeed increases, this generates a correlation between price drop and volatility increase. This of course assumes that the market anticipation of the future volatility is correct, which is dubious in general.

[†] A recent survey of the different models can be found in Bekaert & Wu (2000).

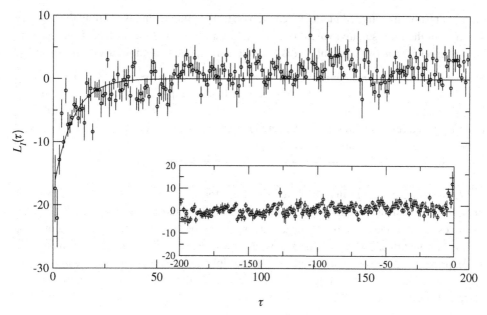

Fig. 8.5. Return-volatility correlation for stock indices. Data points are the empirical correlation averaged over 7 major stock indices, the error bars are two sigma errors bars estimated from the inter-index variability. The full line shows an exponential fit (Eq. (8.17)) with $A_I = 18$ and $T_{\times I} = 9.3$ days. The inset shows the same function for negative times. τ is in days.

8.3.1 Leverage effect for stocks and the retarded model

A simpler interpretation, which accounts rather well for the data on individual stocks, is that the observed 'leverage' correlation is induced by the retardation mechanism discussed above. Because the amplitude of absolute price changes remains constant for a while, the *relative* volatility increases when the price goes down and vice versa. From Eq. (8.14), one expects $L(\ell \to 0)$ to be equal to -2, independently of the detailed shape of \mathcal{K}. This property entirely relies on the normalization of \mathcal{K}. Taking into account the volatility fluctuations would multiply the above result by a factor $\langle \sigma^2(t)\sigma^2(t+\tau)\rangle/\langle\sigma^2(t)\rangle^2 \geq 1$.

As shown in Figure 8.4, $L_S(\ell \to 0)$ is close to (although indeed slightly below) the value -2 for individual stocks. One can give weight to this finding by analyzing a set of 500 European stocks and 300 Japanese stocks, again in the period 1990–2000. One again finds an exponential behaviour for L_S with a time scale on the order of 40 days and, more importantly, and initial values of L_S close to the retarded model value -2: $A_S = 1.96$ and $T_{\times S} = 38$ days for European stocks and $A_S = 1.5$ and $T_{\times S} = 47$ days for Japanese stocks. One should emphasize that these numbers (including the previously quoted U.S. results) carry large uncertainties as they vary significantly with the time period and averaging procedure.

As a more direct test of the retarded model, one can study the correlation between $\delta x_i / x_i$ and $(\delta x_{i+\ell}/x_{i+\ell}^R)^2$. This correlation is found to be zero within error bars, which should be expected if most of the leverage correlations indeed come from a simple retardation effect.

Another test of the retarded model when the volatility is fluctuating is to study the residual variance of the local squared volatility $(\delta x / x_R)^2$ as a function of the horizon T of the averaging kernel. The above value of $T_{\times S}$ is found to correspond to a minimum of this quantity.

> In the retarded model, the leverage effect has a simple interpretation: in a first approximation, price changes are independent of the price. But since the volatility is defined as a relative quantity, it of course increases when the price decreases, simply because the price appears in the denominator.

We therefore conclude that the leverage effect for stocks might not have a very deep economical significance, but can be assigned to a simple 'retarded' effect, where the change of prices are calibrated not on the instantaneous value of the price but rather on a moving average of the price. This means that on short times scales ($< T_{\times S}$), the relevant model for stock price variations is *additive* rather than multiplicative. The observed skewness of the log-price is therefore induced by the 'wrong' choice of the relevant variable in this case – see the discussion of Eq. (8.7).

8.3.2 Leverage effect for indices

Figure 8.5 shows that the 'leverage effect' for indices has a much larger amplitude and tends to decay to zero much faster with the lag ℓ. This is at first sight puzzling, since the stock index is, by definition, an average over stocks. So one can wonder why the strong leverage effect for the index does not show up in stocks and conversely why the long time scale seen in stocks disappears in the index.[†] These seemingly paradoxical features can be rationalized within the framework a simple one factor model (see Chapter 11), where the 'market factor' is responsible for the strong leverage effect.

The empirical leverage effect on indices shown in Figure 8.5 suggests that there exists an index specific 'panic-like' effect, resulting from an increase of activity when the market goes down as a whole. A natural question is why this panic effect does not show up also in the statistics of individual stocks. A detailed analysis of the one factor model actually indicates that, to a good approximation, the index specific leverage effect and the stock leverage effect do not mix in the limit where the volatility of individual stocks is much larger than the index volatility. Since the stock volatility is indeed found to be somewhat larger than the index volatility, one expects the mixing to be small and difficult to detect.

From a practical point of view, the fact that the leverage correlation for indices cannot be explained solely in terms of a retarded effect means that the skewness of the log-price is a real effect rather than an artifact based on a bad choice of the fundamental variable. The strong, genuine negative skewness observed for indices has important consequences for option pricing that will be expanded upon in Chapter 13.

[†] A close look at Fig. 8.4 actually suggests that there is indeed a stronger, short time effect for stocks as well.

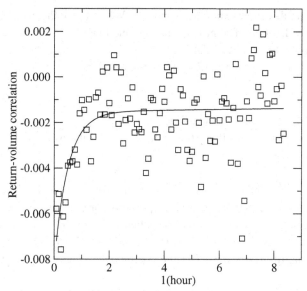

Fig. 8.6. Intraday return-volume correlation for the S&P 500 futures. The plain line corresponds to a fit with the sum of two exponentials, one with a rather short time scale (a few hours), and a second with a correlation time of 9 days, as extracted from the leverage correlation function.

8.3.3 Return-volume correlations

As explained in Section 7.3.2, the sources of volatility are twofold: trading volume on the one hand, amplitude of individual price jumps in the other. In the above 'panic' interpretation of the strong leverage effect for stocks, we have implied that a price drop triggers an increase of activity. It is interesting to check this directly by studying the return-volume correlation function, defined as:

$$C_{rv}(\ell) = \frac{\langle \eta_k \cdot \delta N_{k+\ell} \rangle}{\sqrt{\langle \eta^2 \rangle \langle \delta N^2 \rangle}}, \tag{8.18}$$

where δN_k is the deviation from the average number of trades in a small time interval centered around time $t = k\tau$. Figure 8.6 shows this correlation function for the S&P 500 futures, with $\tau = 5$ minutes. As expected, this correlation shows the same features as the leverage correlation: it is negative, implying that the trading activity indeed increases after a price drop. This correlation function decays on two time scales: one rather short (one hour) corresponding to intraday reactions, and a longer one (several days) similar to the leverage correlation time reported above.

8.4 The Heston model: a model with volatility fluctuations and skew

It is interesting at this stage to discuss a continuous time model for price fluctuations, introduced by Heston (1993), that contains both volatility correlations and the 'leverage' effect, and that is exactly solvable, in the sense that the distribution of price changes for

different time intervals can be explicitly computed. We first give the equations defining the model, then give without proof some exact results, and finally discuss the limitations of this model as compared with empirical data.

The starting point is to write Eq. (7.4) in continuous time:

$$dx_t = x_t(m\,dt + \sigma_t\,dW), \tag{8.19}$$

where σ_t is the volatility at time t and dW is a Brownian noise with unit volatility. Now, let us assume that the volatility itself has random, mean reverting fluctuations such that:

$$d(\sigma_t^2) = -\Omega(\sigma_t^2 - \sigma_0^2)\,dt + \Upsilon\sigma_t\,dW', \tag{8.20}$$

where Ω is the inverse mean reversion time towards the average volatility level σ_0, and dW' is another Brownian white noise of unit variance, possibly correlated with dW. The correlation coefficient between these two noises will be called ρ. When $\rho < 0$, this model describes the leverage effect, since a decrease of price (i.e. $dW < 0$) is typically accompanied by an increase of volatility (i.e. $dW' > 0$). The coefficient Υ measures the 'volatility of the volatility'.

Remarkably, many exact results can be obtained within this framework, due to the particular form of the noise term in the second equation, proportional to σ_t itself.[†] First, one can compute the stationary distribution $P(v)$ of the squared volatility $v = \sigma^2$, which is found to be a gamma distribution (not an inverse gamma distribution):

$$P(v) = \frac{v^{\beta-1}}{\Gamma(\beta)v_0^\beta}e^{-v/v_0}, \tag{8.21}$$

with:

$$v_0 = \frac{\Upsilon^2}{2\Omega}, \qquad \beta = \frac{2\Omega\sigma_0^2}{\Upsilon^2}. \tag{8.22}$$

The unconditional distribution of price returns (averaged over the volatility process) can also be exactly computed, in terms of its characteristic function. The expression is too complicated to be given here; let us only mention three important features of the explicit solution:

- It is strongly non Gaussian for short time scales, and becomes progressively more and more Gaussian for longer time scales. Its skewness and kurtosis can be exactly computed for all time lags τ. For $\rho = 0$, the kurtosis to lowest order in Υ, reads:

$$\kappa(\tau) = \frac{\Upsilon^2}{2\sigma_0^2\Omega}\frac{e^{-\Omega\tau} + \Omega\tau - 1}{(\Omega\tau)^2}, \tag{8.23}$$

 which is plotted in Fig. 8.7. As expected, the kurtosis in this model is induced by the volatility of the volatility Υ. For $\Omega\tau \to 0$, the kurtosis tends to a constant, whereas for $\Omega\tau \to \infty$, the kurtosis has the expected τ^{-1} behaviour.

[†] The results given here are discussed in details in Dragulescu & Yakovlenko (2002).

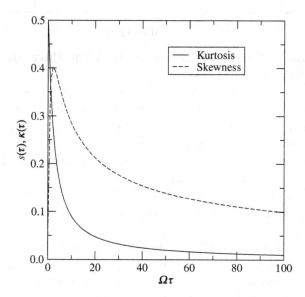

Fig. 8.7. Time structure of the skewness and kurtosis in the Heston model, as given by the non dimensional part of the expressions of $\varsigma(\tau)$ and $\kappa(\tau)$.

- The skewness of the distribution can be tuned by choosing the correlation coefficient ρ. To lowest order in Υ and ρ, the skewness reads:

$$\varsigma(\tau) = \frac{\sqrt{2}\Upsilon\rho}{\sqrt{\Omega}\sigma_0} \frac{e^{-\Omega\tau} - 1 + \Omega\tau}{(\Omega\tau)^{3/2}}, \tag{8.24}$$

which is again plotted in Fig. 8.7. Note that the skewness is negative when $\rho < 0$, and has a non monotonous time structure. When $\Omega\tau \to \infty$, the skewness decays like $\tau^{-1/2}$, as for independent increments.
- The asymptotic tails of the distribution are exponential for all times. The parameter of this exponential fall off becomes time independent in the limit $\Omega\tau \gg 1$.

This model has however three important limitations:

- First, as shown in Section 7.2.2, the empirical distribution of the volatility is closer to an *inverse* gamma distribution than to a gamma distribution. Note that Eq. (8.21) above predicts a Gaussian like decay of the distribution of the volatility for large arguments, at variance with the observed, much slower, power-law.
- Second, the asymptotic decay of the kurtosis is as τ^{-1} for $\Omega\tau \gg 1$, i.e. much faster than the power-law decay mentioned in Section 7.2.1. This is related to the fact that there is a well defined correlation time in the Heston model, given by $1/\Omega$, beyond which the volatility correlation function decays exponentially. Therefore, the multiscale nature of the volatility correlations is missed in the Heston model.
- Finally, the correlation time for the skewness is, in the Heston model, also equal to $1/\Omega$. In other words, the temporal structure of the leverage effect and of the volatility fluctuations are, in this model, strongly interconnected.

8.5 Summary

- Price returns and volatility are negatively correlated. This effect induces a negative skew in the distribution of returns.
- In an additive model for price changes, returns and volatility (defined using *relative* price changes) are negatively correlated for a trivial reason. This simple mechanism, embedded in the retarded volatility model where price increments progressively (rather than instantaneously) adapt to the price level, accounts well for the empirical data on individual stocks. The retarded model interpolates smoothly between an additive and a multiplicative random process.
- For stock indices, the negative correlation is much stronger and cannot be accounted for by the retarded model. A similar negative correlation exists between price changes and volume of activity.
- The Heston model is a continuous time stochastic volatility model that captures both the volatility fluctuations and the skewness effect. However, this model fails to describe the detailed shape of the empirical volatility distribution, and the multi-time scale nature of the volatility correlations.

- FURTHER READING

J. Y. Campbell, A. W. Lo, A. C. McKinley, *The Econometrics of Financial Markets*, Princeton University Press, Priceton, 1997.

A. Crisanti, G. Paladin and A. Vulpiani, *Products of Random Matrices in Statistical Physics*, Springer-Verlag, Berlin, 1993.

- SPECIALIZED PAPERS

G. Bekaert, G. Wu, *Asymmetric volatility and risk in equity markets*, The Review of Financial Studies, **13**, 1 (2000).

J.-P. Bouchaud, M. Potters, A. Matacz, *The leverage effect in financial markets: retarded volatility and market panic*, Phys. Rev. Lett., **87**, 228701 (2001).

A. Dragulescu and V. Yakovlenko, *Probability distribution of returns for a model with stochastic volatility*, Quantitative Finance, **2**, 443 (2002).

C. Harvey, A. Siddique, *Conditional skewness in asset pricing tests*, Journal of Finance, **LV**, 1263 (2000).

S. L. Heston, *A closed-form solution for options with stochastic volatility with applications to bond and currency options*, Review of Financial Studies, **6**, 327 (1993).

J. Masoliver and J. Perelló, *Random diffusion and leverage effect in financial markets*, Physical Review E **67**, 037102 (2003).

J. Perelló, J. Masoliver, and J.-P. Bouchaud, *Multiple time scales in volatility and leverage correlations: An stochastic volatility model*, Applied Mathematical Finance, **11**, 27–50 (2004).

B. Pochart, J.-P. Bouchaud, *The skewed multifractal random walk with applications to option smiles*, Quantitative Finance, **2**, 303 (2002).

- SOME RECENT READS

P. Hagan, D. Kumar, A. Lesniewski, and D. Woodward, *Managing smile risk*, Wilmott magazine, 84–108, September 2002.

9

Cross-correlations

La science est obscure — peut-être parce que la vérité est sombre.[†]

<div align="right">(Victor Hugo)</div>

9.1 Correlation matrices and principal component analysis

9.1.1 Introduction

The previous chapters were concerned with the statistics of individual assets, disregarding any possible cross-correlations between these different assets. However, as we shall see in Chapter 10, an important aspect of risk management is the estimation of the correlations between the price movements of different assets. The probability of large losses for a certain portfolio or option book is dominated by correlated moves of its different constituents – for example, a position which is simultaneously long in stocks and short in bonds is risky because stocks and bonds tend to move in opposite directions in crisis periods. (This is the so-called 'flight to quality' effect.) The study of correlation (or covariance) matrices thus has a long history in finance, and is one of the cornerstone of Markowitz's theory of optimal portfolios (see Chapter 12).

In order to measure these correlations, one often defines the covariance matrix \mathbf{C} between all the assets in the universe one decides to work in. (Other possible definitions of these correlations, in particular for strongly fluctuating assets, will be described in Chapter 11). The number of such assets will be called M. The matrix elements C_{ij} of \mathbf{C} are obtained as:[‡]

$$C_{ij} = \langle \eta_i \eta_j \rangle - m_i m_j, \tag{9.1}$$

where $\eta_i = \delta x_i / x_i$ is the return of asset i, and m_i its average return, that we will, without loss of generality, set to zero in the following. (One can always shift the returns by $-m_i$ in the following formulae if needed.)

Once this matrix is constructed, it is natural to ask for the interpretation of its eigenvalues and eigenvectors. Note that since the matrix is symmetric, the eigenvalues are all real numbers. These numbers are also all positive, otherwise it would be possible to find a linear

[†] Science is obscure — maybe because truth is dark.

[‡] The *correlation* matrix is usually defined in terms of rescaled returns with unit volatility.

combination of the x_i (in other words, a certain portfolio of the different assets) with a negative variance. We will call \mathbf{v}_a the normalized eigenvector corresponding to eigenvalue λ_a, with $a = 1, \ldots, M$. By definition, an eigenvector is a linear combination of the different assets i such that $\mathbf{Cv}_a = \lambda_a \mathbf{v}_a$. The vector \mathbf{v}_a is the list of the weights $v_{a,i}$ in this linear combination of the different assets i. More precisely, let us consider a portfolio such that the dollar amount of stock i is $v_{a,i}$ (the number of stocks is thus $v_{a,i}/x_i$). The variance of the return of such a portfolio is thus:

$$\sigma_a^2 = \left\langle \left(\sum_{i=1}^{M} \frac{v_{a,i}}{x_i} \delta x_i \right)^2 \right\rangle = \sum_{i,j=1}^{M} v_{a,i} v_{a,j} C_{ij} \equiv \mathbf{v}_a \cdot \mathbf{Cv}_a. \tag{9.2}$$

But since \mathbf{v}_a is an eigenvector corresponding to eigenvalue λ_a, we see that λ_a is precisely the variance of the portfolio constructed from the weights $v_{a,i}$. Furthermore, using the fact that different eigenvectors are orthogonal, it is easy to see that the correlation of the returns of two portfolios constructed from two different eigenvectors is zero:

$$\left\langle \left(\sum_{i=1}^{M} \frac{v_{a,i}}{x_i} \delta x_i \right) \left(\sum_{j=1}^{M} \frac{v_{b,j}}{x_j} \delta x_j \right) \right\rangle = \mathbf{v}_b \cdot \mathbf{Cv}_a = 0 \quad (b \neq a). \tag{9.3}$$

We have thus obtained a set of uncorrelated random returns e_a, which are the returns of the portfolios constructed from the weights $v_{a,i}$:

$$e_a = \sum_{i=1}^{M} v_{a,i} \eta_i; \qquad \langle e_a e_b \rangle = \lambda_a \delta_{a,b}. \tag{9.4}$$

Conversely, one can think of the initial returns as a linear combination of the uncorrelated **factors** E_a:

$$\frac{\delta x_i}{x_i} = \eta_i = \sum_{a=1}^{M} v_{a,i} e_a, \tag{9.5}$$

where the last equality holds because the transformation matrix $v_{a,i}$ from the initial vectors to the eigenvectors is orthogonal. This decomposition is called a **principal component analysis**, where the correlated fluctuations of a set of random variables is decomposed in terms of the fluctuations of underlying uncorrelated factors. In the case of financial returns, the principal components E_a often have an economic interpretation in terms of sectors of activity (see Section 9.3.2).

9.1.2 Gaussian correlated variables

When the returns are Gaussian random variables, more precisely, when the joint distribution of the η_i can be written as:[†]

$$
P_G(\eta_1, \eta_2, \ldots, \eta_M) = \frac{1}{\sqrt{(2\pi)^M \det \mathbf{C}}} \exp\left[-\frac{1}{2} \sum_{ij} \eta_i \left(C^{-1}\right)_{ij} \eta_j \right],
\tag{9.6}
$$

where $(C^{-1})_{ij}$ denotes the elements of the matrix inverse of \mathbf{C}, then, the principal components E_a are themselves *independent* Gaussian random variables: any set of correlated Gaussian random variables can always be decomposed as a linear combination of independent Gaussian random variables (the converse is obviously true since the sum of Gaussian random variables is a Gaussian random variable). In other words, correlated Gaussian random variables are fully characterized by their correlation matrix. This is not true in general, much more information is needed to describe the interdependence of non Gaussian correlated variables. This will be discussed further in Section 9.2.

9.1.3 Empirical correlation matrices

A reliable empirical determination of a correlation matrix turns out to be difficult: if one considers M assets, the correlation matrix contains $M(M-1)/2$ entries, which must be determined from M time series of length N; if N is not very large compared to M, one should expect that the determination of the covariances is noisy, and therefore that the empirical correlation matrix is to a large extent random, i.e. the structure of the matrix is dominated by 'measurement' noise. If this is the case, one should be very careful when using this correlation matrix in applications. From this point of view, it is interesting to compare the properties of an empirical correlation matrix \mathbf{C} to a 'null hypothesis' purely *random* matrix as one could obtain from a finite time series of strictly uncorrelated assets. Deviations from the random matrix case might then suggest the presence of true information.

The empirical correlation matrix \mathbf{C} is constructed from the time series of price returns $\eta_i(k)$ (where i labels the asset and k the time) through the equation:

$$
\mathbf{C}_{ij} = \frac{1}{N} \sum_{k=1}^{N} \eta_i(k)\eta_j(k).
\tag{9.7}
$$

In this section we assume that the average value of the η's has been subtracted off, and that the η's are rescaled to have a constant unit volatility. The null hypothesis of independent assets, which we consider now, translates itself in the assumption that the coefficients

[†] The fact that the marginal distribution of individual returns is Gaussian is not sufficient for the following discussion to hold, see Section 9.2.7 for a detailed discussion.

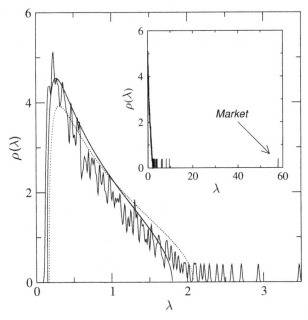

Fig. 9.1. Smoothed density of the eigenvalues of **C**, where the correlation matrix **C** is extracted from $M = 406$ assets of the S&P 500 during the years 1991–96. For comparison we have plotted the density Eq. (9.61) for $Q = 3.22$ and $\sigma^2 = 0.85$: this is the theoretical value obtained assuming that the matrix is purely random except for its highest eigenvalue (dotted line). A better fit can be obtained with a smaller value of $\sigma^2 = 0.74$ (solid line), corresponding to 74% of the total variance. Inset: same plot, but including the highest eigenvalue corresponding to the 'market', which is found to be \sim30 times greater than λ_{\max}.

$\eta_i(k)$ are independent, identically distributed, random variables.[†] The theory of random matrices, briefly expounded in Appendices A and B, allows one to compute the density of eigenvalues of **C**, $\rho_C(\lambda)$, in the limit of very large matrices: it is given by Eq. (9.61), with $Q = N/M$.

Now, we want to compare the empirical distribution of the eigenvalues of the correlation matrix of stocks corresponding to different markets with the theoretical prediction given by Eq. (9.61) in Appendix B, based on the assumption that the correlation matrix is random. We have studied numerically the density of eigenvalues of the correlation matrix of $M = 406$ assets of the S&P 500, based on daily variations during the years 1991–96, for a total of $N = 1309$ days (the corresponding value of Q is 3.22). An immediate observation is that the highest eigenvalue λ_1 is 25 times larger than the predicted λ_{\max} (Fig. 9.1, inset). The corresponding eigenvector is, as expected, the 'market' itself, i.e. it has roughly equal components on all the M stocks. The simplest 'pure noise' hypothesis is therefore inconsistent with the value of λ_1. A more reasonable idea is that the component of the correlation

[†] Note that even if the 'true' correlation matrix $\mathbf{C}_{\mathrm{true}}$ is the identity matrix, its empirical determination from a finite time series will generate non-trivial eigenvectors and eigenvalues.

matrix which are orthogonal to the 'market' are pure noise. This amounts to subtracting the contribution of λ_1 from the nominal value $\sigma^2 = 1$, leading to $\sigma^2 = 1 - \lambda_1/M = 0.85$. The corresponding fit of the empirical distribution is shown as a dotted line in Figure 9.1. Several eigenvalues are still above λ_{max} and contain some information about different economic sectors, thereby reducing the variance of the effectively random part of the correlation matrix. One can therefore treat σ^2 as an adjustable parameter. The best fit is obtained for $\sigma^2 = 0.74$, and corresponds to the dark line in Figure 9.1, which accounts quite satisfactorily for 94% of the spectrum, whereas the 6% highest eigenvalues still exceed the theoretical upper edge by a substantial amount. These 6% highest eigenvalues are however responsible for 26% of the total volatility.

One can repeat the above analysis on different stock markets (e.g. Paris, London, Zurich), or on *volatility correlation matrices*, to find very similar results. In a first approximation, the location of the theoretical edge, determined by fitting the part of the density which contains most of the eigenvalues, allows one to distinguish 'information' from 'noise'.

The conclusion of this section is therefore that a large part of the empirical correlation matrices must be considered as 'noise', and cannot be trusted for risk management. In Chapter 10, we will dwell on Markowitz' portfolio theory, which assumes that the correlation matrix is perfectly known. This theory must therefore be taken with a grain of salt, bearing in mind the results of the present section.

9.2 Non-Gaussian correlated variables

There are many different ways to construct non-Gaussian correlated variables, corresponding to different underlying mechanisms. In the section, we elaborate on three simple ways to generate these multivariate non-Gaussian statistics: a linear model where one sums independent non-Gaussian variables, a non-linear model where one imposes a non-linear transformation to multivariate Gaussian variables, and finally a stochastic volatility model where multivariate Gaussian variables are multiplied by a common random factor.

9.2.1 Sums of non Gaussian variables

We have seen in Section 9.1.1 above that correlated Gaussian variables can always be written as a linear superposition of independent random Gaussian variables. This suggests an immediate generalization in order to construct non Gaussian correlated variables. One can for example consider independent but non Gaussian elementary 'factors' E_a, with an arbitrary probability distribution $P_a(e_a)$, and define the correlated variables of interest as:

$$\eta_i = \sum_{a=1}^{M} v_{a,i} e_a, \tag{9.8}$$

where the $v_{a,i}$ are the elements of a certain rotation matrix that diagonalizes the correlation matrix of the η variables. This means that the E_a can be obtained from the η_i by applying

the inverse transformation:

$$e_a = \sum_{i=1}^{M} v_{a,i} \eta_i. \tag{9.9}$$

A particularly interesting case is when $P_a(e_a)$ has power tails:

$$P_a(e_a) \simeq_{|e_a| \to \pm\infty} \frac{\mu A_a^{\mu}}{|e_a|^{1+\mu}}, \tag{9.10}$$

where the exponent μ does not depend on the factor E_a. In this case, using the results of Appendix C, the η_i also have power-law tails with the same exponent μ, and with a tail amplitude given by:

$$A_i^{\mu} = \sum_{a=1}^{M} |v_{a,i}|^{\mu} A_a^{\mu}. \tag{9.11}$$

A special case is when the E_a are Lévy stable variables, with $\mu < 2$. Then, since the sum of independent Lévy stable random variables is also a Lévy stable variable, the marginal distribution of the η_i's is a Lévy stable distribution. The η_i's then form a set of multivariate Lévy stable random variables, see Section 9.2.6.

9.2.2 Non-linear transformation of correlated Gaussian variables

Another possibility is to start from independent Gaussian variables E_a, and build the η_i's in two steps, as:

$$\tilde{\eta}_i = \sum_{a=1}^{M} v_{a,i} e_a \qquad \eta_i = F_i [\tilde{\eta}_i], \tag{9.12}$$

where F_i is a certain non-linear function, chosen such that the marginal distribution of the η_i matches the empirical distribution $P_i(\eta_i)$. More precisely, since $\tilde{\eta}_i$ is Gaussian, one should have, if F_i is monotonous:

$$P_i(\eta_i) = P_G(\tilde{\eta}_i) \left| \frac{d\tilde{\eta}_i}{d\eta_i} \right|, \tag{9.13}$$

where $P_G(.)$ is the Gaussian distribution. One therefore encodes the correlations is the construction of the intermediate Gaussian random variables $\tilde{\eta}_i$, before converting them into a set of random variables with the correct marginal probability distributions.

9.2.3 Copulas

The above idea of a non linear transformation of correlated Gaussian variables to obtain correlated variables with an arbitrary marginal distribution can be generalized, and leads to the notion of **copulas***. Let us consider N random variables $\eta_1, \eta_2, \ldots, \eta_N$, with a joint distribution $P(\eta_1, \eta_2, \ldots, \eta_N)$. The marginal distributions are $P_i(\eta_i)$, and the corresponding cumulative distributions $\mathcal{P}_{i<}$. By construction, the random variables*

$u_i \equiv \mathcal{P}_{i<}(\eta_i)$ are uniformly distributed in the interval $[0, 1]$. The joint distribution of the u_i defines the copula of the joint distribution P above. It characterizes the structure of the correlations, but is by construction independent of the marginal distributions, in the sense that any non linear transformation of the random variables leaves the copula invariant. More precisely, let $C(u_1, u_2, \ldots, u_N)$ be a function from $[0, 1]^N$ to $[0, 1]$, that is a non decreasing function of all its arguments. The function $C(u_1, u_2, \ldots, u_N)$, called the copula, is the joint cumulative distribution of the u_i. To any joint distribution $P(\eta_1, \eta_2, \ldots, \eta_N)$ one can associate a unique copula C. Conversely, any function C that has the above properties can be used to define a joint distribution $P(\eta_1, \eta_2, \ldots, \eta_N)$ with the desired marginals, by defining the η_i through:

$$\eta_i = \mathcal{P}_{i<}^{-1}(u_i). \tag{9.14}$$

The 'non-linear' model described in the previous subsection therefore corresponds to a Gaussian copula.

9.2.4 Comparison of the two models

Given a set of non Gaussian correlated variables, which of the above two models is most faithful in describing the data? In order to test the linear model where each variable is the sum of independent non Gaussian variables, one can first diagonalize the correlation matrix and construct the factors E_a using Eq. (9.9). One can then transform each of these E_a as $\tilde{e}_a = F_a^{-1}(e_a)$ such that the marginal distribution of each \tilde{E}_a is Gaussian of unit variance. The assumption of the linear model is that the E_a (and thus the \tilde{E}_a) are independent. One can therefore study the following correlation function (which is a generalized kurtosis, see Section 9.2.7):

$$K_{ab} = \langle \tilde{e}_a^2 \tilde{e}_b^2 \rangle - \langle \tilde{e}_a^2 \rangle \langle \tilde{e}_b^2 \rangle - 2\langle \tilde{e}_a \tilde{e}_b \rangle^2, \tag{9.15}$$

which should be very small if the linear model holds. Note that, because the individual \tilde{E}_a are Gaussian by construction, one has $K_{aa} = 0$. A global measure of the merit of the model could be defined as:

$$\overline{K}_l = \frac{1}{M(M-1)} \sum_{a \neq b} K_{ab}. \tag{9.16}$$

As for the non linear model of correlations, one should first construct from the η_i the intermediate Gaussian variables such that $\tilde{\eta}_i = F_i^{-1}(\eta_i)$, then determine the correlation matrix \tilde{C} of the $\tilde{\eta}_i$ in order to find the e_a from:

$$e_a = \sum_{i=1}^{M} \tilde{v}_{a,i} \tilde{\eta}_i, \tag{9.17}$$

where $\tilde{v}_{a,i}$ is the rotation matrix that diagonalizes \tilde{C}. The assumption of the non-linear model is that the resulting e_a are independent Gaussian random variables. By construction, the e_a are uncorrelated, and can be normalized to have unit variance. The first non trivial

test for independence is thus to consider again the above non linear correlation function:

$$K_{ab} = \langle e_a^2 e_b^2 \rangle - \langle e_a^2 \rangle \langle e_b^2 \rangle - 2 \langle e_a e_b \rangle^2 , \tag{9.18}$$

(where now the last term $\langle e_a e_b \rangle^2 = 0$ by construction), and the global merit of the non-linear model:

$$\overline{K}_{nl} = \frac{1}{M(M-1)} \sum_{a \neq b} K_{ab}. \tag{9.19}$$

The best of the two descriptions corresponds to one with \overline{K} closest to zero. These numbers are directly comparable since in both cases, the role of the tail events has been regularized by the change of variable that transforms them into Gaussian variables. We have computed these quantities averaged over all 50×99 pairs made of 100 U.S. stocks from 1993 to 1999. We find $\overline{K}_{nl} = 0.118$ and $\overline{K}_l = 0.182$, which shows that the non-linear, Gaussian copula model is better than a linear model of non-Gaussian factors. However, the residual 'cross-kurtosis' \overline{K}_{nl} is still significant, showing that the non linear model is not capturing all the correlations. As discussed at length in Chapter 7 and in the next section, Section 9.2.5, volatility fluctuations appears to be an important source of non linear correlations in financial markets. We thus expect, and will indeed find, that a model which explicitly accounts for these volatility fluctuations should be more accurate.

The matrix K_{ab} can be understood as the correlation matrix of the square volatility of the factors. This matrix can itself be diagonalized, thereby defining principal components of volatility fluctuations. This suggests another, financially motivated, model of correlated returns:

$$\eta_i = \sum_{a=1}^{M} v_{a,i} \left(\sum_{c=1}^{M} w_{c,a} f_c^2 \right)^{1/2} e_a, \tag{9.20}$$

where the f_c^2 are now the principal components of the square volatility, and $w_{c,a}$ the rotation matrix that diagonalizes K_{ab}. A particular case of interest is when one volatility component, say $c = 1$, dominates all the others. This means, intuitively, that the volatilities of all economic factors tend to increase or decrease in a strongly correlated way. In that case, one can write:

$$\eta_i = f \sum_{a=1}^{M} v_{a,i} \tilde{e}_a, \qquad \tilde{e}_a \equiv \sqrt{w_{1,a}} e_a, \tag{9.21}$$

where $f = f_1$ is the common fluctuating volatility. When the E_a are independent Gaussian factors and f has an inverse gamma distribution, this model generates a set of multivariate Student variables η_i, which we discuss in the next section.

This idea can be directly tested on empirical data by dividing the returns of all stocks on a given day by a proxy of the fluctuating volatility f_1. Such a proxy can be constructed by computing the root mean square of the returns of all the stocks on a given day (a quantity that we will call the 'variety' and discuss in Section 11.2.1, see Eq. (11.24)). All the returns of a given day are divided by the variety, and the same test on K_{ab} is performed on these rescaled returns. One now finds $\overline{K}_{nl} = 0.0169$ and $\overline{K}_l = 0.0107$. These numbers are much smaller than the ones quoted above and very small. This shows that indeed, volatility fluctuations is

the major source of non linear correlations between stocks. Furthermore, once the fluctuating volatility effect is removed, the linear model appears better than the non-linear, Gaussian copula model.

9.2.5 Multivariate Student distributions

Let us make the above remarks more precise. Consider M correlated Gaussian variables $\tilde{\eta}_i$, such that their multivariate distribution is given by Eq. (9.6). Now, multiply all $\tilde{\eta}_i$ by the same random factor $f = \sqrt{\mu/s}$, where s is a positive random variable with a gamma distribution:

$$P(s) = \frac{1}{\Gamma(\frac{\mu}{2})} s^{\frac{\mu}{2}-1} e^{-s}. \tag{9.22}$$

The joint distribution P_S of the resulting variables $\eta_i = \sqrt{\mu/s}\,\tilde{\eta}_i$ is clearly given by:

$$P_S(\eta_1, \eta_2, \ldots, \eta_M)$$
$$= \int_0^\infty \left(\frac{s}{\mu}\right)^{M/2} P(s) P_G(\sqrt{s/\mu}\,\eta_1, \sqrt{s/\mu}\,\eta_2, \ldots, \sqrt{s/\mu}\,\eta_M) ds, \tag{9.23}$$

where P_G is the multivariate Gaussian (9.6) and the factor $(s/\mu)^{M/2}$ comes from the change of variables. Using the explicit form of P_G with a correlation matrix $\tilde{\mathbf{C}}$, one sees that the integral over s can be explicitly performed, finally leading to:

$$P_S(\eta_1, \eta_2, \ldots, \eta_M) = \frac{\Gamma(\frac{M+\mu}{2})}{\Gamma(\frac{\mu}{2})\sqrt{(\mu\pi)^M \det \tilde{\mathbf{C}}}} \frac{1}{\left(1 + \frac{1}{\mu}\sum_{ij}\eta_i(\tilde{C}^{-1})_{ij}\eta_j\right)^{\frac{M+\mu}{2}}}, \tag{9.24}$$

which is called the multivariate Student distribution. Let us list a few useful properties of this distribution, that can easily be shown using the representation (9.23) above:

- The marginal distribution of any η_i is a monovariate Student distribution of parameter μ:

$$P_S(\eta_i) = \frac{1}{\sqrt{\pi\mu}} \frac{\Gamma(\frac{1+\mu}{2})}{\Gamma(\frac{\mu}{2})} \frac{\tilde{C}_{ii}^{\mu/2}}{\left(\tilde{C}_{ii} + \frac{\eta_i^2}{\mu}\right)^{\frac{1+\mu}{2}}}. \tag{9.25}$$

- In the limit $\mu \to \infty$, one can show that the multivariate Student distribution P_S tends towards the multivariate Gaussian P_G. This is expected, since in this limit, $P(s)$ tends to a delta function $\delta(s - \mu/2)$, and therefore the random volatility $f = \sqrt{\mu/s}$ does not fluctuate anymore and is equal to $f = \sqrt{2}$.
- The correlation matrix of the η_i is given, for $\mu > 2$, by:

$$C_{ij} = \langle \eta_i \eta_j \rangle = \frac{2\mu}{\mu - 2} \tilde{C}_{ij}. \tag{9.26}$$

Note that, as expected, the above result tends to the Gaussian result $2\tilde{C}_{ij}$ when $\mu \to \infty$ and diverges when $\mu \to 2$.

- Wick's theorem for Gaussian variables[†] can be extended to Student variables. For example, one can show that:

$$\langle \eta_i \eta_j \eta_k \eta_l \rangle = \frac{\mu - 2}{\mu - 4} [C_{ij} C_{kl} + C_{ik} C_{jl} + C_{il} C_{jk}], \tag{9.28}$$

which again reproduces Wick's contractions in the limit $\mu \to \infty$, where the prefactor tends to unity.

Finally, note that the matrix \tilde{C}_{ij} can be estimated from empirical data using a maximum likelihood procedure. Given a time series of stock returns $\eta_i(k)$, with $i = 1, \ldots, M$ and $k = 1, \ldots, N$, the most likely matrix \tilde{C}_{ij} is given by the solution of the following equation:

$$\tilde{C}_{ij} = \frac{M + \mu}{N} \sum_{k=1}^{N} \frac{\eta_i(k) \eta_j(k)}{\mu + \sum_{mn} \eta_m(k)(\tilde{C}^{-1})_{mn} \eta_n(k)}. \tag{9.29}$$

Note that in the Gaussian limit $\mu \to \infty$, the denominator of the above expression is simply given by μ, and the final expression is simply:

$$\tilde{C}_{ij} = C_{ij} = \frac{1}{N} \sum_{k=1}^{N} \eta_i(k) \eta_j(k), \tag{9.30}$$

as it should.

9.2.6 Multivariate Lévy variables (*)

Consider finally the case where the distribution of the squared volatility f^2 is a totally asymmetric Lévy distribution of index $\mu < 1$. When f multiplies a single Gaussian random variable, the resulting product has, as mentioned in Section 7.1, a symmetric Lévy distribution of index 2μ. When one multiplies a set correlated Gaussian variables by the same f (such that f^2 has a totally asymmetric Lévy distribution), one generates a multivariate symmetric Lévy distribution of index 2μ, that reads:

$$P_L(\eta_1, \eta_2, \ldots, \eta_M) = \int \int .. \int \prod_{j=1}^{M} \frac{dz_j}{2\pi} \exp \left[-i \sum_j z_j \eta_j - \frac{1}{2} \left| \sum_{ij} z_i C_{ij} z_j \right|^\mu \right]. \tag{9.31}$$

Note that this multivariate distribution is different from the one obtained in the case of a linear combination of independent symmetric Lévy variables of index 2μ, discussed in Section 9.2.1. If the η_i's read:

$$\eta_i = \sum_{b=1}^{N} v_{b,i} e_b, \tag{9.32}$$

[†] Wick's theorem states that the expectation value of a product of any number of multivariate Gaussian variables (of zero mean) can be obtained by replacing every pair of variables by the two-point correlation matrix with the same indices and summing over all possible pairings. For example,

$$\langle \eta_i \eta_j \eta_k \eta_l \rangle = [C_{ij} C_{kl} + C_{ik} C_{jl} + C_{il} C_{jk}], \tag{9.27}$$

when $\{\eta_i\}$ are multivariate Gaussian variables of zero mean.

then one finds:

$$P(\eta_1, \eta_2, \ldots, \eta_M) =$$

$$\int \int \cdots \int \prod_{j=1}^{M} \frac{dz_j}{2\pi} \exp\left(-i \sum_j z_j \eta_j - \sum_{b=1}^{N} a_{2\mu,b} \left[\sum_{ij} z_i v_{b,i} v_{b,j} z_j\right]^\mu\right), \tag{9.33}$$

where $a_{2\mu,b}$ are related to the tail amplitude of the variables e_b (see Eq. (1.20)).

Both Eqs (9.31) and (9.33) correspond to a *stable* multivariate process, in the sense that if $\{\eta_1, \eta_2, \ldots, \eta_M\}$ and $\{\eta_1', \eta_2', \ldots, \eta_M'\}$ are two realizations of the process, then the sum $\{\eta_1 + \eta_1', \eta_2 + \eta_2', \ldots, \eta_M + \eta_M'\}$ has (up to a rescaling) the same multivariate distribution. This is particularly obvious in this last case, since:

$$\eta_i + \eta_i' = \sum_{b=1}^{N} v_{b,i}(e_b + e_b'), \tag{9.34}$$

where $e_b + e_b'$ still has a symmetric Lévy distribution, due to the stability of these under addition.

The most general symmetric multivariate Lévy distribution of index 2μ in fact reads:

$$L_{2\mu}(\eta_1, \eta_2, \ldots, \eta_M) =$$

$$\int \int \cdots \int \prod_{j=1}^{M} \frac{dz_j}{2\pi} \exp\left(-i \sum_j z_j \eta_j - \int \gamma(\vec{n}) \left[\sum_{ij} z_i n_i n_j z_j\right]^\mu d\vec{n}\right), \tag{9.35}$$

where \vec{n} is a unit vector on the M-dimensional sphere, and $\gamma(\vec{n})$ an arbitrary function that satisfies $\gamma(\vec{n}) = \gamma(-\vec{n})$. Equation (9.33) corresponds to the case where $\gamma(\vec{n})$ is concentrated in a finite number of directions:

$$\gamma(\vec{n}) = \frac{1}{2} \sum_{b=1}^{N} a_{2\mu,b} \left[\delta(\vec{n} - \vec{v}_b) + \delta(\vec{n} + \vec{v}_b)\right], \tag{9.36}$$

whereas Eq. (9.31) with $C_{ij} = \sigma^2 \delta_{i,j}$ corresponds to a uniform function:

$$\gamma(\vec{n}) \propto \sigma^\mu. \tag{9.37}$$

A full account of these multivariate Lévy distributions, together with their asymmetric generalizations, can be found in Samorodnitsky & Taqqu (1994).

9.2.7 Weakly non Gaussian correlated variables (*)

Another way to characterize non Gaussian correlated variables is to introduce, from the joint distribution of the asset fluctuations, a generalized kurtosis K_{ijkl} that measures the first correction to Gaussian statistics. One writes*:

$$P(\eta_1, \eta_2, \ldots, \eta_M) = \int \int \cdots \int \prod_{j=1}^{M} \frac{dz_j}{2\pi} \exp\left[-i \sum_j z_j(\eta_j - m_j)\right.$$

$$\left. -\frac{1}{2} \sum_{ij} z_i C_{ij} z_j + \frac{1}{4!} \sum_{ijkl} K_{ijkl} z_i z_j z_k z_l + \cdots\right]. \tag{9.38}$$

* One could obviously also take into account any generalized skewness S_{ijk} in the following formula.

For $K_{ijkl} = 0$ (and also all higher order cumulants), this formula reduces to the multivariate Gaussian distribution, Eq. (9.6). The marginal distribution of, say, η_i, obtained by integrating $P(\eta_1, \eta_2, \ldots, \eta_M)$ over the $M - 1$ other variables η_j, $j \neq i$, and is given by:

$$P(\eta_i) = \int \frac{dz_i}{2\pi} \exp\left[-iz_i(\eta_i - m_i) - \frac{1}{2}C_{ii}z_i^2 + \frac{1}{4!}K_{iiii}z_i^4 + \cdots\right]. \tag{9.39}$$

This shows that whenever all 'diagonal' cumulants are zero (e.g. $K_{iiii} = 0$), the marginal distribution of η_i is Gaussian; however as soon as, say K_{ijij}, is non zero, the joint distribution of all η_i's is not a multivariate Gaussian.

From a simple manipulation of Fourier variables, one establishes that the four-point correlation function is given by:

$$\langle \eta_i \eta_j \eta_k \eta_l \rangle = C_{ij}C_{kl} + C_{ik}C_{jl} + C_{il}C_{jk} + K_{ijkl}. \tag{9.40}$$

The first three terms correspond to Wick's theorem for Gaussian variables. In the case $i = k$, $j = l$, one finds:

$$\langle \eta_i^2 \eta_j^2 \rangle - \langle \eta_i^2 \rangle \langle \eta_j^2 \rangle = 2\langle \eta_i \eta_j \rangle^2 + K_{ijij}. \tag{9.41}$$

Suppose that the η_i are uncorrelated, such that $\langle \eta_i \eta_j \rangle = 0$ for $i \neq j$. The above formula means, as we have discussed at length in Chapter 7 and in Section 9.2.5 above, that correlated volatility fluctuations between different assets create non Gaussian effects. Note also that the indicators \overline{K}_l and \overline{K}_{nl} are indeed generalized kurtosis, and are natural candidates to measure non Gaussian multivariate effects, that can exist even if the monovariate kurtosis is zero.

Finally, the generalized kurtosis can be computed for multivariate Student variables. One finds:

$$K_{ijkl} = \frac{2}{\mu - 4}[C_{ij}C_{kl} + C_{ik}C_{jl} + C_{il}C_{jk}]. \tag{9.42}$$

9.3 Factors and clusters

9.3.1 One factor models

The empirical results of Section 9.1.3 show that the correlation matrix has one dominant eigenvalue λ_1, much larger than all the others. This suggests that each return η_i can be decomposed as:

$$\eta_i = \beta_i \phi_0 + e_i, \tag{9.43}$$

where ϕ_0 and e_i are $M + 1$ uncorrelated random variables, of variance respectively equal to Σ^2 and σ_i^2, and β_i are fixed parameters. The same random variable ϕ_0 is common to all returns and corresponds to a 'factor' that affects all stocks, albeit with different 'impacts' β_i. The e_i are random variables, specific to each individual stock, and are often called the idiosyncratic contributions, or residuals. The correlations between stocks are, in this so called

one factor model, entirely induced by the common factor ϕ_0. In reality, as we will discuss below, the residuals are themselves correlated, for example when two companies belong to the same industrial sector.

The correlation matrix of the one factor model is easily computed to be:

$$C_{ij} = \beta_i \beta_j \Sigma^2 + \sigma_i^2 \delta_{i,j}. \tag{9.44}$$

If all the σ_i were equal to σ_0, this matrix would be very easy to diagonalize: one would have one dominant eigenvalue $\lambda_1 = \Sigma^2 \sum_i \beta_i^2 + \sigma_0^2$, corresponding to the eigenvector $v_{1,i} = \beta_i$, and $M - 1$ eigenvalues all equal to σ_0^2. For M large, $\lambda_1 \sim M \Sigma^2 \overline{\beta^2} \gg \sigma_0^2$. When M is large and the σ_i's not very different from one another, the spectrum of \mathbf{C} still has one dominant eigenvalue of order M and a sea of $M - 1$ other 'small' eigenvalues, much as the empirical correlation matrix.

In reality, several other eigenvalues emerge from the noise level (see Fig. 9.1), suggesting that one factor is not sufficient to reproduce quantitatively the form of the empirical correlation matrix. The most important factor ϕ_0 is traditionally called the 'market', whereas other relevant factors correspond to major activity sectors (for example, high technology stocks) and leads to a more complex structure of the correlation matrix.

Furthermore, as discussed in Section 11.2.3 below, there exist important non linear correlations between the volatility of the market Σ and that of the residuals σ_i that, as discussed above, cannot be described within a linear factor model, even if the factors are non Gaussian.

9.3.2 Multi-factor models

Once the 'market' factor has been subtracted, the idiosyncratic parts e_i still have non trivial correlations. A natural parametrization of these correlations that accounts for the existence of economic sectors starts by assuming the existence of uncorrelated factors ϕ_α, with $\alpha = 1, \ldots, N_s$ labels the N_s different sectors. One then writes:

$$e_i = \sum_{\alpha=1}^{N_s} \beta_{i\alpha} \phi_\alpha + \epsilon_i, \tag{9.45}$$

where the new residuals ϵ_i are assumed to be uncorrelated.[†] A natural choice for the $\{\beta_{i\alpha}\}$ is the N_s most significant eigenvectors (or principal components) of the correlation matrix. If the different ϕ_α and ϵ_i are not only uncorrelated but independent, this multi-factor model is in fact the linear model of correlated random variables described in Section 9.2.1.

A simple case is when each company is only sensitive to one particular sector: $\beta_{i\alpha} = 1$ if company i belongs to sector α, and $\beta_{i\alpha} = 0$ otherwise (remember that the market has been subtracted). If all the residuals ϵ_i belonging to the same sector have the same variance σ_α^2, the correlation matrix of the e_i can readily be diagonalized, since it is block diagonal, with each block representing a one-factor model for which the results of the above section can be transcribed. Calling M_α the size of the sector α, to each sector is associated

[†] In the financial textbooks multi-factor models are usually introduced in the context of the Arbitrage Pricing Theory (APT) which relates the expected gain of an asset to its factor exposure.

Table 9.1. Bloomberg sectors with size (number of companies in S&P 500) and example (member with largest market capitalization).

Sector Name	Size	Example
Basic materials	30	Du Pont
Communications	44	Cisco Systems
Consumer, cyclical	73	Wal-Mart Stores
Consumer, non-cyclical	89	Pfizer
Diversified	0	LVMH
Energy	27	Exxon Mobil
Financial	78	Citigroup
Industrial	67	General Electric
Technology	59	Microsoft
Utilities	33	Southern Co.

one large eigenvalue $\lambda_\alpha = M_\alpha \Sigma_\alpha^2 + \sigma_\alpha^2$, (where Σ_α^2 is the variance of the factor ϕ_α), and $M_\alpha - 1$ eigenvalues all equal to σ_α^2. In this model, the 'sea' of small eigenvalues seen in Fig. 9.1 corresponds to the latter eigenvalues, whereas the large eigenvalues that emerge from the noise level corresponds to the λ_α. If, as reasonable, the Σ_α's corresponding to different sectors have similar magnitudes, one sees that large λ_α's correspond to large sectors (large M_α). From the results of Fig. 9.1, one sees that there are roughly twenty sectors that stand out. This can be compared to the classification of economic sectors given by Bloomberg, see Table 9.1.

A natural question is then: how does one identify the sectors using empirical correlation matrices? The statistical identification of the sectors corresponds to a clustering problem that we now briefly discuss.

9.3.3 Partition around medoids

A traditional method of clustering is to choose *a priori* a certain number N_c of centres (**medoids**), and to assign each point of the set to be partitioned to its nearest medoid. Clusters \mathcal{C} are then the set of points closest to a given medoid. The cost function of this clustering is simply the total distance between all the points and their corresponding medoid, and the optimal choice of the medoid positions is such that this cost function is minimized. In the context of stocks, a medoid would be a certain portfolio, i.e. a linear combination of the stocks, and the quadratic distance equal to one minus the correlation coefficient. The clusters can be seen as statistical sectors. Once clusters are identified, one can create an 'index' associated to each cluster \mathcal{C}, which one can call a 'factor':

$$\eta_{\mathcal{C}} = \frac{1}{S(\mathcal{C})} \sum_{i \in \mathcal{C}} \eta_i, \tag{9.46}$$

where $S(\mathcal{C})$ is the size of the cluster \mathcal{C}, and η_i are the returns of the stocks belonging to \mathcal{C}. For the chosen (quadratic) distance, the centre of mass $\eta_{\mathcal{C}}$ is nothing but the medoid itself. When \mathcal{C} is the whole market, $\eta_{\mathcal{C}}$ is the market return $\eta_m \propto \phi_0$, or the market 'factor'.

It is often very useful to analyze the performance of a given portfolio in terms of its exposure to the different factors, by computing the correlation (traditionally called the β's) between the portfolio and the returns $\eta_{\mathcal{C}}$.

The drawback of the above method is that the choice of the number of clusters is fixed a priori. The optimal choice, that would minimize the cost function, is that the number of clusters equals the number of stocks so that each stock is its own medoid, is obviously absurd. An interesting idea to create clusters without imposing the number of clusters, is to assume a block diagonal correlation matrix (as discussed in the previous section), and to use a maximum likelihood method to determine the most probable clusters.[†]

9.3.4 Eigenvector clustering

Since the largest eigenvectors seem to carry some useful interpretation, one can also try to define clusters on the basis of the structure of these eigenvectors. The largest one, \mathbf{v}_1, has positive (and comparable) components on all stocks, and defines the largest cluster, containing all stocks. The second one \mathbf{v}_2, which by construction has to be orthogonal to \mathbf{v}_1, has roughly half of its components positive, and half negative. This means that a probable move of the cloud of stocks around the average (market) fluctuations is one where half of the stocks overperforms the market, and the other half underperforms it, or vice-versa. Therefore, the sign of the components of \mathbf{v}_2 can be used to group the stocks in two families. Each family can then be divided further, using the relative signs of \mathbf{v}_3, \mathbf{v}_4, etc. Using, say, the five largest eigenvectors leads to $2^4 = 16$ different clusters (since \mathbf{v}_1 cannot be used to distinguish different stocks).

9.3.5 Maximum spanning tree

Another simple idea, due to Mantegna, is to create a tree of stocks in the following way: first rank by decreasing order the $M(M-1)/2$ values of the correlations C_{ij} between all pair of stocks.[‡] Pick the pair corresponding to the largest C_{ij} and create a link between these two stocks. Take the second largest pair, and create a link between these two. Repeat the operation, *unless* adding a link between the pair under consideration creates a loop in the graph, in which case skip that value of C_{ij}. Once all stocks have been linked at least once without creating loops, one finds a tree which only contains the strongest possible correlations, called the maximum spanning tree. An example of this construction for U.S. stocks is shown in Fig. 9.2.

Now, clusters can easily be created by removing all links of the maximum spanning tree such that $C_{ij} < C^*$. Since the tree contains no loop, removing one link cuts the tree into

[†] On this method, see Marsili (2002). Alternative methods, based on 'fuzzy clusters', where each stock can belong to different clusters, can be found in the papers cited at the end of the chapter.
[‡] In principle, one should consider $|C_{ij}|$ rather than C_{ij}. In practice, most stocks have positive correlations.

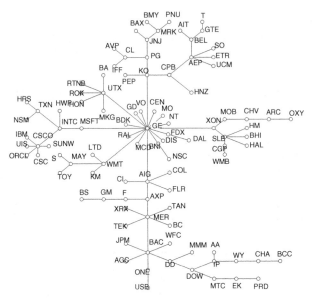

Fig. 9.2. Maximum spanning tree for the U.S. stock market. Each node is a stock with its code name. The bonds correspond to the largest $|C_{ij}|$'s that have been retained in the construction of the tree. See how economic sectors emerge in this representation, with 'leaders' such as General Electric (GE) appearing as multi-connected nodes. Courtesy of R. Mantegna.

disconnected pieces. The remaining pieces are clusters within which all remaining links are 'strong', i.e., such that $C_{ij} \geq C^*$, which can be considered as strongly correlated. The number of disconnected pieces grows as the correlation threshold C^* increases.

The problem with the above methods is that one often ends up with clusters of very dissimilar sizes. For example, progressively dismantling the maximum spanning tree might create singlets, which are clusters of their own. However, the fact that clusters have dissimilar sizes is perhaps a reality, related to the organization of the economic activity as a whole.

9.4 Summary

- Cross correlations between assets are captured, at the simplest level, by the correlation matrix. The eigenvectors of this matrix can be used to express the returns as a linear combination of uncorrelated (but not necessarily independent) random variables, called principal components. The (positive) eigenvalues are the squared volatilities of these components. In the case of Gaussian correlated returns, the principal components are in fact *independent*.
- The empirical determination of the full correlation matrix is difficult. This is illustrated by comparing the spectrum of eigenvalues of the empirical correlation matrix with that of a purely random correlation matrix, which is known analytically for large matrices. Most eigenvalues can be explained in terms of a purely random

correlation matrix, except for the largest ones, which correspond to the fluctuations of the market as a whole, and of several industrial sectors. These correlations can be described within factor models, which however fail to describe non linear (volatility) correlations.

- There are many ways to construct correlated non Gaussian random variables, that lead to different multivariate distributions. Three particularly important cases correspond to: (a) linear combination of independent, but non Gaussian, random variables; (b) non linear transformations of correlated Gaussian variables; and (c) correlated Gaussian variables multiplied by a common, random volatility. This last case gives rise to two interesting families of multivariate distributions: Student and Lévy.

9.5 Appendix A: central limit theorem for random matrices

One interesting application of the CLT concerns the spectral properties of 'random matrices'. The theory of random matrices has made enormous progress during the past 30 years, with many applications in physical sciences and elsewhere. As illustrated in the previous section, it has been recently suggested that random matrices might also play an important role in finance. It is therefore appropriate to give a cursory discussion of some salient properties of random matrices. The simplest ensemble of random matrices is one where all elements of the matrix \mathbf{H} are iid random variables, with the only constraint that the matrix be symmetrical ($H_{ij} = H_{ji}$). One interesting result is that in the limit of very large matrices, the distribution of its eigenvalues has universal properties, which are to a large extent independent of the distribution of the elements of the matrix. This is actually the consequence of the CLT, as we will show below. Let us introduce first some notation. The matrix \mathbf{H} is a square, $M \times M$ symmetric matrix. Its eigenvalues are λ_a, with $a = 1, \ldots, M$. The *density* of eigenvalues is defined as:

$$\rho(\lambda) = \frac{1}{M} \sum_{a=1}^{M} \delta(\lambda - \lambda_a),$$

(9.47)

where δ is the Dirac function. We shall also need the so-called 'resolvent' $\mathbf{G}(\lambda)$ of the matrix \mathbf{H}, defined as:

$$G_{ij}(\lambda) \equiv \left(\frac{1}{\lambda \mathbf{1} - \mathbf{H}} \right)_{ij},$$

(9.48)

where $\mathbf{1}$ is the identity matrix. The trace of $\mathbf{G}(\lambda)$ can be expressed using the eigenvalues of \mathbf{H} as:

$$\mathrm{Tr}\mathbf{G}(\lambda) = \sum_{a=1}^{M} \frac{1}{\lambda - \lambda_a}.$$

(9.49)

The 'trick' that allows one to calculate $\rho(\lambda)$ in the large M limit is the following representation of the δ function:

$$\frac{1}{x - i\epsilon} = PP\frac{1}{x} + i\pi\delta(x) \qquad (\epsilon \to 0),$$

(9.50)

where PP means the principal part. Therefore, $\rho(\lambda)$ can be expressed as:

$$\rho(\lambda) = \lim_{\epsilon \to 0} \frac{1}{M\pi} \Im \left(\mathrm{Tr}\mathbf{G}(\lambda - i\epsilon) \right). \tag{9.51}$$

Our task is therefore to obtain an expression for the resolvent $\mathbf{G}(\lambda)$. This can be done by establishing a recursion relation, allowing one to compute $\mathbf{G}(\lambda)$ for a matrix \mathbf{H} with one extra row and one extra column, the elements of which being H_{0i}. One then computes $G_{00}^{M+1}(\lambda)$ (the superscript stands for the size of the matrix \mathbf{H}) using the standard formula for matrix inversion:

$$G_{00}^{M+1}(\lambda) = \frac{\mathrm{minor}(\lambda\mathbf{1} - \mathbf{H})_{00}}{\det(\lambda\mathbf{1} - \mathbf{H})}. \tag{9.52}$$

Now, one expands the determinant appearing in the denominator in minors along the first row, and then each minor is itself expanded in subminors along their first column. After a little thought, this finally leads to the following expression for $G_{00}^{M+1}(\lambda)$:

$$\frac{1}{G_{00}^{M+1}(\lambda)} = \lambda - H_{00} - \sum_{i,j=1}^{M} H_{0i} H_{0j} G_{ij}^{M}(\lambda). \tag{9.53}$$

This relation is general, without any assumption on the H_{ij}. Now, we assume that the H_{ij}'s are iid random variables, of zero mean and variance equal to $\langle H_{ij}^2 \rangle = \sigma^2/M$. This scaling with M can be understood as follows: when the matrix H acts on a certain vector, each component of the image vector is a sum of M random variables. In order to keep the image vector (and thus the corresponding eigenvalue) finite when $M \to \infty$, one should scale the elements of the matrix with the factor $1/\sqrt{M}$.

One could also write a recursion relation for G_{0i}^{M+1}, and establish self-consistently that $G_{ij} \sim 1/\sqrt{M}$ for $i \neq j$. On the other hand, due to the diagonal term λ, G_{ii} remains finite for $M \to \infty$. This scaling allows us to discard all the terms with $i \neq j$ in the sum appearing in the right-hand side of Eq. (9.53). Furthermore, since $H_{00} \sim 1/\sqrt{M}$, this term can be neglected compared to λ. This finally leads to a simplified recursion relation, valid in the limit $M \to \infty$:

$$\frac{1}{G_{00}^{M+1}(\lambda)} \simeq \lambda - \sum_{i=1}^{M} H_{0i}^2 G_{ii}^{M}(\lambda). \tag{9.54}$$

Now, using the CLT, we know that the last sum converges, for large M, towards $\sigma^2/M \sum_{i=1}^{M} G_{ii}^{M}(\lambda)$. This result is independent of the precise statistics of the H_{0i}, provided their variance is finite.[†] This shows that G_{00} converges for large M towards a well-defined limit G_{∞}, which obeys the following limit equation:

$$\frac{1}{G_{\infty}(\lambda)} = \lambda - \sigma^2 G_{\infty}(\lambda). \tag{9.55}$$

The solution to this second-order equation reads:

$$G_{\infty}(\lambda) = \frac{1}{2\sigma^2}[\lambda - \sqrt{\lambda^2 - 4\sigma^2}]. \tag{9.56}$$

[†] The case of Lévy distributed H_{ij}'s with infinite variance has been investigated in Cizeau & Bouchaud (1994).

(The correct solution is chosen to recover the right limit for $\sigma = 0$.) Now, the only way for this quantity to have a non-zero imaginary part when one adds to λ a small imaginary term $i\epsilon$ which tends to zero is that the square root itself is imaginary. The final result for the density of eigenvalues is therefore:

$$\rho(\lambda) = \frac{1}{2\pi\sigma^2}\sqrt{4\sigma^2 - \lambda^2} \qquad \text{for } |\lambda| \leq 2\sigma, \tag{9.57}$$

and zero elsewhere. This is the well-known 'semi-circle' law for the density of states, first derived by Wigner. This result can be obtained by a variety of other methods if the distribution of matrix elements is Gaussian. In finance, one often encounters *correlation matrices* \mathbf{C}, which have the special property of being positive definite. \mathbf{C} can be written as $\mathbf{C} = \mathbf{H}\mathbf{H}^\dagger$, where \mathbf{H}^\dagger is the matrix transpose of \mathbf{H}. In general, \mathbf{H} is a rectangular matrix of size $M \times N$, so \mathbf{C} is $M \times M$. In Section 9.1.3, M was the number of assets, and N, the number of observations (days). In the particular case where $N = M$, the eigenvalues of \mathbf{C} are simply obtained from those of \mathbf{H} by squaring them:

$$\lambda_C = \lambda_H^2. \tag{9.58}$$

If one assumes that the elements of \mathbf{H} are random variables, the density of eigenvalues of \mathbf{C} can be obtained from:

$$\rho(\lambda_C)\,d\lambda_C = 2\rho(\lambda_H)\,d\lambda_H \quad \text{for } \lambda_H > 0, \tag{9.59}$$

where the factor of 2 comes from the two solutions $\lambda_H = \pm\sqrt{\lambda_C}$; this then leads to:

$$\rho(\lambda_C) = \frac{1}{2\pi\sigma^2}\sqrt{\frac{4\sigma^2 - \lambda_C}{\lambda_C}} \qquad \text{for } 0 \leq \lambda_C \leq 4\sigma^2, \tag{9.60}$$

and zero elsewhere. For $N \neq M$, a similar formula exists, which we shall use in the following. In the limit $N, M \to \infty$, with a fixed ratio $Q = N/M \geq 1$, one has:[†]

$$\rho(\lambda_C) = \frac{Q}{2\pi\sigma^2}\frac{\sqrt{(\lambda_{\max} - \lambda_C)(\lambda_C - \lambda_{\min})}}{\lambda_C},$$

$$\lambda_{\min}^{\max} = \sigma^2(1 + 1/Q \pm 2\sqrt{1/Q}), \tag{9.61}$$

with $\lambda_C \in [\lambda_{\min}, \lambda_{\max}]$ and where σ^2/N is the variance of the elements of \mathbf{H}, equivalently σ^2 is the average eigenvalue of \mathbf{C}. This form is actually also valid for $Q < 1$, except that there appears a finite fraction of strictly zero eigenvalues, of weight $1 - Q$ (Fig. 9.3).

The most important features predicted by Eq. (9.61) are:

- The fact that the lower 'edge' of the spectrum is positive (except for $Q = 1$); there is therefore no eigenvalue between 0 and λ_{\min}. Near this edge, the density of eigenvalues exhibits a sharp maximum, except in the limit $Q = 1$ ($\lambda_{\min} = 0$) where it diverges as $\sim 1/\sqrt{\lambda}$.
- The density of eigenvalues also vanishes above a certain upper edge λ_{\max}.

[†] A derivation of Eq. (9.61) is given in Appendix B.

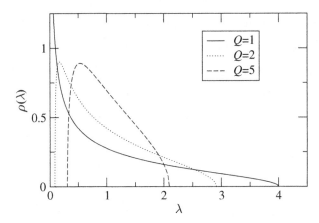

Fig. 9.3. Graph of Eq. (9.61) for $Q = 1, 2$ and 5.

Note that all the above results are only valid in the limit $N \to \infty$. For finite N, the singularities present at both edges are smoothed: the edges become somewhat blurred, with a small probability of finding eigenvalues above λ_{\max} and below λ_{\min}, which goes to zero when N becomes large, see Bowick and Brézin (1991).

9.6 Appendix B: density of eigenvalues for random correlation matrices

This very technical appendix aims at giving a few of the steps of the computation needed to establish Eq. (9.61). One starts from the following representation of the resolvent $G(\lambda) = Tr\mathbf{G}(\lambda)$:

$$G(\lambda) = \sum_a \frac{1}{\lambda - \lambda_a} = \frac{\partial}{\partial \lambda} \log \prod_a (\lambda - \lambda_a) = \frac{\partial}{\partial \lambda} \log \det(\lambda \mathbf{1} - \mathbf{C}) \equiv \frac{\partial}{\partial \lambda} Z(\lambda). \tag{9.62}$$

Using the following representation for the determinant of a symmetrical matrix \mathbf{A}:

$$[\det \mathbf{A}]^{-1/2} = \left(\frac{1}{\sqrt{2\pi}} \right)^M \int \exp\left[-\frac{1}{2} \sum_{i,j=1}^M \varphi_i \varphi_j A_{ij} \right] \prod_{i=1}^M \mathrm{d}\varphi_i, \tag{9.63}$$

we find, in the case where $\mathbf{C} = \mathbf{H}\mathbf{H}^\dagger$:

$$Z(\lambda) = -2 \log \int \exp\left[-\frac{\lambda}{2} \sum_{i=1}^M \varphi_i^2 - \frac{1}{2} \sum_{i,j=1}^M \sum_{k=1}^N \varphi_i \varphi_j H_{ik} H_{jk} \right] \prod_{i=1}^M \left(\frac{\mathrm{d}\varphi_i}{\sqrt{2\pi}} \right). \tag{9.64}$$

The claim is that the quantity $G(\lambda)$ is self-averaging in the large M limit. So in order to perform the computation we can replace $G(\lambda)$ for a specific realization of \mathbf{H} by the average over an ensemble of \mathbf{H}. In fact, one can in the present case show that the average of the logarithm is equal, in the large N limit, to the logarithm of the average. This is not true in general, and one has to use the so-called 'replica trick' to deal with this problem: this amounts to taking the nth power of the quantity to be averaged (corresponding to n copies (replicas) of the system) and to let n go to zero at the end of the computation.[†]

[†] For more details on this technique, see, for example, Mézard et al. (1987).

We assume that the $M \times N$ variables H_{ik} are iid Gaussian variable with zero mean and variance σ^2/N. To perform the average of the H_{ik}, we notice that the integration measure generates a term $\exp - M \sum H_{ik}^2/(2\sigma^2)$ that combines with the $H_{ik}H_{jk}$ term above. The summation over the index k doesn't play any role and we get N copies of the result with the index k dropped. The Gaussian integral over $H_{i\circ}$ gives the square-root of the ratio of the determinant of $[M\delta_{ij}/\sigma^2]$ and $[M\delta_{ij}/\sigma^2 - \varphi_i\varphi_j]$:

$$\left\langle \exp\left[-\frac{1}{2}\sum_{i,j=1}^{M}\sum_{k=1}^{N}\varphi_i\varphi_j H_{ik}H_{jk}\right]\right\rangle = \left(1 - \frac{\sigma^2}{N}\sum\varphi_i^2\right)^{-N/2}. \tag{9.65}$$

We then introduce a variable $q \equiv \sigma^2\sum\varphi_i^2/N$ which we fix using an integral representation of the delta function:

$$\delta\left(q - \sigma^2\sum\varphi_i^2/N\right) = \int \frac{1}{2\pi}\exp\left[i\zeta\left(q - \sigma^2\sum\varphi_i^2/N\right)\right]d\zeta. \tag{9.66}$$

After performing the integral over the φ_i's and writing $z = 2i\zeta/N$, we find:

$$Z(\lambda) = -2\log\frac{N}{4\pi}\int_{-i\infty}^{i\infty}\int_{-\infty}^{\infty}\exp\left[-\frac{M}{2}(\log(\lambda - \sigma^2 z) + Q\log(1-q) + Qqz)\right]dq\,dz, \tag{9.67}$$

where $Q = N/M$. The integrals over z and q are performed by the saddle point method, leading to the following equations:

$$Qq = \frac{\sigma^2}{\lambda - \sigma^2 z} \quad \text{and} \quad z = \frac{1}{1-q}. \tag{9.68}$$

The solution in terms of $q(\lambda)$ is:

$$q(\lambda) = \frac{\sigma^2(1-Q) + Q\lambda \pm \sqrt{(\sigma^2(1-Q) + Q\lambda)^2 - 4\sigma^2 Q\lambda}}{2Q\lambda}. \tag{9.69}$$

We find $G(\lambda)$ by differentiating Eq. (9.67) with respect to λ. The computation is greatly simplified if we notice that at the saddle point the partial derivatives with respect to the functions $q(\lambda)$ and $z(\lambda)$ are zero by construction. One finally finds:

$$G(\lambda) = \frac{M}{\lambda - \sigma^2 z(\lambda)} = \frac{MQq(\lambda)}{\sigma^2}. \tag{9.70}$$

We can now use Eq. (9.51) and take the imaginary part of $G(\lambda)$ to find the density of eigenvalues:

$$\rho(\lambda) = \frac{\sqrt{4\sigma^2 Q\lambda - (\sigma^2(1-Q) + Q\lambda)^2}}{2\pi\lambda\sigma^2}, \tag{9.71}$$

which is identical to Eq. (9.61).

- FURTHER READING

O. Bohigas, M. J. Giannoni, *Random Matrix Theory and applications*, in *Mathematical and Computational Methods in Nuclear Physics*, Lecture Notes in Physics, vol. 209, Springer-Verlag, New York, 1983.

A. Crisanti, G. Paladin and A. Vulpiani, *Products of Random Matrices in Statistical Physics*, Springer-Verlag, Berlin 1993.

N. L. Johnson, S. Kotz, *Distribution in Statistics: Continuous Multivariate Distributions*, John Wiley & Sons, 1972.

M. L. Mehta, *Random Matrix Theory*, Academic Press, New York, 1995.

M. Mézard, G. Parisi, M. A. Virasoro, *Spin Glasses and Beyond*, World Scientific, 1987.

G. Samorodnitsky, M. S. Taqqu, *Stable Non-Gaussian Random Processes*, Chapman & Hall, New York, 1994.

See also Chapter 11.

- SPECIALIZED PAPERS: MULTIVARIATE STATISTICS

Y. Malevergne, D. Sornette, *Tail Dependence of Factor Models*, e-print cond-mat/0202356; *Minimizing Extremes*, Risk, **15**, 129 (2002).

Y. Malevergne, D. Sornette, *Testing the Gaussian copula hypothesis for financial assets dependence*, Quantitative Finance, **3**, 231–250 (2003).

- SPECIALIZED PAPERS: RANDOM MATRICES

M. J. Bowick, E. Brézin, *Universal scaling of the tails of the density of eigenvalues in random matrix models*, Physics Letters, **B 268**, 21 (1991).

Z. Burda, J. Jurkiewicz, M. A. Nowak, G. Papp, I. Zahed, *Free Random Levy Variables and Financial Probabilities*, e-print cond-mat/0103140.

P. Cizeau, J.-P. Bouchaud, *Theory of Lévy matrices*, Physical Review, **E 50**, 1810 (1994).

A. Edelmann, *Eigenvalues and condition numbers of random matrices*, SIAM Journal of Matrix Analysis and Applications, **9**, 543 (1988).

L. Laloux, P. Cizeau, J.-P. Bouchaud and M. Potters, *Noise dressing of financial correlation matrices*, Phys. Rev. Lett., **83**, 1467 (1999).

L. Laloux, P. Cizeau, J.-P. Bouchaud and M. Potters, *Noise dressing of financial correlation matrices*, Int. J. Theo. Appl. Fin., **3**, 391 (2000).

V. Plerou, P. Gopikrishnan, B. Rosenow, L. A. N. Amaral, H. E. Stanley, *Universal and non-universal properties of cross-correlations in financial time series*, Phys. Rev. Lett., **83**, 1471 (1999).

V. Plerou, P. Gopikrishnan, B. Rosenow, L. A. N. Amaral, T. Guhr, H. E. Stanley, *A random matrix approach to financial cross-correlations*, Phys. Rev., **E 65**, 066126 (2002).

B. Rosenow, V. Plerou, P. Gopikrishnan, H. E. Stanley, *Portfolio Optimization and the Random Magnet Problem*, Europhys. Lett., **59**, 500–506 (2002).

A. N. Sengupta, P. Mitra, *Distributions of singular values for some random matrices*, Physical Review, **E 80**, 3389 (1999).

- SPECIALIZED PAPERS: CLUSTERING

P. A. Barès, R. Gibson, S. Gyger, *Style consistency and survival probability in the Hedge Funds industry*, submitted to Financial Analysts Journal, (2000).

G. Bonanno, N. Vandewalle, R. N. Mantegna, *Taxonomy of stock market indices*, Physical Review, **E 62**, R7615 (2000).

E. Domany, *Super-paramagnetic clustering of data*, Physica A, **263**, 158 (1999).

P. Gopikrishnan, B. Rosenow, V. Plerou, H. E. Stanley, *Identifying business sectors from stock price fluctuations*, Physical Review, **E 64** 035106(R) (2001).

R. Mantegna, *Hierarchical structure in financial markets*, Eur., J. Phys. B **11**, 193 (1999).

M. Marsili, *Dissecting financial markets: sectors and states*, Quantitative Finance, **2**, 297 (2002).

J. Noh, *A model for correlations in stock markets*, Physical Review, **E 61**, 5981 (2000).

• SOME RECENT READS

J. Baik, G. Ben Arous, S. Péché, *Phase transition of the largest eigenvalue for non-null complex sample covariance matrices*, Ann. Probab., **33**, 1643 (2005).

G. Biroli, J. P. Bouchaud, M. Potters, *On the top eigenvalue of heavy-tailed random matrices*, Europhys. Lett., **78**, 10001 (2007).

G. Biroli, J.-P. Bouchaud, M. Potters, *The Student Ensemble of Correlation Matrices: Eigenvalue Spectrum and Kullback–Leibler Entropy*, Acta Physica Polonica B, **38**, 4009 (2007).

J.-P. Bouchaud, L. Laloux, M. A. Miceli, M. Potters, *Large dimension forecasting models and random singular value spectra*, Eur. Phys. J. B (2006), b06036.

Y. Malevergne, D. Sornette, *Extreme Financial Risks: From Dependence to Risk Management*, Springer, 2005.

M. Potters, J.-P. Bouchaud, L. Laloux, *Financial Applications of Random Matrix Theory: Old laces and new pieces*, Acta Physica Polonica B, **36** 2767 (2005).

A. Tulino, S. Verdù, *Random Matrix Theory and Wireless Communications*, Foundations and Trends in Communication and Information Theory, **1**, 1 (2004).

10

Risk measures

Il faut entrer dans le discours du patient et non tenter de lui imposer le nôtre.[†]
(Gérard Haddad, *Le jour où Lacan m'a adopté*.)

10.1 Risk measurement and diversification

Measuring and controlling risks is now a major concern in many modern human activities. The financial markets, which act as highly sensitive (probably over sensitive) economical and political thermometers, are no exception. One of their rôles is actually to allow the different actors in the economic world to *trade* their risks, to which a price must therefore be given.

The very essence of the financial markets is to fix thousands of prices all day long, thereby generating enormous quantities of data that can be analysed statistically. An objective measure of risk therefore appears to be easier to achieve in finance than in most other human activities, where the definition of risk is vaguer, and the available data often very poor. Even if a purely statistical approach to financial risks is itself a dangerous scientists' dream (see e.g. Fig. 1.1), it is fair to say that this approach has not been fully exploited until the very recent years, and that many improvements can be still expected in the future, in particular concerning the control of extreme risks. The aim of this chapter is to introduce some classical ideas on financial risks, to illustrate their weaknesses, and to propose several theoretical ideas devised to handle more adequately the 'rare events' where the true financial risk resides.

10.2 Risk and volatility

Financial risk has been traditionally associated with the statistical uncertainty on the final outcome. Its traditional measure is the RMS, or, in financial terms, the *volatility*. We will note by $R(T)$ the logarithmic return on the time interval T, defined by:

$$R(T) = \log \left[\frac{x(T)}{x_0} \right], \tag{10.1}$$

[†] One should accept the patient's language, and not try to impose our own.

where $x(T)$ is the price of the asset X at time T, knowing that it is equal to x_0 today ($t = 0$). When $|x(T) - x_0| \ll x_0$, this definition is equivalent to $R(T) = x(T)/x_0 - 1$.

If $P(x, T|x_0, 0)\,\mathrm{d}x$ is the conditional probability of finding $x(T) = x$ within $\mathrm{d}x$, the volatility σ of the investment is the standard deviation of $R(T)$, defined by:

$$\sigma^2 = \frac{1}{T}\left[\int P(x, T|x_0, 0)R^2(T)\,\mathrm{d}x - \left(\int P(x, T|x_0, 0)R(T)\,\mathrm{d}x\right)^2\right]. \tag{10.2}$$

The volatility is still now often chosen as the adequate measure of risk associated to a given investment. We notice however that this definition includes in a symmetrical way both abnormal gains and abnormal losses. This fact is *a priori* curious. The theoretical foundations behind this particular definition of risk are numerous:

- First, operational; the computations involving the variance (volatility squared) are relatively simple and can be generalized easily to multi-asset portfolios.
- Second, the central limit theorem (CLT) presented in Chapter 2 seems to provide a general and solid justification: by decomposing the motion from x_0 to $x(T)$ in $N = T/\tau$ increments, one can write:

$$x(T) = x_0 \prod_{k=0}^{N-1}(1 + \eta_k), \tag{10.3}$$

where η_k is by definition the instantaneous return. Therefore, we have:

$$R(T) = \log\left[\frac{x(T)}{x_0}\right] = \sum_{k=0}^{N-1}\log(1 + \eta_k). \tag{10.4}$$

In the classical approach one assumes that the returns η_k are independent variables. From the CLT we learn that *in the limit where $N \to \infty$*, $R(T)$ becomes a Gaussian random variable centred on a given average return $\tilde{m}T$, with $\tilde{m} = \langle\log(1 + \eta)\rangle/\tau$, and whose standard deviation is given by $\sigma\sqrt{T}$.[†] Therefore, in this limit, the entire probability distribution of $R(T)$ is parameterized by two quantities only, \tilde{m} and σ: any reasonable measure of risk must therefore be based on σ.

However, as we have discussed at length in Chapter 2, this is not true for finite N (which corresponds to the financial reality: for example, there are only roughly $N \approx 320$ half-hour intervals in a working month), especially in the 'tails' of the distribution, that correspond precisely to the extreme risks. We will discuss this point in detail below. Furthermore, as explained in Chapter 7, the presence of long-ranged correlations in the way the volatility itself fluctuates can slow down dramatically the asymptotic convergence towards the Gaussian limit: non Gaussian effects are in fact important after several weeks, sometimes even after several months.

One can give to σ the following intuitive meaning: after a long enough time T, the price of asset X is given by:

$$x(T) = x_0 \exp[\tilde{m}T + \sigma\sqrt{T}\,\xi], \tag{10.5}$$

[†] To second order in $\eta_k \ll 1$, we find: $\sigma^2 = \frac{1}{\tau}\langle\eta^2\rangle$ and $\tilde{m} = \frac{1}{\tau}\langle\eta\rangle - \frac{1}{2}\sigma^2$.

where ξ is a Gaussian random variable with zero mean and unit variance. The quantity $\sigma \sqrt{T}$ gives us the order of magnitude of the deviation from the expected return. By comparing the two terms in the exponential, one finds that when $T \gg \hat{T} \equiv \sigma^2/\tilde{m}^2$, the expected return becomes more important than the fluctuations. This means that the probability that $x(T)$ turns out to be less than x_0 (and that the investment leads to losses) becomes small. The **security horizon** \hat{T} increases with σ. For a rather good individual stock, one has – say – $\tilde{m} = 10\%$ per year and $\sigma = 20\%$ per year, which gives a \hat{T} as long as 4 years!

In fact, one should not compare $x(T)$ to x_0, because a risk free investment would have produced $x_0 e^{rT}$. The investor should therefore only be concerned with the *excess return* over the risk free rate, $m^* = \tilde{m} - r$. (Note however that the existence of guaranteed capital products shows that many people think in terms of raw returns and not in terms of excess returns!)

The quality of an investment is often measured by its **Sharpe ratio** \mathcal{S}, that is, the 'signal-to-noise' ratio of the excess return m^*T to the fluctuations $\sigma\sqrt{T}$. The Sharpe ratio is in fact directly related to the 'security horizon' \hat{T}:

$$\mathcal{S} = \frac{m^*\sqrt{T}}{\sigma} \equiv \sqrt{\frac{T}{\hat{T}}}. \tag{10.6}$$

The Sharpe ratio increases with the investment horizon and is equal to 1 precisely when $T = \hat{T}$. Practitioners usually define the Sharpe ratio for a 1-year horizon. In the case of stocks mentioned above, the Sharpe ratio is equal to 0.25 if one subtracts a risk-free rate of 5% annual. An investment with a Sharpe ratio equal to one is considered as extremely good. Note however that the Sharpe ratio itself is only known approximately. Suppose that one has ten years of daily returns, corresponding to $N = 2500$ observations. We neglect volatility correlations, which leads to a substantial underestimation of the error. The error on \tilde{m} is σ/\sqrt{N}, whereas the relative error on σ is $\sqrt{(2 + \kappa)/N}$, where κ is the daily kurtosis. The error on the Sharpe ratio therefore reads:

$$\Delta\mathcal{S} = \frac{1}{\sqrt{N}}[\sqrt{T} + \mathcal{S}\sqrt{(2 + \kappa)}], \tag{10.7}$$

where T is expressed in days. For the one year Sharpe ratio, $T = 250$, one finds $\Delta\mathcal{S} \approx 0.31 + 0.045\mathcal{S}$ with $\kappa = 3$. Take the case of stocks, for which $\mathcal{S} \approx 0.25$: the error on the Sharpe is of the same order as the Sharpe itself! This illustrates the difficulty of establishing the quality of a given investment, especially based on a few months of performance.

As a rule of thumb, for the one-year Sharpe ratio, the error is:

$$\Delta\mathcal{S} \approx \frac{1}{\sqrt{n}}, \tag{10.8}$$

where n is the number of available years of data.

Note that the most probable value of $x(T)$, as given by Eq. (10.5), is equal to $x_0 \exp(\tilde{m}T)$, whereas the mean value of $x(T)$ is higher: assuming that ξ is Gaussian, one finds: $x_0 \exp[(\tilde{m}T + \sigma^2 T/2)]$. This difference is due to the fact that the *returns* η_k, rather than the absolute increments δx_k, are taken to be iid random variables. However, if T is short (say up to a few months), the difference between the two descriptions is hard to detect. As explained in details in Chapter 8, a purely additive description is actually more adequate at short times. We shall therefore often write in the following $x(T)$ as:

$$x(T) = x_0 \exp[\tilde{m}T + \sigma\sqrt{T}\xi] \simeq x_0 + mT + \sqrt{DT}\xi, \tag{10.9}$$

where we have introduced the following notations, which we shall use throughout the following: $m \equiv \tilde{m}x_0$ for the absolute return per unit time, and $D \equiv \sigma^2 x_0^2$ for the absolute variance per unit time (usually called the diffusion constant of the stochastic process). As we now discuss, the non-Gaussian nature of the random variable ξ is the most important factor determining the probability for extreme risks.

10.3 Risk of loss, 'value-at-risk' (VaR) and expected shortfall

10.3.1 Introduction

The fact that financial risks are often described using the volatility is actually intimately related to the idea that the distribution of price changes is Gaussian. In the extreme case of 'Lévy fluctuations', for which the variance is infinite, this definition of risk would obviously be meaningless. Even in a less risky world where the variance is well defined, this measure of risk has three major drawbacks:

- The financial risk is obviously associated to losses and not to profits. A definition of risk where both events play symmetrical roles is thus not in conformity with the intuitive notion of risk, as perceived by professionals.
- As discussed at length in Chapter 2, a Gaussian model for the price fluctuations is *never justified* for the extreme events, since the CLT only applies in the *centre* of the distributions. Now, it is precisely these extreme risks that are of most concern for all financial houses, and thus those which need to be controlled in priority. In recent years, international regulators have tried to impose some rules to limit the exposure of banks to these extreme risks.
- The presence of extreme events in the financial time series can actually lead to a very bad empirical determination of the variance: its value can be substantially changed by a few 'big days'. A bold solution to this problem is simply to remove the contribution of these so-called aberrant events! This rather absurd solution was actually quite commonly used in the past.

Both from a fundamental point of view, and for a better control of financial risks, another definition of risk is thus needed. An interesting notion that we shall present now is the *probability of extreme losses*, or, equivalently, the **value-at-risk** (VaR).

10.3.2 Value-at-risk

The probability of losing an amount $-\delta x$ larger than a certain threshold Λ on a given time interval τ is defined as:

$$\mathcal{P}[\delta x < -\Lambda] = \mathcal{P}_<[-\Lambda] = \int_{-\infty}^{-\Lambda} P_\tau(\delta x)\,d\delta x, \tag{10.10}$$

where $P_\tau(\delta x)$ is the probability density for a price change on the time interval τ. One can alternatively define the risk as a level of loss (the 'VaR') Λ^* corresponding to a certain probability of loss \mathcal{P}^* over the time interval τ (for example, $\mathcal{P}^* = 1\%$):

$$\int_{-\infty}^{-\Lambda^*} P_\tau(\delta x)\,d\delta x = \mathcal{P}^*. \tag{10.11}$$

This definition means that a loss greater than Λ^* over a time interval of $\tau = 1$ day (for example) happens only every 100 days on average (for $\mathcal{P}^* = 1\%$). Let us note that:

- This definition does not take into account the fact that losses can accumulate on consecutive time intervals τ, leading to an overall cumulated loss which might substantially exceed Λ^*.
- Similarly, this definition does not take into account the value of the maximal loss 'inside' the period τ. In other words, only the closing price over the period $[k\tau, (k+1)\tau]$ is considered, and not the lowest point reached during this time interval: we shall come back on these temporal aspects of risk in Section 10.4.

More precisely, one can discuss the probability distribution $P(\Lambda, N)$ for the worst daily loss Λ (we choose $\tau = 1$ day to be specific) on a temporal horizon $T = N\tau$. Using the results of Section 2.1, one has:

$$P(\Lambda, N) = N[\mathcal{P}_>(-\Lambda)]^{N-1} P_\tau(-\Lambda). \tag{10.12}$$

For N large, this distribution takes a universal shape that only depends on the asymptotic behaviour of $P_\tau(\delta x)$ for $\delta x \to -\infty$. In the important case where $P_\tau(\delta x)$ decays faster than any power-law, one finds that $P(\Lambda, N)$ is given by Eq. (2.7), which is represented in Figure 10.1.

Choosing the temporal horizon to be $T^* = \tau/\mathcal{P}^*$ such that on average one event beyond Λ^* takes place, we find that the distribution of the worst daily loss has a maximum precisely for $\Lambda = \Lambda^*$. The intuitive meaning of Λ^* is thus the value of the most probable worst day over a time interval T^*.[†]

Note that the probability for Λ to be even worse ($\Lambda > \Lambda^*$) is equal to 63% (Fig. 10.1). One could define Λ^* in such a way that this exceedance probability is smaller, by requiring a higher confidence level, for example 95%. This would mean that on a given horizon (for example 100 days), the probability that the worst day is found to be beyond Λ^* is equal to

[†] If the distribution $P_\tau(\delta x)$ is a power-law with an index $\mu = 3$, one finds that the most probable worst day over a time interval T^* is $(3/4)^{1/3}\Lambda^* \approx 0.91\Lambda^*$.

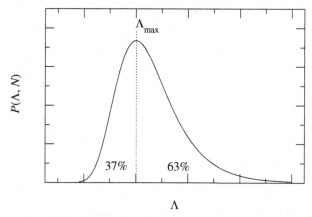

Fig. 10.1. Extreme value distribution (the so-called Gumbel distribution) $P(\Lambda, N)$ when $P_\tau(\delta x)$ decreases faster than any power-law. The most probable value, Λ_{max}, has a probability equal to 0.63 to be exceeded.

5%, instead of 63%. This new Λ^* then corresponds to the most probable worst day on a time period equal to $100/0.05 = 2000$ days.

The standard in the financial industry is to choose $\tau = 10$ trading days (two weeks) and $\mathcal{P}^* = 0.05$, corresponding to the 95% VaR over two weeks. In this case, the time interval and the time horizon are identical.

In the Gaussian case, the VaR is directly related to the volatility σ. Indeed, for a Gaussian distribution of RMS equal to $\sigma_1 x_0 = \sigma x_0 \sqrt{\tau}$, and of mean m_1, one finds that Λ^* is given by:

$$\mathcal{P}_{G<}\left(-\frac{\Lambda^* + m_1}{\sigma_1 x_0}\right) = \mathcal{P}^* \longrightarrow \Lambda^* = \sqrt{2}\sigma_1 x_0 \mathrm{erfc}^{-1}[2\mathcal{P}^*] - m_1 \tag{10.13}$$

(cf. Eq. (2.36)). When m_1 is small, minimizing Λ^* is thus equivalent to minimizing σ. It is furthermore important to note the very slow growth of Λ^* as a function of the time horizon in a Gaussian framework. For example, for $T_{VaR} = 250\tau$ (1 market year), corresponding to $\mathcal{P}^* = 0.004$, one finds that $\Lambda^* \approx 2.65\sigma_1 x_0$. Typically, for $\tau = 1$ day, $\sigma_1 = 1\%$, and therefore, the most probable worst day on a market year is equal to -2.65%, and grows only to -3.35% over a 10-year horizon!

In the general case, however, there is no useful link between σ and Λ^*. In some cases, decreasing one actually increases the other (see below, Sections 12.1.3 and 12.3). Let us nevertheless mention the Chebyshev inequality, often invoked by the volatility fans, which states that if σ exists,

$$\Lambda^{*2} \leq \frac{\sigma_1^2 x_0^2}{2\mathcal{P}^*}. \tag{10.14}$$

This inequality suggests that in general, the knowledge of σ is tantamount to that of Λ^*. This is however completely wrong, as illustrated in Figure 10.2. We have represented Λ^* as a function of the time horizon $T^* = N\tau$ for three distributions $P_\tau(\delta x)$ which all have exactly the same variance, but decay as a Gaussian, as an exponential, or as a power-law

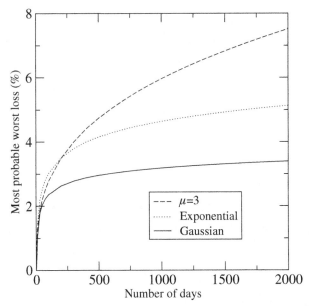

Fig. 10.2. Growth of Λ^* as a function of the number of days N, for an asset of daily volatility equal to 1%, with different distribution of price increments: Gaussian, (symmetric) exponential, or power-law with $\mu = 3$ (cf. Eq. (2.55)). Note that for intermediate N, the exponential distribution leads to a larger VaR than the power-law; this relation is however inverted for $N \to \infty$.

with an exponent $\mu = 3$ (cf. Eq. (2.55)). Of course, the slower the decay of the distribution, the faster the growth of Λ^* with time.

Table 10.1 shows, in the case of international bond indices, a comparison between the prediction of the most probable worst day using a Gaussian model, or using the observed quasi-exponential character of the tail of the distribution (TLD), and the actual worst day observed the following year. It is clear that the Gaussian prediction is systematically over-optimistic. The exponential model leads to a number which is seven times out of 11 below the observed result, which is indeed the expected result (Fig. 10.1).

Note finally that the measure of risk as a loss probability keeps its meaning even if the variance is infinite, as for a Lévy process. Suppose indeed that $P_\tau(\delta x)$ decays very slowly when δx is very large, as:

$$P_\tau(\delta x) \underset{\delta x \to -\infty}{\simeq} \frac{\mu A^\mu}{|\delta x|^{1+\mu}}, \tag{10.15}$$

with $\mu < 2$, such that $\langle \delta x^2 \rangle = \infty$. A^μ is the 'tail amplitude' of the distribution P_τ; A gives the order of magnitude of the probable values of δx. The calculation of Λ^* is immediate and leads to:

$$\Lambda^* = A \mathcal{P}^{*-1/\mu}, \tag{10.16}$$

which shows that in order to minimize the VaR one should minimize A^μ, independently of the probability level \mathcal{P}^*.

Table 10.1. J. P. Morgan international bond indices (expressed in French francs), analysed over the period 1989–93, and worst day observed day in 1994. The predicted numbers correspond to the most probable worst day Λ_{max}. The amplitude of the worst day with 95% confidence level is easily obtained, in the case of exponential tails, by multiplying Λ_{max} by a factor 1.53. The last line corresponds to a portfolio made of the 11 bonds with equal weight. All numbers are in per cent.

Country	Worst day Log-normal	Worst day TLD	Worst day Observed
Belgium	0.92	1.14	1.11
Canada	2.07	2.78	2.76
Denmark	0.92	1.08	1.64
France	0.59	0.74	1.24
Germany	0.60	0.79	1.44
Great Britain	1.59	2.08	2.08
Italy	1.31	2.60	4.18
Japan	0.65	0.82	1.08
Netherlands	0.57	0.70	1.10
Spain	1.22	1.72	1.98
United States	1.85	2.31	2.26
Portfolio	0.61	0.80	1.23

10.3.3 Expected shortfall

Although more directly related to large risks than the variance, the concept of Value-at-Risk still suffers from some inconsistencies. For example, Λ^* is the amount (in dollar) that has a certain probability \mathcal{P}^* to be exceeded, but if exceeded, what is the typical loss incurred? The second problem is that value-at-risk is not subadditive in general, which means that if one mixes two assets X_1 and X_2 with weights p_1 and p_2, one does not necessarily have that:

$$\Lambda^*(p_1 X_1 + p_2 X_2) \leq p_1 \Lambda^*(X_1) + p_2 \Lambda^*(X_2). \tag{10.17}$$

This inequality means that diversification should never increase the total risk, which is indeed a required property of any consistent measure of risk. Note that the volatility is indeed subadditive, since:

$$p_1^2 \sigma_1^2 + p_2^2 \sigma_2^2 + 2\rho_{12}\, p_1 p_2 \sigma_1 \sigma_2 \leq (p_1 \sigma_1 + p_2 \sigma_2)^2, \tag{10.18}$$

which follows because the correlation coefficient $\rho_{12} \leq 1$.

A quantity which is subadditive, and gives a more faithful estimate of the amplitude of large losses, is the average loss conditioned to the fact that Λ^* is exceeded, called the

expected shortfall (or conditional value-at-risk) E^*:

$$E^* = \frac{\int_{-\infty}^{-\Lambda^*} (-\delta x) P_\tau(\delta x)\, d\delta x}{\mathcal{P}^*}. \tag{10.19}$$

Note that in this definition, \mathcal{P}^* is fixed, and Λ^* deduced from Eq. (10.11).

We should nevertheless mention that for typical (well behaved) distributions $P_\tau(\delta x)$, E^* and Λ^* are, in the small \mathcal{P}^* limit, proportional to each other. For example, if $P_\tau(\delta x)$ is a power-law as in Eq. (10.15), one finds (see Eq. (2.16)):

$$E^* = \frac{\mu}{\mu - 1} \Lambda^* \qquad (\mu > 1). \tag{10.20}$$

Typically, for stocks μ is in the range 3 to 5, and thus $E^* \sim 1.25 - 1.5 \Lambda^*$.

In some special cases, however, the distribution of profit and losses contains discrete components, or has a non monotone behaviour, that make E^* very different from Λ^*.

10.4 Temporal aspects: drawdown and cumulated loss

Worst low

A first problem which arises is the following: we have defined \mathcal{P} as the probability for the loss observed at the end of the period $[k\tau, (k+1)\tau]$ to be at least equal to Λ. However, in general, a worse loss still has been reached *within* this time interval. What is then the probability that the worst point reached within the interval (the 'low') is a least equal to Λ? The answer is easy for symmetrically distributed increments (we thus neglect the average return $m\tau \ll \Lambda$, which is justified for small enough time intervals): this probability is simply equal to $2\mathcal{P}$,

$$\mathcal{P}[X_{\text{lo}} - X_{\text{op}} < -\Lambda] = 2\mathcal{P}[X_{\text{cl}} - X_{\text{op}} < -\Lambda], \tag{10.21}$$

where X_{op} is the value at the beginning of the interval (open), X_{cl} at the end (close) and X_{lo}, the lowest value in the interval (low). The reason for this is that for each trajectory just reaching $-\Lambda$ between $k\tau$ and $(k+1)\tau$, followed by a path which ends up above $-\Lambda$ at the end of the period, there is a mirror path with precisely the same weight which reaches a 'close' value lower than $-\Lambda$.[†] This factor of 2 in cumulative probability is illustrated in Figure 10.3. Therefore, if one wants to take into account the possibility of further loss within the time interval of interest in the computation of the VaR, one should simply divide by a factor 2 the probability level \mathcal{P}^* appearing in Eq. (10.11).

A simple consequence of this 'factor 2' rule is the following: the average value of the maximal excursion of the price during time τ, given by the high minus the low over that period, is equal to twice the average of the absolute value of the variation from the beginning

[†] In fact, this argument – which dates back to Bachelier himself (1900)! – assumes that the moment where the trajectory reaches the point $-\Lambda$ can be precisely identified. This is not the case for a discontinuous process, for which the doubling of \mathcal{P} is only an approximate result.

Table 10.2. Average value of the absolute value of the open/close daily returns and maximum daily range (high-low) over open for S&P 500, DEM/$ and Bund. Note that the ratio of these two quantities is indeed close to 2. Data from 1989 to 1998.

| | $A = \left\langle \frac{|X_{cl} - X_{op}|}{X_{op}} \right\rangle$ | $B = \left\langle \frac{X_{hi} - X_{lo}}{X_{op}} \right\rangle$ | B/A |
|---|---|---|---|
| S&P 500 | 0.472% | 0.924% | 1.96 |
| DEM/$ | 0.392% | 0.804% | 2.05 |
| Bund | 0.250% | 0.496% | 1.98 |

Fig. 10.3. Cumulative probability distribution of losses (Rank histogram) for the S&P 500 for the period 1989–98. Shown is the daily loss $(X_{cl} - X_{op})/X_{op}$ (thin line, left axis) and the intraday loss $(X_{lo} - X_{op})/X_{op}$ (thick line, right axis). Note that the right axis is shifted downwards by a factor of two with respect to the left one, so that in theory the two lines should fall on top of one another.

to the end of the period ($\langle |\text{open-close}| \rangle$). This relation can be tested on real data; the corresponding empirical factor is reported in Table 10.2.

Cumulated losses

Another very important aspect of the problem is to understand how losses can accumulate over successive time periods. For example, a bad day can be followed by several other bad days, leading to a large overall loss. One would thus like to estimate the most probable value of the worst week, month, etc. In other words, one would like to construct the graph of $\Lambda^*(M)$, where M is the number of successive days over which the risk must be estimated, given a fixed overall investment horizon $T = N\tau$.

The answer is straightforward in the case where the price increments are independent random variables. When the elementary distribution P_τ is Gaussian, $P_{M\tau}$ is also Gaussian, with a variance multiplied by M. One must also take into account the fact that the number of different intervals of size $M\tau$ for a fixed investment horizon T decreases by a factor M. For large enough N/M, one then finds, using the asymptotic behaviour of the error function:

$$\Lambda^*(M)\big|_T \simeq \sigma_1 x_0 \sqrt{2M \log\left(\frac{N}{\sqrt{2\pi}M}\right)}, \tag{10.22}$$

where the notation $|_T$ means that the investment period is fixed.[†] The main effect is then the \sqrt{M} increase of the volatility, up to a small logarithmic correction.

The case where $P_\tau(\delta x)$ decreases as a power-law has been discussed in Chapter 2: for any M, the far tail remains a power-law (that is progressively 'eaten up' by the Gaussian central part of the distribution if the exponent μ of the power-law is greater than 2, but keeps its integrity whenever $\mu < 2$). For finite M, the *largest moves* will always be described by the power-law tail. Its amplitude A^μ is simply multiplied by a factor M (cf. Section 2.2.2). Since the number of independent intervals is divided by the same factor M, one finds:[‡]

$$\Lambda^*(M)\big|_T = AM^{1/\mu}\left(\frac{N}{M\tau}\right)^{\frac{1}{\mu}} = AN^{1/\mu}, \tag{10.23}$$

independently of M. Note however that for $\mu > 2$, the above result is only valid if $\Lambda^*(M)$ is located outside of the Gaussian central part, the size of which growing as $\sigma\sqrt{M}$ (Fig. 10.4). In the opposite case, the Gaussian formula (10.22) should be used.

One can ask a slightly different question, by fixing not the investment horizon T but rather the total probability of occurrence \mathcal{P}^*. This amounts to multiplying simultaneously the period over which the loss is measured and the investment horizon by the same factor M. In this case, one finds that:

$$\Lambda^*(M) = M^{\frac{1}{\mu}}\Lambda^*(1). \tag{10.24}$$

The growth of the value-at-risk with M is thus faster for small μ, as expected. The case of Gaussian fluctuations corresponds to $\mu = 2$.

Drawdowns

One can finally analyze the amplitude of a cumulated loss over a period that is not *a priori* limited. More precisely, the question is the following: knowing that the price of the asset today is equal to x_0, what is the probability that the lowest price ever reached in the future will be x_{min}; and how long such a *drawdown* will last? This is a classical problem in probability theory (see Feller (1971)). We shall assume a multiplicative model of independent returns, and denote as $\Delta = \log(x_0/x_{min}) > 0$ the maximum amplitude of the loss. If the $P_\tau(\eta)$ decays

[†] One can also take into account the average return, $m_1 = \langle \delta x \rangle$. In this case, one must subtract to $\Lambda^*(M)|_T$ the quantity $-m_1 N$ (the potential losses are indeed smaller if the average return is positive).

[‡] cf. previous footnote.

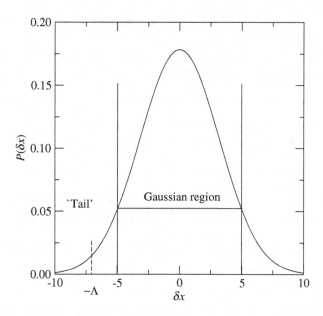

Fig. 10.4. Distribution of cumulated losses for finite M: the central region, of width $\sim \sigma M^{1/2}$, is Gaussian. The tails, however, remember the specific nature of the elementary distribution $P_\tau(\delta x)$, and are in general fatter than Gaussian. The point is then to determine whether the confidence level \mathcal{P}^* puts $\Lambda^*(M)$ within the Gaussian part or far in the tails.

sufficiently fast for large negative η, the result is that the tail of the distribution of Δ behaves as an exponential:

$$P(\Delta) \underset{\Delta \to \infty}{\propto} \exp\left(-\frac{\Delta}{\Delta_0}\right), \tag{10.25}$$

where $\Delta_0 > 0$ is the finite solution of the following equation:

$$\int (1+\eta)^{-\frac{1}{\Delta_0}} P_\tau(\eta)\, d\eta = 1. \tag{10.26}$$

If this equation only has $\Delta_0 = \infty$ as a solution, the decay of the distribution of drawdowns is slower than exponential.

It is interesting to show how the above result can be obtained. Let us introduce the cumulative distribution $\mathcal{P}_>(\Delta) = \int_\Delta^\infty P(\Delta')\, d\Delta'$. The first step of the 'walk' of the logarithm of the price, of size $\log(1+\eta)$, can either exceed $-\Delta$, or stay above it. In the first case, the level Δ is immediately exceeded, and the event contributes to $\mathcal{P}_>(\Delta)$. In the second case, one recovers the very same problem as initially, but with Δ shifted to $\Delta + \log(1+\eta)$. Therefore, $\mathcal{P}_>(\Delta)$ obeys the following equation:

$$\mathcal{P}_>(\Delta) = \int_{-1}^{\eta^*} P_\tau(\eta)\, d\eta + \int_{\eta^*}^{+\infty} P_\tau(\eta)\mathcal{P}_>(\Delta + \log(1+\eta))\, d\eta, \tag{10.27}$$

where $\log(1+\eta^) = -\Delta$. If one can neglect the first term in the right-hand side for large*

Δ (which should be self-consistently checked), then, asymptotically, $\mathcal{P}_>(\Delta)$ should obey the following equation:

$$\mathcal{P}_>(\Delta) \approx \int P_\tau(\eta)\mathcal{P}_>(\Delta + \log(1+\eta))\, d\eta. \tag{10.28}$$

Inserting an exponential ansatz for $\mathcal{P}_>(\Delta)$ then leads to Eq. (10.26) for Δ_0, in the limit $\Delta \to \infty$.

Let us study the simple case of Gaussian fluctuations, for which:

$$P_\tau(\eta) = \frac{1}{\sqrt{2\pi\sigma^2\tau}} \exp -\frac{(\eta - m^*\tau)^2}{2\sigma^2\tau}. \tag{10.29}$$

where m^* is the excess return, which means that we are defining drawdowns with respect to a risk free investment. For $m^*\tau$, $\sigma^2\tau \ll 1$, equation (10.26) becomes:

$$-\frac{m^*}{\Delta_0} + \frac{\sigma^2}{2\Delta_0^2} = 0 \to \Delta_0 = \frac{\sigma^2}{2m^*}. \tag{10.30}$$

The quantity Δ_0 gives the order of magnitude of the worst drawdown. One can understand the above result as follows: Δ_0 is the amplitude of the probable fluctuation over the characteristic time scale $\hat{T} = \sigma^2/m^{*2}$ introduced above. By definition, for times shorter than \hat{T}, the average return m^* is negligible. Therefore, the order of typical drawdowns is: $\propto \sqrt{\sigma^2\hat{T}} = \sigma^2/m^*$.

If $m^* = 0$, the very idea of worst drawdown loses its meaning: if one waits a time long enough, the price can reach arbitrarily low values (again, compared to the risk free investment). It is natural to introduce a quality factor that compares the average excess return m^*T on a time T to the amplitude of the worst drawdown Δ_0. One thus has $m^*T/\Delta_0 = 2m^{*2}T/\sigma^2 \equiv 2T/\hat{T}$. This quality factor is therefore twice the square of the Sharpe ratio. The larger the Sharpe ratio, the smaller the time needed to end a drawdown period.

More generally, if the width of $P_\tau(\eta)$ is much smaller than Δ_0, one can expand the exponential and find that the above equation, $-(m^*/\Delta_0) + (\sigma^2/2\Delta_0^2) = 0$ is still valid, independently of the detailed shape of the distribution of returns.

How long will these drawdowns last? The answer can only be statistical. The probability for the time of first return T to the initial investment level x_0, which one can take as the definition of a drawdown (although it could of course be immediately followed by a second drawdown), has, for large T, the following form:

$$P(T) \simeq \frac{\tau^{1/2}}{T^{3/2}} \exp\left(-\frac{T}{\hat{T}}\right). \tag{10.31}$$

The probability for a drawdown to last much longer than \hat{T} is thus very small. In this sense, \hat{T} appears as the characteristic drawdown time. However, \hat{T} is *not* equal to the *average* drawdown time, which is on the order of $\sqrt{\tau\hat{T}}$, and thus much smaller than \hat{T}. This is

related to the fact that *short* drawdowns have a large probability: as τ decreases, a large number of drawdowns of order τ appear, thereby reducing their average size.

10.5 Diversification and utility – satisfaction thresholds

It is intuitively clear that one should not put all his eggs in the same basket. A *diversified* portfolio, composed of different assets with small mutual correlations, is less risky because the gains of some of the assets more or less compensate the loss of the others. Now, an investment with a small risk and small return must sometimes be preferred to a high yield, but very risky, investment.

The theoretical justification for the idea of diversification comes again from the CLT. A portfolio made up of M uncorrelated assets and of equal volatility, with weight $1/M$, has an overall volatility reduced by a factor \sqrt{M}. Correspondingly, the amplitude (and duration) of the worst drawdown is divided by a factor M (cf. Eq. (10.30)), which is obviously satisfying.

This qualitative idea stumbles over several difficulties. First of all, the fluctuations of financial assets are in general *strongly correlated*; this substantially decreases the possibility of true diversification, and requires a suitable theory to deal with these correlations and construct an 'optimal' portfolio: this is Markowitz's theory, to be detailed in Chapter 12. Furthermore, since price fluctuations can be strongly non-Gaussian, a volatility-based measure of risk might be unadapted: one should rather try to minimize the value-at-risk, or the expected shortfall, of the portfolio. It is therefore interesting to look for an extension of the classical formalism, allowing one to devise *minimum VaR* portfolios. This will also be presented in Chapter 12.

One is thus in general confronted with the problem of defining properly an 'optimal' portfolio. Usually, one invokes the rather abstract concept of 'utility functions', on which we shall briefly comment in this section, in particular to show that it does not naturally accommodate for the notion of value-at-risk or expected shortfall.

We will call W_T the wealth of a given operator at time $t = T$. If one argues that the level of satisfaction of this operator is quantified by a certain function of W_T only, which one usually calls the **utility function** $U(W_T)$. Note that one assumes that the satisfaction does *not* depend on the whole 'history' of the wealth between $t = 0$ and T. One thus assumes that the operator is insensitive to what can happen between these two dates. This is not very realistic and a generalization to path dependent utility *functionals* $U(\{W(t)\}_{0 \leq t \leq T})$ could change significantly some of the following discussion.

The utility function $U(W_T)$ is furthermore taken to be continuous and even twice differentiable. The postulated 'rational' behaviour for the operator is then to look for investments which maximize his expected utility, averaged over all possible histories of price changes:

$$\langle U(W_T) \rangle = \int P(W_T) U(W_T) \, dW_T. \tag{10.32}$$

The utility function should be non-decreasing: a larger profit is clearly always more satisfying. One can furthermore consider the case where the distribution $P(W_T)$ is sharply peaked around its mean value $\langle W_T \rangle = W_0 + mT$. Performing a Taylor expansion of $\langle U(W_T) \rangle$

around $U(W_0 + mT)$ to second order, one finds:

$$\langle U(W_T)\rangle = \int U(W_0 + mT + \varepsilon)P(\varepsilon)\,d\varepsilon \approx U(W_0 + mT) + \frac{1}{2}\left.\frac{d^2U}{dW^2}\right|_{W_0+mT}\langle\varepsilon^2\rangle + \dots,$$

(10.33)

where we have used that $\langle\varepsilon\rangle = 0$. On the other hand $\langle\varepsilon^2\rangle$ measures the uncertainty on the final wealth, which should also degrade the satisfaction of the agent. One deduces that the utility function must be such that:

$$\frac{d^2U}{dW^2} < 0.$$

(10.34)

This property reflects the fact that for the same average return, a less risky investment should always be preferred.[†]

A simple example of utility function compatible with the above constraints is the exponential function $U(W_T) = -\exp[-W_T/w_0]$. Note that w_0 has the dimensions of a *wealth*, thereby fixing a wealth scale in the problem. A natural candidate is the initial wealth of the operator, $w_0 \propto W_0$. If $P(W_T)$ is Gaussian, of mean $W_0 + mT$ and variance DT, one finds that the expected utility is given by:

$$\langle U\rangle = -\exp\left[-\frac{W_0}{w_0} - \frac{T}{w_0}\left(m - \frac{D}{2w_0}\right)\right].$$

(10.35)

One could think of constructing a utility function with no intrinsic wealth scale by choosing a power-law: $U(W_T) = (W_T/w_0)^\alpha$ with $\alpha < 1$ to ensure the correct convexity. Indeed, in this case a change of w_0 can be reabsorbed in a change of scale of the utility function itself, which is arbitrary. The choice $\alpha = 0$ corresponds to Bernouilli's logarithmic utility function $U(W_T) = \log(W_T)$. However, these definitions cannot allow for negative final wealths, and is thus sometimes problematic.

Despite the fact that these postulates sound reasonable, and despite the very large number of academic studies based on the concept of utility function, this axiomatic approach suffers from a certain number of fundamental flaws. For example, it is not clear that one could ever measure the utility function used by a given agent.[‡] The theoretical results are thus:

- Either relatively weak, because independent of the special form of the utility function, and only based on its general properties.
- Or rather arbitrary, because based on a specific, but unjustified, form for $U(W_T)$.

On the other hand, the idea that the utility function is regular is probably not always realistic. The satisfaction of an operator is often governed by rather sharp *thresholds*, separated

[†] This is however not obvious when one considers path dependent utilities, since a more volatile stock leads to the possibility of larger potential profits if the reselling time is not a priori determined.

[‡] Even the idea that an operator would really optimize his expected utility, and not take decisions partly based on 'non-rational', 'random' motivations, is far from obvious. On this point, see e.g.: Kahnemann & Tversky (1979), Marsili & Zhang (1997), Shleifer (2000).

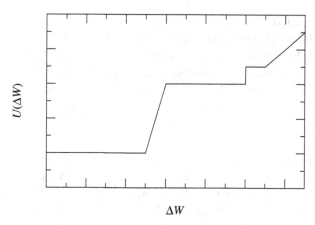

Fig. 10.5. Example of a utility function with thresholds, where the utility function is non-continuous. These thresholds correspond to special values for the profit, or the loss, and are often of purely psychological origin.

by regions of indifference (Fig. 10.5).[†] For example, one can be in a situation where a specific project can only be funded if the profit $\Delta W = W_T - W_0$ exceeds a certain amount. Symmetrically, the clients of a fund manager will take their money away as soon as the losses exceed a certain value: this is the strategy of 'stop-losses', which fix a level for acceptable losses, beyond which the position is closed. The existence of option markets (which allow one to limit the potential losses below a certain level – see Chapter 13), or of items the price of which is \$99 rather than \$100, are concrete examples of the existence of these psychological thresholds where 'satisfaction' changes abruptly. Therefore, the utility function U is not necessarily continuous. This remark is actually intimately related to the fact that the value-at-risk is often a better measure of risk than the variance. Let us indeed assume that the operator is 'happy' if $\Delta W > -\Lambda$ and 'unhappy' whenever $\Delta W < -\Lambda$. This corresponds formally to the following utility function:

$$U_\Lambda(\Delta W) = \begin{cases} U_1 \ (\Delta W > -\Lambda) \\ U_2 \ (\Delta W < -\Lambda) \end{cases}, \tag{10.36}$$

with $U_2 - U_1 < 0$.

The expected utility is then simply related to the loss probability:

$$\langle U_\Lambda \rangle = U_1 + (U_2 - U_1) \int_{-\infty}^{-\Lambda} P(\Delta W)\,\mathrm{d}\Delta W$$
$$= U_1 - |U_2 - U_1|\mathcal{P}. \tag{10.37}$$

Therefore, optimizing the expected utility is in this case tantamount to minimizing the probability of losing more that Λ. Despite this rather appealing property, which certainly corresponds to the behaviour of some market operators, the function $U_\Lambda(\Delta W)$ does not

[†] It is possible that the presence of these thresholds actually plays an important role in the fact that the price fluctuations are strongly non-Gaussian.

satisfy the above criteria (continuity and negative curvature). The same remark would hold for the expected shortfall, that corresponds to replacing U_2 by $U_2 \times \Delta W$ in Eq. (10.36).

Confronted to the problem of choosing between risk (as measured by the variance) and return, a very natural strategy for those not acquainted with utility functions would be to compare the average return to the potential loss \sqrt{DT}. This can thus be thought of as defining a risk-corrected, 'pessimistic' estimate of the profit, as:

$$m_\lambda T = mT - \lambda \sqrt{DT}, \tag{10.38}$$

where λ is an arbitrary coefficient that measures the pessimism (or the risk aversion) of the operator. A rather natural criterion would then be to look for the optimal portfolio that maximizes the risk adjusted return m_λ. However, this optimal portfolio cannot be obtained using the standard utility function formalism. For example, Eq. (10.35) shows that the object which should be maximized in the context this particular utility function is *not* $mT - \lambda\sqrt{DT}$ but rather $mT - DT/2w_0$. This amounts to comparing the average profit to the *square* of the potential losses, divided by the reference wealth scale w_0 – a quantity that depends *a priori* on the operator.[†] On the other hand, the quantity $mT - \lambda\sqrt{DT}$ is directly related (at least in a Gaussian world) to the value-at-risk Λ^*, cf. Eq. (10.22).

This argument can be expressed slightly differently: a reasonable objective could be to maximize the value of the **probable gain** \mathcal{G}_p, such that the probability of earning more is equal to a certain probability p:[‡]

$$\int_{\mathcal{G}_p}^{+\infty} P(\Delta W)\,d\Delta W = p. \tag{10.39}$$

In the case where $P(\Delta W)$ is Gaussian, this amounts again to maximizing $mT - \lambda\sqrt{DT}$, where λ is related to p in a simple manner. Now, one can show that it is impossible to construct a utility function such that, in the general case, different strategies can be ordered according to their probable gain \mathcal{G}_p. Therefore, the concepts of loss probability, value-at-risk or probable gain cannot be accommodated naturally within the framework of utility functions.

10.6 Summary

- The simplest measure of risk is the volatility σ. Compared to the average return \tilde{m}, this defines a 'security' time horizon $\hat{T} = \sigma^2/\tilde{m}^2$ beyond which the probability of loss becomes small. Correspondingly, the typical drawdown amplitude is σ^2/\tilde{m}.

- The volatility is not always adapted to the real world. The tails of the distributions, where the large events lie, are very badly described by a Gaussian law: this leads to a systematic underestimation of the extreme risks. Sometimes, the measurement of volatility using historical data is difficult, precisely because of the presence of these large fluctuations.

[†] This comparison is actually meaningful, since it corresponds to comparing the reference wealth w_0 to the order of magnitude of the worst drawdown D/m, cf. Eq. (10.30).

[‡] Maximizing \mathcal{G}_p is thus equivalent to minimizing Λ^* such that $\mathcal{P}^* = 1 - p$. Note that one could alternatively try to maximize the probability p that a certain gain \mathcal{G} is achieved.

- The measure of risk through the probability of loss, value-at-risk, or expected shortfall on the other hand, focus specifically on the tails. Extreme events are considered as the true source of risk, whereas the small fluctuations that contribute to the 'centre' of the distributions (and contribute to the volatility) can be seen as background noise, inherent to the very activity of financial markets, but not relevant for risk assessment.
- From a theoretical point of view, these definitions of risk (based on extreme events) do not easily fit into the classical utility function framework. The minimization of a loss probability rather assumes that there exist well-defined thresholds (possibly different for each operator) where the utility function is discontinuous. The concept of value-at-risk, or probable gain, cannot be naturally dealt with using classical utility functions.

- FURTHER READING

C. Alexander, *Market Models: A Guide to Financial Data Analysis*, John Wiley, 2001.

L. Bachelier, *Théorie de la Spéculation*, Gabay, Paris, 1995.

M. Dempster, *Value at Risk and Beyond*, Cambridge University Press, 2001.

P. Embrechts, C. Klüppelberg, T. Mikosch, *Modelling Extremal Events for Insurance and Finance*, Springer-Verlag, Berlin, 1997.

W. Feller, *An Introduction to Probability Theory and its Applications*, Wiley, New York, 1971.

J. Galambos, *The Asymptotic Theory of Extreme Order Statistics*, R. E. Krieger Publishing Co., Malabar, Florida, 1987.

E. J. Gumbel, *Statistics of Extremes*, Columbia University Press, New York, 1958.

P. Jorion, *Value at Risk: The New Benchmark for Managing Financial Risk*, McGraw-Hill, 2nd edition, 2000.

A. Shleifer, *Inefficient Markets, An Introduction to Behavioral Finance*, Oxford University Press, 2000.

- SPECIALIZED PAPERS

C. Acerbi and D. Tasche, *On the coherence of expected shortfall*, Journal of Banking and Finance, **26**, 1487 (2002).

P. Artzner, F. Delbaen, J.-M. Eber, D. Heath, *Coherent measures of risk*, Mathematical Finance, **9**, 203 (1999).

V. Bawa, *Optimal rules for ordering uncertain prospects*, J. Fin. Eco., **2**, 95 (1975).

S. Galluccio, J.-P. Bouchaud, M. Potters, *Rational decisions, random matrices and spin glasses*, Physica A, **259**, 449 (1998).

D. Kahnemann, A. Tversky, *Prospect theory: An analysis of decision under risk*, Econometrica, **47**, 263 (1979).

M. Marsili, Y. C. Zhang, *Fluctuations around Nash equilibria in game theory*, Physica A, **245**, 181 (1997).

R. T. Rockafellar, S. Uryasev, *Conditional value-at-risk for general loss distributions*, Journal of Banking & Finance, **26**, 1443 (2001).

D. Tasche, *Expected Shortfall and Beyond*. Journal of Banking and Finance, **26**, 1519 (2002).

- SOME RECENT READS

Y. Malevergne, D. Sornette, *Extreme Financial Risks: From Dependence to Risk Management*, Springer, 2005.

11

Extreme correlations and variety

Etait-ce parce que je ne l'avais qu'entr-aperçue que je l'avais trouvée si belle?[†]
(Marcel Proust, *A l'ombre des jeunes filles en fleurs*.)

A proper understanding of cross-correlations between different stocks, or more generally between different financial assets, is of crucial importance for risk management and, as will be discussed in the Chapter 12, for portfolio optimization. In fact, much as the tail of the distributions of individual returns is of particular relevance for risk management, it is the correlations between extreme events that must be faithfully described.

As we have discussed in depth in Chapter 7, the volatility of each asset is in general not constant in time, but fluctuates randomly. It is therefore tempting to think that the whole covariance matrix (of which the diagonal contains the squared volatilities) also randomly fluctuates in time. The simplest case would be that the correlation matrix has a fixed structure, and that the variations of the covariance matrix only reflects that of the volatilities. However, it is a common belief in the financial industry that cross-correlations between stocks themselves fluctuate in time, and in fact increase substantially in a period of high market volatility. If true, this would mean that the possibility of risk diversification is much reduced, since in the periods that really matter as far as risk is concerned, all stocks would behave similarly.

The aim of this chapter is first to introduce several possible measures of correlations of extreme events, such as conditional correlations, exceedance correlations, tail dependence and tail covariance. We show that some of these indicators are misguided in the sense that they can lead to an apparent increase of correlations in extreme market conditions whereas the underlying stochastic process has fixed, time independent correlations. We illustrate this point using a simple one-factor model with a time dependent random volatility, that we compare with data using stocks from the S&P 500. We then discuss the limitations of the one-factor description, that fails to describe the time dependence of the *variety*, a new quantity akin to the volatility, that measures the variability of the returns of the different stocks for a given day. Similarly, the skewness of the distribution of returns of the different stocks for a given day is also time dependent.

[†] Was it because I only had a glimpse of her that I thought she was so beautiful?

11.1 Extreme event correlations

11.1.1 Correlations conditioned on large market moves

The simplest idea is to measure correlations between stocks conditioned on certain extreme events, for example an extreme market return. It is indeed commonly believed that cross-correlations between stocks increase in such "high-volatility" periods. Introducing the return of stock i, η_i, and the return of the market η_m simply defined as the average of all η_i's, a natural measure of these correlations is given by the following coefficient:

$$\rho_>(\lambda) = \frac{\frac{1}{N^2} \sum_{i,j} (\langle \eta_i \eta_j \rangle_{>\lambda} - \langle \eta_i \rangle_{>\lambda} \langle \eta_j \rangle_{>\lambda})}{\frac{1}{N} \sum_i (\langle \eta_i^2 \rangle_{>\lambda} - \langle \eta_i \rangle_{>\lambda}^2)}, \tag{11.1}$$

where the subscript $> \lambda$ indicates that the averaging is restricted to market returns η_m in absolute value larger than a certain λ. For $\lambda = 0$ the conditioning disappears. Note that the quantity $\rho_>$ is the average covariance divided by the average variance, and therefore differs from the average correlation coefficient. The two objects however behave very similarly with λ, and the theoretical discussion is much easier in terms of $\rho_>$.

In a first approximation, the distribution of individual stocks returns can be taken to be symmetrical (see Section 6.3.2), leading $\langle \eta_i \rangle_{>\lambda} \approx 0$. Using the fact that $\eta_m = \sum_i \eta_i / N$, the above equation can be transformed into:

$$\rho_>(\lambda) \approx \frac{\sigma_m^2(\lambda)}{\frac{1}{N} \sum_{i=1}^N \sigma_i^2(\lambda)}, \tag{11.2}$$

where $\sigma_m^2(\lambda)$ is the market volatility conditioned to market returns in absolute value larger than λ, and $\sigma_i^2(\lambda) = \langle r_i^2 \rangle_{>\lambda}$.

Now, assume the following decomposition to hold:

$$\eta_i(t) = \beta_i \eta_m(t) + \epsilon_i(t), \tag{11.3}$$

where β_i are fixed, time independent coefficients such that $\sum_i \beta_i / N = 1$, and the ϵ_i are uncorrelated with the market and are the so-called idiosyncratic parts, or residuals. (The **one-factor** model would correspond to the case where all idiosyncrasies are uncorrelated, which we do not need to assume here.) Then one obtains for the average conditional correlation $\rho_>(\lambda)$:

$$\rho_>(\lambda) = \frac{\sigma_m^2(\lambda)}{\left(\frac{1}{N} \sum_{i=1}^N \beta_i^2\right) \sigma_m^2(\lambda) + \frac{1}{N} \sum_{i=1}^N \sigma_{\epsilon_i}^2}. \tag{11.4}$$

The quantity $\sigma_m^2(\lambda)$ is obviously an increasing function of λ. If the residual volatilities $\sigma_{\epsilon_i}^2$ are independent of η_m (and therefore of λ) then the coefficient $\rho_>(\lambda)$ is an increasing function of λ, see Fig. 11.1.

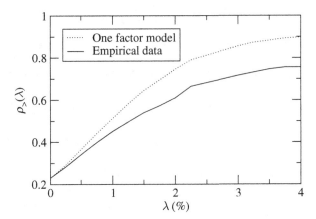

Fig. 11.1. Correlation measure $\rho_>(\lambda)$ conditional to the absolute market return to be larger than λ, both for the empirical data and the one-factor model. Note that both show a similar apparent increase of correlations with λ. This effect is actually overestimated by the one-factor model with fixed residual volatilities. λ is in percents.

Equation (11.4) naturally predicts an increase of the correlations (as measured by $\rho_>(\lambda)$) in high volatility periods. This conclusion is quite general: the very fact of conditioning the correlation on large market returns leads to an increase of the measured correlation, even if the true structure of correlations is strictly time independent. Equation (11.4) actually even *overestimates* the correlations for large λ, as shown in Fig. 11.1. This overestimation can be understood qualitatively as a result of neglecting the positive correlation between the amplitude of the market return $|r_m|$ and the residual volatilities σ_{ϵ_i}, which we discuss in more details in Section 11.2.1 below in terms of the 'variety'. For large values of λ, σ_{ϵ_i} is indeed found to be larger than its average value. From Eq. (11.4), this effect lowers the correlation $\rho_>(\lambda)$ for a given value of $\sigma_m^2(\lambda)$.

11.1.2 Real data and surrogate data

The data set we consider in this chapter is composed of the daily returns of 450 U.S. equities among the most liquid ones from 1993 up to 1999. In order to test the relevance of a simple one-factor model, we also generated surrogate data compatible with this model. Very importantly, the one-factor model we consider is not based on Gaussian distributions, but rather on fat-tailed distributions that match the empirical observations for both the market and the stocks daily returns – see Chapter 6.

The procedure used to generate the surrogate data is the following:

- *Compute the β_i's using*

$$\beta_i = \frac{\langle \eta_i \eta_m \rangle - \langle \eta_i \rangle \langle \eta_m \rangle}{\langle \eta_m^2 \rangle - \langle \eta_m \rangle^2}, \tag{11.5}$$

where the brackets refer to time averages over the whole time period. These β_i's are found to be rather narrowly distributed around 1, with $(1/N) \sum_{i=1}^{N} \beta_i^2 = 1.05$.

- *Compute the variance of the residuals $\sigma_{\epsilon_i}^2 = \sigma_i^2 - \beta_i^2 \sigma_m^2$. On the dataset we used σ_m was 0.91% (per day) whereas the volatility σ_{ϵ_i} was 1.66%.*
- *Generate the residual $\epsilon_i(t) = \sigma_{\epsilon_i} u_i(t)$, where the $u_i(t)$ are independent random variables of unit variance with a Student distribution with an exponent $\mu = 4$:*

$$P(u) = \frac{3}{(2 + u^2)^{5/2}},\qquad(11.6)$$

see Sections 1.9 and 18.6.

- *Compute the surrogate return as $\eta_i^{surr}(t) = \beta_i \eta_m^t(t) + \epsilon_i(t)$, where $\eta_m^t(t)$ is the true (historical) market return at day t.*

Therefore, within this method, both the empirical and surrogate returns are based on the very same realization of the market statistics. This allows one to compare meaningfully the results of the surrogate model with real data, without further averaging. It also short-cuts the precise parameterization of the distribution of market returns, in particular its correct negative skewness, which turns out to be crucial.

11.1.3 Conditioning on large individual stock returns: exceedance correlations

We now study more specifically how extreme stock returns are correlated between themselves. A first possibility is to study the so-called **exceedance correlation** function, introduced in Longin & Solnik (2001). One first defines from the η_i's normalized centered returns $\tilde{\eta}_i$ with zero mean and unit variance. The positive exceedance correlation between i and j is defined as:

$$\rho_{ij}^+(\theta) = \frac{\langle \tilde{\eta}_i \tilde{\eta}_j \rangle_{>\theta} - \langle \tilde{\eta}_i \rangle_{>\theta} \langle \tilde{\eta}_j \rangle_{>\theta}}{\sqrt{\left(\langle \tilde{\eta}_i^2 \rangle_{>\theta} - \langle \tilde{\eta}_i \rangle_{>\theta}^2\right)\left(\langle \tilde{\eta}_j^2 \rangle_{>\theta} - \langle \tilde{\eta}_j \rangle_{>\theta}^2\right)}},\qquad(11.7)$$

where the subscript $> \theta$ means that both normalized returns $\tilde{\eta}_{i,j}$ are larger than a certain θ. For $\theta \to -\infty$, ρ_{ij}^+ becomes the usual correlation coefficient, whereas large positive θ's correspond to correlations between extreme positive moves. The negative exceedance correlation $\rho_{ij}^-(\theta)$ is defined similarly, the conditioning being now on normalized returns smaller than θ.

Figure 11.2 shows the exceedance correlation function as a function of θ, when $\tilde{\eta}_{i,j}$ are obtained as the linear combination of two independent random variables $\xi_{1,2}$, such that their correlation coefficient is equal to $\rho = 0.5$, but when (a) $\xi_{1,2}$ are Gaussian random variables and (b) $\xi_{1,2}$ are Student variables with a tail index $\mu = 4$. Interestingly, the shape of $\rho_{ij}^{\pm}(\theta)$ is completely different in the two cases. For correlated Gaussian variables, $\rho_{ij}^{\pm}(\theta)$ decreases with $|\theta|$; in this sense, extreme events become uncorrelated for Gaussian returns. On the contrary, $\rho_{ij}^{\pm}(\theta)$ increases with $|\theta|$ in the Student case and becomes unity for large $|\theta|$. More generally, the growth of the exceedance correlation with $|\theta|$ is associated to distribution tails fatter than exponential (as for the Student case).

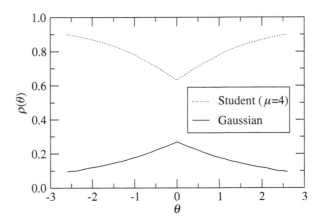

Fig. 11.2. Exceedance correlation functions for two Gaussian variables (full line) and two Student variables (dotted line). We have shown $\rho_{ij}^+(\theta)$ for positive θ and $\rho_{ij}^-(\theta)$ for negative θ. In both cases the correlation coefficient is set to $\rho = 0.5$. Note the important difference of shape in the two cases.

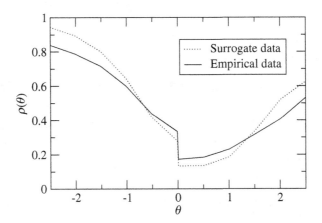

Fig. 11.3. Average exceedance correlation functions between stocks as a function of the level parameter θ, both for real data and the surrogate one-factor model. We have shown $\rho_{ij}^+(\theta)$ for positive θ and $\rho_{ij}^-(\theta)$ for negative θ. Note that this quantity *grows* with $|\theta|$ and is strongly asymmetric.

Figure 11.3 shows the exceedance correlation function, averaged over all the possible pairs of stocks i and j, both for real data and for the surrogate one-factor model data. As in other studies, the empirical $\rho^\pm(\theta)$ is found to grow with $|\theta|$ and is larger for large negative moves than for large positive moves.

Figure 11.3 however clearly shows that a *fixed* correlation non Gaussian one-factor model is enough to explain quantitatively the level and asymmetry of the exceedance correlation function,

without having to invoke time dependent correlations (such as, for example, 'regime switching' scenarios, see Beckaert & Wu (2000)). In particular, the asymmetry is entirely induced by the large negative skewness in the distribution of index returns.

11.1.4 Tail dependence

An interesting way to measure the correlation between extreme events is to compute the so-called **coefficient of tail dependence**, defined as follows. Consider two random variables η_i, η_j, with marginal distributions $P_i(\eta_i)$, $P_j(\eta_j)$, and the corresponding cumulative distributions $\mathcal{P}_{>,i,j}$. If we choose a certain probability level p, one can define the values of η_i^*, η_j^* such that:

$$\mathcal{P}_{>,i}(\eta_i^*) = \mathcal{P}_{>,j}(\eta_j^*) = p. \tag{11.8}$$

The coefficient of tail dependence $\rho_{t,ij}^+$ measures the probability that η_i exceeds η_i^*, conditional to the fact that η_j itself exceeds η_j^*, in the limit of extreme events, i.e. $p \to 0$:[†]

$$\rho_{t,ij}^+ = \lim_{p \to 0} \mathcal{P}(\eta_i > \eta_i^* | \eta_j > \eta_j^*). \tag{11.9}$$

In other words, ρ_t^+ measures the tendency for extreme events to occur simultaneously. As defined, ρ_t^+ is between 0 and 1 but appears not to be symmetrical in the permutation of i and j. However, the conditional probability is by definition given by:

$$\mathcal{P}(\eta_i > \eta_i^* | \eta_j > \eta_j^*) = \frac{\mathcal{P}(\eta_i > \eta_i^* \wedge \eta_j > \eta_j^*)}{\mathcal{P}_>(\eta_j^*)}, \tag{11.10}$$

and since we have chosen η_i^* and η_j^* such that Eq. (11.8) holds, one sees that $\rho_{t,ij}^+$ is indeed symmetrical.

If η_i and η_j are independent, then $\mathcal{P}(\eta_i > \eta_i^* | \eta_j > \eta_j^*) = p$ and therefore $\rho_{t,ij}^+ = 0$. Interestingly, if η_i and η_j are correlated Gaussian variables with a correlation coefficient ρ then, unless $\rho = 1$, one also has $\rho_{t,ij}^+ = 0$. Note that the coefficient of tail dependence is by construction invariant under any change of variables $\tilde{\eta}_i = f_i(\eta_i)$, $\tilde{\eta}_j = f_j(\eta_j)$, which changes the marginal distributions but not $\rho_{t,ij}^+$. The coefficient of tail dependence therefore only reflects the correlation structure of the underlying 'copula' (see Section 9.2.3).

More interesting is the case where η_i and η_j are obtained as the sum of power-law distributed independent 'factors':

$$\eta_i = \sum_{a=1}^{M} v_{a,i} e_a, \tag{11.11}$$

where

$$P(e_a) \underset{e_a \to \pm\infty}{\simeq} \frac{\mu A_a^\mu}{|e_a|^{1+\mu}}, \tag{11.12}$$

[†] We define here the coefficient of tail dependence for the right tail. The left tail coefficient ρ_t^- can be defined similarly and is a priori different from ρ_t^+.

the $v_{a,i}$ are coefficients. Since power-law variables are (asymptotically) stable under addition, the decomposition Eq. (11.11) leads for all μ to correlated power-law variables η_i with, as discussed in Chapter 2, a tail amplitude given by:

$$A_i^\mu = \sum_{a=1}^M |v_{a,i}|^\mu A_a^\mu. \tag{11.13}$$

The crucial remark is that for power law variables, η_i is asymptotically large whenever one of the factor e_a is individually large (this property is also useful for the analysis of the risk of complex portfolios, see Section 12.4). The coefficient of tail dependence can then be computed to be:

$$\rho_{t,ij}^+ = \sum_{a/v_{a,i}v_{a,j}>0} A_a^\mu \min \left(\frac{|v_{a,i}|^\mu}{\sum_b |v_{b,i}|^\mu A_b^\mu}, \frac{|v_{a,j}|^\mu}{\sum_b |v_{b,j}|^\mu A_b^\mu} \right). \tag{11.14}$$

Note that $0 \le \rho_{t,ij}^+ \le 1$ from the above formula. In the simple case of a one-factor model given by Eq. (11.3), where both the market and residual have a power-law distribution and $\beta_i > 0$, the coefficient of tail dependence between asset i and the market can be computed using the above formula, and reads:[†]

$$\rho_{t,im}^\pm = \frac{\beta_i^\mu A_m^\mu}{\beta_i^\mu A_m^\mu + A_{\epsilon_i}^\mu}. \tag{11.15}$$

As intuitively expected, $\rho_{t,im}^\pm$ grows from zero to one as β_i increases from zero to infinity.

Another case of interest is when the η_i, η_j are multivariate Student variables. As explained in Section 9.2.5, one can obtain such variables by first choosing two random Gaussian variables ξ_i, ξ_j, of unit variance and correlation coefficient ρ, and multiplying both of them by a common factor $f = \sqrt{\mu/s}$, where s is a positive random variable with a Gamma distribution. The coefficient of tail dependence in this case can be computed, but the final formula is not very helpful. A plot of $\rho_{t,ij}^+$ as a function of ρ for $\mu = 4$ is shown in Fig. 11.4.

On a more practical side, one might wonder whether the tail dependence coefficient and the usual correlation coefficient really grasp different information about a given pair of assets. To say it bluntly: is the tail dependence useful in real life?

To compute an empirical tail dependence, one needs to set a value for p. This value should be small since in theory we should let p tend towards zero. On the other hand the smaller the value of p, the fewer the events that determine $\rho_{t,ij}^\pm$ and therefore the more noise in its determination. The relative standard error on $\rho_{t,ij}^\pm$ is given by $1/\sqrt{\rho_{t,ij}^\pm pN}$, where N is the number of data points (days) used. This is found to be 28% for the values given below, which is much larger than the relative error on the usual correlation coefficient: $1/\sqrt{N}$ (2.3%). Note also that the possible values of $\rho_{t,ij}^\pm$ are discrete and come in increments of $1/pN$.

We have computed $\rho_{t,ij}^\pm$ and the standard correlation ρ_{ij} for a set of 388 large cap. US stocks using daily data from Jan. 1993 to June 2000. Figure 11.5 shows a scatter plot of $\rho_{t,ij}^-$

[†] This formula was also obtained in Malevergne & Sornette (2002b).

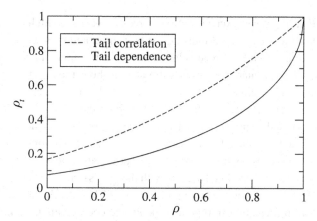

Fig. 11.4. Tail dependence coefficient ρ_t^+ and tail correlation coefficient ρ_t as a function of the correlation coefficient ρ of the two Gaussian variables used to generate a bivariate Student distribution with $\mu = 4$. Note that these coefficients are non zero even if $\rho = 0$, since the correlation is induced by the strong common 'volatility' factor.

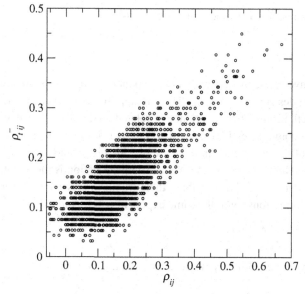

Fig. 11.5. Scatter plot of $\rho_{t,ij}^-$ versus the usual correlation ρ_{ij}. The empirical $\rho_{t,ij}^-$ was defined using $p = 5\%$. The dataset consists of 7.5 years of daily data for 388 US large cap. stocks. The discreteness of $\rho_{t,ij}^-$ arises from the empirical determination of probabilities, the minimum difference between two values is given by $1/(0.05T) \approx 0.01$.

vs. ρ_{ij}. We notice that the two quantities are very correlated ($\rho = 76\%$) with a fairly linear relationship between them. On our dataset, the average $\rho_{t,ij}^-$ is 0.154 while the average ρ_{ij} is 0.168. Results for $\rho_{t,ij}^+$ (not shown) were very similar, with a lower average value (0.123).[†]

[†] The difference between $\langle \rho_t^+ \rangle$ and $\langle \rho_t^- \rangle$ is a manifestation of the large skewness of the market returns.

Note however that $\rho^-_{t,ij}$ never exceeds 0.45, whereas ρ_{ij} can be as high as 0.7. Interestingly, from that point of view, extreme events appear *less* correlated than naively expected.

Are the deviations from the linear relationship seen in Figure 11.5 signal or noise? In other words are the two quantities noisy estimates of the same thing or are they carrying different information? To answer that question one can divide the dataset between odd days and even days and correlate the coefficients obtained on the two non-overlapping datasets. The first finding is that the empirical tail dependences are much more noisy than the standard correlation coefficients. Values for the same pair of stocks from non-overlapping datasets are correlated to a mere 34% while the standard coefficients have an 80% correlation. More interestingly the ρ_{ij}'s are correlated at 54% with the $\rho^-_{t,ij}$'s computed on a different dataset, i.e. more than the tail dependence among themselves! This clearly shows that on real data the potential extra information in the tail dependence coefficients is completely obscured by measurement noise.

11.1.5 Tail covariance (*)

Let us again assume a linear model of correlations:

$$\eta_i = \sum_{a=1}^{M} v_{a,i} e_a, \tag{11.16}$$

where the e_a have a distribution with a tail exponent μ and, for simplicity, equal left and right tail amplitudes $A_a^+ = A_a^- = A_a$. The tail amplitude of η_i is denoted, as usual, as A_i.

The usual definition of the covariance is related to the average value of $\eta_i \eta_j$, but this quantity can in some cases be divergent (i.e. when $\mu < 2$). In order to find a sensible generalization of the covariance in this case, the idea is to study directly the behaviour of the tail of the product variable $\pi_{ij} = \eta_i \eta_j$.[†] The justification is the following: if e_a and e_b are two independent power-law variables, their product π_{ab} is also a power-law variable (up to logarithmic corrections) with the same exponent μ (cf. Appendix C):

$$P(\pi_{ab}) \underset{\pi_{ab} \to \infty}{\simeq} \frac{\mu^2 \left(A_a^\mu A_b^\mu \right) \log(|\pi_{ab}|)}{|\pi_{ab}|^{1+\mu}} \qquad (a \neq b). \tag{11.17}$$

On the contrary, the quantity π_{aa} is distributed as a power-law *with an exponent* $\mu/2$ (cf. Appendix C):

$$P(\pi_{aa}) \underset{\pi_{aa} \to \infty}{\simeq} \frac{\mu A_a^\mu}{2|\pi_{aa}|^{1+\frac{\mu}{2}}}. \tag{11.18}$$

Hence, the variable π_{ij} gets both 'non-diagonal' contributions π_{ab} and 'diagonal' ones π_{aa}. For $\pi_{ij} \to \infty$, however, only the latter survive. Therefore π_{ij} is a power-law variable of exponent $\mu/2$, with a tail amplitude that we will note $A_{ij}^{\mu/2}$, and an asymmetry coefficient β_{ij} (see Section 1.8). Using the additivity of the tail amplitudes, and the results of Appendix

[†] Other generalizations of the covariance have been proposed in the context of Lévy processes, such as the 'covariation' (Samorodnitsky & Taqqu, 1994).

C, one finds:

$$A_{ij}^{\mu/2} = \sum_{a=1}^{M} |v_{a,i} v_{a,j}|^{\frac{\mu}{2}} A_a^\mu,$$

(11.19)

and

$$C_{t,ij} \equiv \beta_{ij} A_{ij}^{\mu/2} = \sum_{a=1}^{M} \mathrm{sign}(v_{a,i} v_{a,j}) |v_{a,i} v_{a,j}|^{\frac{\mu}{2}} A_a^\mu.$$

(11.20)

In the limit $\mu = 2$, one sees that the quantity $C_{t,ij}$ reduces to the standard covariance, provided one identifies A_a^2 with the variance of the explicative factors, as should be. This suggests that $C_{t,ij}$ is the suitable generalization of the covariance for power-law variables, which is constructed using extreme events only. The expression Eq. (11.20) furthermore shows that the matrix $O_{ia} = \mathrm{sign}(v_{a,i})|v_{a,i}|^{\mu/2}$ allows one to diagonalize the 'tail covariance matrix' $C_{t,ij}$;[†] its eigenvalues are given by the A_a^μ's. Correspondingly, a natural generalization of the correlation coefficient to tail events is:

$$\rho_{t,ij} = \frac{\beta_{ij} A_{ij}^{\mu/2}}{\sqrt{A_i^\mu A_j^\mu}}.$$

(11.21)

This definition is similar to, but different from, the tail dependence coefficient of the previous section. For example, in the case of the linear model studied here, one finds:

$$\rho_{t,ij} = \frac{\sum_{a=1}^{M} \mathrm{sign}(v_{a,i} v_{a,j}) |v_{a,i} v_{a,j}|^{\frac{\mu}{2}} A_a^\mu}{\sqrt{\sum_{a=1}^{M} |v_{a,i}|^\mu A_a^\mu} \sqrt{\sum_{a=1}^{M} |v_{a,j}|^\mu A_a^\mu}},$$

(11.22)

to be compared with Eq. (11.14). In the case of the simple one factor model, one finds:

$$\rho_{t,im} = \frac{\beta_i^{\mu/2} A_m^{\mu/2}}{\sqrt{\beta_i^\mu A_m^\mu + A_{\epsilon_i}^\mu}}.$$

(11.23)

One can also study the tail correlation coefficient of two correlated Student variables. The result is shown in Figure 11.4 for $\mu = 4$ and compared to the tail dependence coefficient.

In summary, the tail covariance matrix $C_{t,ij}$ can be obtained by studying the asymptotic tail of the product variable $\eta_i \eta_j$, which is a power-law variable of exponent $\mu/2$. The product of its tail amplitude and of its asymmetry coefficient is the natural definition of the tail covariance $C_{t,ij}$. This quantity will be useful in the context of portfolio optimization.

11.2 Variety and conditional statistics of the residuals

11.2.1 The variety

On any given day, every stock of the S&P 500 performs differently. The market return $\eta_m(t)$ is the average return over the population of stocks. A measure of the heterogeneity of the performance of the different stocks on a given day can be defined as the standard

[†] Note that $O_{ia} = v_{a,i}$ for $\mu = 2$, as it should.

deviation of the returns over the population of stocks, that we will call the **variety** $V(t)$:

$$\eta_m(t) = \frac{1}{M} \sum_{i=1}^{M} \eta_i(t); \qquad V(t) = \sqrt{\frac{1}{M} \sum_{i=1}^{M} (\eta_i(t) - \eta_m(t))^2}. \qquad (11.24)$$

Suppose the market return is $\eta_m(t) = 3\%$. If the variety is, say, 0.1%, then most stocks have indeed made between 2.9% and 3.1%. But if the variety is 10%, then stocks followed rather different trends during the day and their average happened to be positive. The variety is not the volatility of the index. The latter is a measure of the temporal fluctuations of the market return η_m, whereas the variety measures the instantaneous heterogeneity of stocks. Consider a day where the market has gone down 5% with a variety of 0.1% – that is, all stocks have gone down by nearly 5%. This is a very volatile day, but with a low variety. Note that low variety means that it is hard to diversify: all stocks behave the same way.

The intuition is however that there should be a positive correlation between volatility and variety: when the market makes big swings, stocks are expected to be all over the place. This is indeed observed empirically: for example, the correlation coefficient between $V(t)$ and $|\eta_m|$, which is a simple proxy for the volatility, is 0.68. A study of the temporal correlations of the variety shows a similar long-ranged dependence as the one discussed in details in the case of the volatility. More precisely, the variety variogram $\langle (V(t + \tau) - V(t))^2 \rangle$ can also be fitted as an inverse power-law of τ with a small exponent, as in Section 7.2.2.

11.2.2 The variety in the one-factor model

A theoretical relation between variety and market average return can be obtained within the framework of the one-factor model, that suggests that the variety increases when the market volatility increases. We assume that Eq. (11.3) holds with constant (in time) β_i's. In the standard one-factor model the distributions of η_m and ϵ_i are chosen to be Gaussian with constant variances. We do not make this assumption here and let these distributions be completely general including possible volatility fluctuations.

In the study of the properties of the one-factor model it is useful to consider the variety $v(t)$ of the residuals, defined as

$$v^2(t) = \frac{1}{M} \sum_{i=1}^{M} [\epsilon_i(t)]^2. \qquad (11.25)$$

Under the above assumptions the relation between the variety and the market average return is well approximated by (see below):

$$V^2(t) \approx v^2(t) + \Delta\beta^2 \, \eta_m^2(t), \qquad (11.26)$$

where $\Delta\beta^2$ is the variance of the β's.

If the number of stocks M is large, then up to terms of order $1/\sqrt{M}$, Eq. (11.26) indeed holds. Summing Eq. (11.3) over $i = 1, \ldots, M$ one finds:

$$\eta_m(t) = \eta_m(t) \frac{1}{M} \sum_{i=1}^{M} \beta_i + \frac{1}{M} \sum_{i=1}^{M} \epsilon_i(t) \qquad (11.27)$$

Since for a given day t the idiosyncratic factors are uncorrelated from stock to stock, the second term on the right hand side is of order $1/\sqrt{M}$, and can thus be neglected in a first approximation giving

$$\overline{\beta} = 1, \qquad (11.28)$$

where $\overline{\beta} \equiv \sum_{i=1}^{M} \beta_i / M$. Now square Eq. (11.3) and sum again over $i = 1, \ldots, M$. This leads to:

$$\frac{1}{M} \sum_{i=1}^{M} \eta_i^2(t) = \eta_m^2(t) \frac{1}{M} \sum_{i=1}^{M} \beta_i^2 + \frac{1}{M} \sum_{i=1}^{M} \epsilon_i^2(t) + 2 \eta_m(t) \frac{1}{M} \sum_{i=1}^{M} \beta_i \epsilon_i(t) \qquad (11.29)$$

Under the assumption that $\epsilon_i(t)$ and β_i are uncorrelated the last term can be neglected and the variety $V(t)$, using (11.24), is given by Eq. (11.26).

Therefore, even if the idiosyncratic variety v is constant, Eq. (11.26) predicts an increase of the variety with η_m^2, which is a proxy of the market volatility. Because $\Delta\beta^2 \approx 0.05$ is small, however, the increase predicted by this simple model is too small to account for the empirical data. As we shall now discuss, the effect is enhanced by the fact that v itself increases with the market volatility.

11.2.3 Conditional variety of the residuals

In its simplest version, the one-factor model assumes that the idiosyncratic part ϵ_i is *independent* of the market return. In this case, the variety of idiosyncratic terms $v(t)$ is constant in time and independent from η_m. In Fig. (11.6), we show the variety of idiosyncratic terms as a function of the market return. In contrast with these predictions, the empirical results show that a significant correlation between $v(t)$ and $\eta_m(t)$ indeed exists. The degree of correlation is in fact different for positive and negative values of the market average. In fact, the best linear least-squares fit between $v(t)$ and $\eta_m(t)$ provides different slopes when the fit is performed for positive (slope $+0.55$) or negative (slope -0.30) value of the market average. Therefore, from Eq. (11.26) we find that the increase of variety in highly volatile periods is stronger than what is expected from the simplest one-factor model, although not as strong for negative (crashes) than it is for positive (rally) days.

The fact that the idiosyncratic variety $v(t)$ increases when the market return is large means that Eq. (11.4) grows less rapidly with $|\eta_m|$ than anticipated by the simplest one factor model. This allows one to understand qualitatively the effect shown in Fig. 11.1.

Hence, as anticipated in Section 9.2.4, one of the most serious drawback of linear correlation models such as the one-factor model is their inability to account for a common volatility factor.

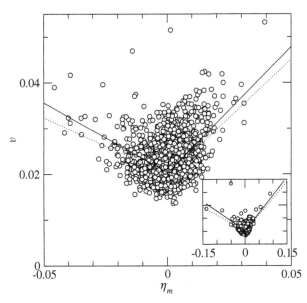

Fig. 11.6. Daily variety v of the residuals terms as a function of the market return η_m of the 1071 NYSE stocks continuously traded from January 1987 to December 1998. Each circle refers to one trading day of the investigated period. Main panel: trading days with η_m belonging to the interval $[-0.05, 0.05]$. Inset: the whole data set including five most extreme days. The two solid lines are linear fits over all days of positive (right line) and negative (left line) market average. The dotted lines are 'robust' linear fits. (Courtesy of F. Lillo and R. Mantegna.)

11.2.4 Conditional skewness of the residuals

Therefore, the idiosyncrasies are by construction uncorrelated, but not independent of the market. This shows up in the variety, does it also appear in different quantities? A natural question is: what fraction f of stocks did actually better than the market? A balanced market would have $f = 50\%$. If f is larger than 50%, then the majority of the stocks beat the market, but a few ones lagging behind rather badly, and vice versa. A closely related measure is the **asymmetry** \mathcal{A}, defined as $\mathcal{A}(t) = \eta_m(t) - \eta^*(t)$, where the median η^* is, by definition, the return such that 50% of the stocks are above, 50% below. If f is larger than 50%, then the median is larger than the average, and vice versa. Is the asymmetry \mathcal{A} also correlated with the market factor? Figure 11.7 shows that it is indeed the case: large positive days show a positive skewness in the distribution of returns – that is, a few stocks do exceptionally well – whereas large negative days show the opposite behaviour. In the figure each day is represented by a circle and all the circles cluster in a pattern which has a sigmoidal shape. The asymmetrical behaviour observed during two extreme market events is shown in the insets of Figure 11.7. This empirical observation cannot be explained by a one-factor model, which predicts that \mathcal{A} is a constant. Intuitively, one possible explanation of this anomalous skewness (and a corresponding increase of variety) might be related to the existence of different sectors which strongly separate from each other during volatile days.

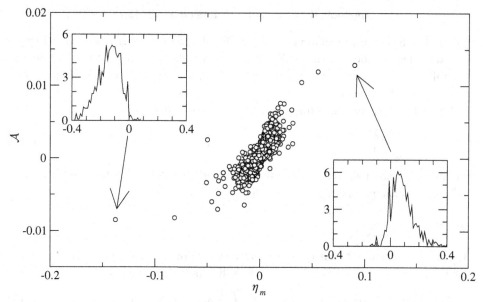

Fig. 11.7. Daily asymmetry \mathcal{A} of the probability density function of daily returns of a set of 1071 NYSE stocks continuously traded from January 1987 to December 1998 as a function of the market return η_m. Each circle refers to one trading day of the investigated period. Insets: the probability density function of daily returns observed for the two most extreme market days, both in October 1987. (Courtesy of F. Lillo and R. Mantegna).

11.3 Summary

- Many indicators can be devised to measure correlations specific to extreme events. However, several of them, such as exceedance correlations, are misleading since their increase as one probes more extreme events does not reflect a genuine increase of correlations, but simply reflects the existence of volatility fluctuations and fat tails.
- Tail dependence coefficients measure the propensity of extreme events to occur simultaneously, and can be of interest for risk management purposes. However, the empirical determination of these coefficients is quite noisy and they carry less information than the standard correlation coefficient.
- The variety is a measure of the dispersion of the returns of all the stocks of a given market on a given day. The variety is strongly correlated with the amplitude of the market return, an effect that would be absent in a simple (linear) factor model. This shows that the variance of the residuals is strongly correlated with the market return. Similarly, the asymmetry of the residuals is also strongly correlated with the market return.

11.4 Appendix C: some useful results on power-law variables

Let us assume that the random variable X is distributed as a power-law, with an exponent μ and a tail amplitude A_X^μ. The random variable λX is then also distributed as a power-law, with the same exponent μ and a tail amplitude equal to $\lambda^\mu A_X^\mu$. This can easily be found using the rule for changing variables in probability densities, recalled in Section 1.2.1: if y is a monotonic function of x, then $P(y)\,dy = P(x)\,dx$. Therefore:

$$P(\lambda x) = \frac{1}{\lambda} \frac{\mu A_X^\mu \lambda^{1+\mu}}{(\lambda x)^{1+\mu}}. \tag{11.30}$$

For the same reason, the random variable X^2 is distributed as a power-law, with an exponent $\mu/2$ and a tail amplitude $(A_X^2)^{\mu/2}$. Indeed:

$$P(x^2) = \frac{P(x)}{2x} \underset{x \to \infty}{\simeq} \frac{\mu A_X^\mu}{2(x^2)^{1+\mu/2}}. \tag{11.31}$$

On the other hand, if X and Y are two independent power-law random variables with the same exponent μ, the variable $Z = X \times Y$ is distributed according to:

$$P(z) = \int \int P(x)P(y)\delta(xy - z)\,dx\,dy = \int \frac{1}{x}P(x)P\left(y = \frac{z}{x}\right)dx. \tag{11.32}$$

When z is large, one can replace $P(y = z/x)$ by $\mu A_Y^\mu x^{1+\mu}/z^{1+\mu}$ as long as $x \ll z$. Hence, in the limit of large z's:

$$P(z) \simeq \int^{az} x^\mu P(x) \frac{\mu A_Y^\mu}{z^{1+\mu}}\,dx \tag{11.33}$$

where a is a certain constant. Now, since the integral x diverges logarithmically for large x's (since $P(x) \propto x^{-1-\mu}$), one can, to leading order, replace $P(x)$ in the integral by its asymptotic behaviour, and find:

$$P(z) \simeq \frac{\mu^2 A_X^\mu A_Y^\mu}{z^{1+\mu}} \log z. \tag{11.34}$$

- FURTHER READING

P. Embrechts, C. Klüppelberg, T. Mikosch, *Modelling Extremal Events for Insurance and Finance*, Springer-Verlag, Berlin, 1997.

G. Samorodnitsky, M. S. Taqqu, *Stable Non-Gaussian Random Processes*, Chapman & Hall, New York, 1994.

- SPECIALIZED PAPERS

A. Ang, G. Bekaert, *International Asset Allocation with time varying correlations*, NBER working paper (1999).

A. Ang, G. Bekaert, *International Asset Allocation with Regime Shifts*, working paper (2000).

G. Bekaert, G. Wu, *Asymmetric volatility and risk in equity markets*, The Review of Financial Studies, **13**, 1 (2000).

P. Embrechts, A. J. Mc Neil, D. Straumann, *Correlation and Dependency in Risk Management: Properties and Pitfalls*, in *Value at risk and Beyond*, M. Dempster Edt., Cambridge University Press, 2001.

B. H. Boyer, M. S. Gibson, M. Lauretan, *Pitfalls in tests for changes in correlations*, international Finance Discussion paper 597, Board of the Governors of the Federal Reserve (1988).

C. B. Erb, C. R. Harvey, T. E. Viskanta, *Forecasting International correlations*, Financial Analysts Journal, **50**, 32 (1994).

P. Cizeau, M. Potters, J.-P. Bouchaud, *Correlation structure of extreme stock returns*, Quantitative Finance, **1**, 217 (2001).

W. L. Lin, R. F. Engle, T. Ito, *Do bulls and bears move across borders? International transmission of stock returns and volatility*, The review of Financial Studies, **7**, 507 (1994).

F. Lillo, R. N. Mantegna, *Symmetry alteration of ensemble return distribution in crash and rally days of financial markets*, e-print cond-mat/0002438 (2000).

F. Lillo, R. N. Mantegna, J.-P. Bouchaud, M. Potters, *Introducing variety in risk management*, Wilmott Magazine, (December 2002).

F. Longin, B. Solnik, *Correlation structure of international equity markets during extremely volatile periods*, Journal of Finance, **LVI**, 649 (2001).

Y. Malevergne, D. Sornette, *Tail Dependence of Factor Models*, e-print cond-mat/0202356; *Minimizing Extremes*, Risk, **15**, 129 (2002).

Y. Malevergne, D. Sornette, *Testing the Gaussian copula hypothesis for financial assets dependence*, Quantitative Finance, **3**, 231–250 (2003).

B. Solnik, C. Boucrelle, Y. Le Fur, *International market correlations and volatility*, Financial Analysts Journal, **52**, 17 (1996).

C. Starica, *Multivariate extremes for models with constant conditional correlations*, Journal of Empirical Finance, **6**, 515 (1999).

- SOME RECENT READS

Y. Malevergne, D. Sornette, *Extreme Financial Risks: From Dependence to Risk Management*, Springer, 2005.

12

Optimal portfolios

Il n'est plus grande folie que de placer son salut dans l'incertitude.[†]

<div align="right">(Madame de Sévigné, Lettres.)</div>

12.1 Portfolios of uncorrelated assets

The aim of this section is to explain, first in the very simple case where all assets that can be mixed in a portfolio are uncorrelated, how the trade-off between risk and return can be dealt with. (The case where some correlations between the asset fluctuations exist will be considered in the next section.) One thus considers a set of M different risky assets X_i, $i = 1, \ldots, M$ and one risk-less asset X_0. The quantity of asset i in the portfolio is n_i, and its present value is x_i^0. If the total wealth to be invested in the portfolio is W, then the n_i's are constrained to be such that $\sum_{i=0}^{M} n_i x_i^0 = W$. We shall rather use the weight of asset i in the portfolio, defined as: $p_i = n_i x_i^0 / W$, which therefore must be normalized to one: $\sum_{i=0}^{M} p_i = 1$. The p_i's can be negative (short positions). The value of the portfolio at time T is given by: $S = \sum_{i=0}^{M} n_i x_i(T) = W \sum_{i=0}^{M} p_i x_i(T)/x_i^0$. In the following, we will set the initial wealth W to 1, and redefine each asset i in such a way that all initial prices are equal to $x_i^0 = 1$. (Therefore, the all returns that we shall consider below are relative, rather than absolute, returns.)

One furthermore *assumes* that the average return m_i is known. A possibility to determine the value of the m_i's is to argue that past returns can be used as estimators of future returns, i.e. that time series are to some extent stationary. However, this is very far from the truth: the life of a company (in particular high-tech ones) is very clearly non-stationary; a whole sector of activity can be booming or collapsing, depending upon global factors, not graspable within a purely statistical framework. Furthermore, the markets themselves evolve with time, and it is clear that some statistical parameters do depend on time, and have significantly shifted over the past 20 years. This means that the empirical determination of the average return is difficult: volatilities are such that at least several years are needed to obtain a reasonable signal-to-noise ratio, this time must indeed be large compared to the 'security time' \hat{T} defined in Chapter 10. But as discussed above, several years is also the time scale over which the intrinsically non-stationary nature of the markets starts to be important.

[†] Nothing is more foolish than betting on uncertainty for salvation.

One should thus rather understand m_i as an 'expected' (or anticipated) future return, which includes some extra information (or intuition) available to the investor. These m_i's can therefore vary from one investor to the next. The relevant question is then to determine the composition of an optimal portfolio compatible with the information contained in the knowledge of the different m_i's.

The determination of the risk parameters is *a priori* subject to the same *caveat*. We have actually seen in Chapter 9 that the empirical determination of the correlation matrices contains a large amount of noise, which blur the true information. However, the statistical nature of the *fluctuations* seems to be more robust in time than the average returns. Since the volatility is correlated in time, the analysis of past price changes distributions is, to a certain extent, predictive for future price fluctuations. However, it sometimes happens that the correlations between two assets change abruptly.

12.1.1 Uncorrelated Gaussian assets

Let us suppose that the variation of the value of the ith asset X_i over the time interval T is Gaussian, centred around $m_i T$ and of variance $\sigma_i^2 T$. The portfolio $\mathbf{p} = \{p_0, p_1, \ldots, p_M\}$ as a whole also obeys Gaussian statistics (since the Gaussian is stable). The average return m_p of the portfolio is given by:

$$m_p = \sum_{i=0}^{M} p_i m_i = m_0 + \sum_{i=1}^{M} p_i (m_i - m_0), \tag{12.1}$$

where we have used the constraint $\sum_{i=0}^{M} p_i = 1$ to introduce the excess return $m_i - m_0$, as compared to the risk-free asset ($i = 0$). If the X_i's are all independent, the total variance of the portfolio is given by:

$$\sigma_p^2 = \sum_{i=1}^{M} p_i^2 \sigma_i^2, \tag{12.2}$$

(since σ_0^2 is zero by assumption). If one tries to minimize the variance without any constraint on the average return m_p, one obviously finds the trivial solution where all the weight is concentrated on the risk-free asset:

$$p_i = 0 \quad (i \neq 0); \quad p_0 = 1. \tag{12.3}$$

On the contrary, the maximization of the return without any risk constraint but such that $p_i \geq 0$, leads to a full concentration of the portfolio on the asset with the highest return.

More realistically, one can look for a trade-off between risk and return, by imposing a certain average return m_p, and by looking for the less risky portfolio (for Gaussian assets, risk and variance are identical). This can be achieved by introducing a Lagrange multiplier in order to enforce the constraint on the average return:

$$\left. \frac{\partial \left(\sigma_p^2 - \zeta m_p \right)}{\partial p_i} \right|_{p_i = p_i^*} = 0 \qquad (i \neq 0), \tag{12.4}$$

while the weight of the risk-free asset p_0^* is determined *via* the equation $\sum_i p_i^* = 1$. The value of ζ is ultimately fixed such that the average value of the return is precisely m_p. Therefore:

$$2p_i^* \sigma_i^2 = \zeta(m_i - m_0),\tag{12.5}$$

and the equation for ζ:

$$m_p - m_0 = \frac{\zeta}{2} \sum_{i=1}^{M} \frac{(m_i - m_0)^2}{\sigma_i^2}.\tag{12.6}$$

The variance of this optimal portfolio is therefore given by:

$$\sigma_p^{2*} = \frac{\zeta^2}{4} \sum_{i=1}^{M} \frac{(m_i - m_0)^2}{\sigma_i^2}.\tag{12.7}$$

The case $\zeta = 0$ corresponds to the risk-free portfolio $p_0^* = 1$. The set of all optimal portfolios is thus described by the parameter ζ, and defines a *parabola* in the σ_p^2, m_p plane (compare the last two equations above, and see Fig. 12.1). This line is called the 'efficient frontier'; all portfolios must lie below this line. Finally, the Lagrange multiplier ζ itself has a direct interpretation: Equation (10.30) tells us that the worst drawdown is of order $\sigma_p^{2*}/2m_p$, which is, using the above equations, equal to $\zeta/4$.

The case where the portfolio only contains risky assets (i.e. $p_0 \equiv 0$) can be treated in a similar fashion. One introduces a second Lagrange multiplier ζ' to deal with the normalization constraint $\sum_{i=1}^{M} p_i = 1$. Therefore, one finds:

$$p_i^* = \frac{\zeta m_i + \zeta'}{2\sigma_i^2}.\tag{12.8}$$

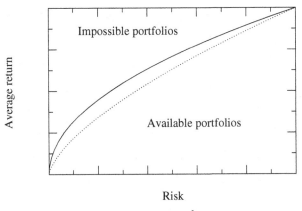

Fig. 12.1. 'Efficient frontier' in the return/risk plane m_p, σ_p^2. In the absence of constraints, this line is a parabola (dark line). If some constraints are imposed (for example, that all the weights p_i should be positive), the boundary moves downwards (dotted line).

The least risky portfolio corresponds to the one such that $\zeta = 0$ (no constraint on the average return):

$$p_i^* = \frac{1}{Z\sigma_i^2} \qquad Z = \sum_{j=1}^{M} \frac{1}{\sigma_j^2}. \tag{12.9}$$

Its total variance is given by $\sigma_p^{2*} = 1/Z$. If all the σ_i^2's are of the same order of magnitude, one has $Z \sim M/\sigma^2$; therefore, one finds the result, expected from the CLT, that the variance of the portfolio is M times smaller than the variance of the individual assets.

In practice, one often adds extra constraints to the weights p_i^* in the form of linear inequalities, such as $p_i^* \geq 0$ (no short positions). The solution is then more involved, but is still unique. Geometrically, this amounts to looking for the restriction of a paraboloid to an hyperplane, which remains a paraboloid. The efficient border is then shifted downwards (Fig. 12.1). A much richer case is when the constraint is *non-linear* – see Section 12.2.2 below.

Effective asset number in a portfolio

It is useful to introduce an objective way to measure the diversification, or the asset concentration, in a given portfolio. Once such an indicator is available, one can actually use it as a constraint to construct portfolios with a minimum degree of diversification. Consider the quantity Y_2 defined as:

$$Y_2 = \sum_{i=1}^{M} (p_i^*)^2. \tag{12.10}$$

If a subset $M' \leq M$ of all p_i^* are equal to $1/M'$, while the others are zero, one finds $Y_2 = 1/M'$. More generally, Y_2 represents the average weight of an asset in the portfolio, since it is constructed as the average of p_i^* itself. It is thus natural to define the 'effective' number of assets in the portfolio as $M_{\mathrm{eff}} = 1/Y_2$. In order to avoid an overconcentration of the portfolio on very few assets (a problem often encountered in practice), one can look for the optimal portfolio with a given value for Y_2. This amounts to introducing another Lagrange multiplier ζ'', associated to Y_2. The equation for p_i^* then becomes:

$$p_i^* = \frac{\zeta m_i + \zeta'}{2(\sigma_i^2 + \zeta'')}. \tag{12.11}$$

An example of the modified efficient border is given in Figure 12.2.

More generally, one could have considered the quantity Y_q defined as:

$$Y_q = \sum_{i=1}^{M} (p_i^*)^q, \tag{12.12}$$

and used it to define the effective number of assets via $Y_q = M_{\mathrm{eff}}^{1-q}$. It is interesting to note that the quantity of missing information (or entropy) \mathcal{I} associated to the very choice of the

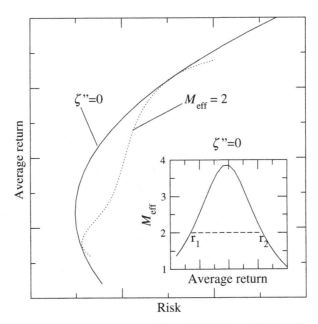

Fig. 12.2. Example of a standard efficient border $\zeta'' = 0$ (thick line) with four risky assets. If one imposes that the effective number of assets is equal to 2, one finds the sub-efficient border drawn in dotted line, which touches the efficient border at r_1, r_2. The inset shows the effective asset number of the unconstrained optimal portfolio ($\zeta'' = 0$) as a function of average return. The optimal portfolio satisfying $M_{\text{eff}} \geq 2$ is therefore given by the standard portfolio for returns between r_1 and r_2 and by the $M_{\text{eff}} = 2$ portfolios otherwise.

p_i^*'s is related to Y_q when $q \rightarrow 1$. Indeed, one has:

$$\mathcal{I} = -\sum_{i=1}^{M} p_i^* \log p_i^* \equiv -\left.\frac{\partial Y_q}{\partial q}\right|_{q=1}. \tag{12.13}$$

Approximating Y_q as a function of q by a straight line thus leads to $\mathcal{I} \approx -Y_2$.

12.1.2 Uncorrelated 'power-law' assets

As we have already underlined, tails of distributions are often non-Gaussian. In this case, the minimization of the variance is not necessarily equivalent to an optimal control of the large fluctuations. The case where these distribution tails are power-laws is interesting because one can then explicitly solve the problem of the minimization of the value-at-risk of the full portfolio. Let us thus assume that the fluctuations of each asset X_i are described, in the region of large losses, by a probability density that decays as a power-law:

$$P_T(\eta_i) \underset{\eta_i \rightarrow -\infty}{\simeq} \frac{\mu A_i^\mu}{|\eta_i|^{1+\mu}}, \tag{12.14}$$

with an arbitrary exponent μ, restricted however to be larger than 1, so that the average return is well defined. (The distinction between the cases $\mu < 2$, for which the variance diverges, and $\mu > 2$ will be considered below.) The coefficient A_i provides an order of magnitude for the extreme losses associated with the asset i (cf. Eq. (10.16)).

As we have mentioned in Section 1.5.2, power-law tails are interesting because they are stable upon addition: the tail amplitudes A_i^μ (that generalize the variance) simply add to describe the far-tail of the distribution of the sum. Using the results of Appendix C, one can show that if the asset X_i is characterized by a tail amplitude A_i^μ, the quantity $p_i X_i$ has a tail amplitude equal to $p_i^\mu A_i^\mu$. The tail amplitude of the global portfolio **p** is thus given by:

$$A_p^\mu = \sum_{i=1}^{M} p_i^\mu A_i^\mu, \tag{12.15}$$

and the probability that the loss exceeds a certain level Λ is given by $\mathcal{P} = A_p^\mu / \Lambda^\mu$. Hence, independently of the chosen loss level Λ, the minimization of the loss probability \mathcal{P} requires the minimization of the tail amplitude A_p^μ; the optimal portfolio is therefore independent of Λ. (This is *not* true in general: see Section 12.1.4.) The minimization of A_p^μ for a fixed average return m_p leads to the following equations (valid if $\mu > 1$):

$$\mu p_i^{*\mu-1} A_i^\mu = \zeta (m_i - m_0), \tag{12.16}$$

with an equation to fix ζ:

$$\left(\frac{\zeta}{\mu} \right)^{\frac{1}{\mu-1}} \sum_{i=1}^{M} \frac{(m_i - m_0)^{\frac{1}{\mu-1}}}{A_i^{\frac{\mu}{\mu-1}}} = m_p - m_0. \tag{12.17}$$

The optimal loss probability is then given by:

$$\mathcal{P}^* = \frac{1}{\Lambda^\mu} \left(\frac{\zeta}{\mu} \right)^{\frac{\mu}{\mu-1}} \sum_{i=1}^{M} \frac{(m_i - m_0)^{\frac{\mu}{\mu-1}}}{A_i^{\frac{\mu}{\mu-1}}}. \tag{12.18}$$

Therefore, the concept of 'efficient border' is still valid in this case: in the plane return/probability of loss, it is similar to the dotted line of Figure 12.1. Eliminating ζ from the above two equations, one finds that the shape of this line is given by $\mathcal{P}^* \propto (m_p - m_0)^\mu$. The parabola is recovered in the limit $\mu = 2$.

In the case where the risk-free asset cannot be included in the portfolio, the optimal portfolio which minimizes extreme risks with no constraint on the average return is given by:

$$p_i^* = \frac{1}{Z A_i^{\frac{\mu}{\mu-1}}} \qquad Z = \sum_{j=1}^{M} A_i^{-\frac{\mu}{\mu-1}}, \tag{12.19}$$

and the corresponding loss probability is equal to:

$$\mathcal{P}^* = \frac{1}{\Lambda^\mu} Z^{1-\mu}. \tag{12.20}$$

If all assets have comparable tail amplitudes $A_i \sim A$, one finds that $Z \sim M A^{-\mu/(\mu-1)}$. Therefore, the probability of large losses for the optimal portfolio is a factor $M^{\mu-1}$ smaller than the individual probability of loss.

> *Note again that this result is only valid if $\mu > 1$. If $\mu < 1$, one finds that the risk increases with the number of assets M. In this case, when the number of assets is increased, the probability of an unfavourable event also increases – indeed, for $\mu < 1$ this largest event is so large that it dominates over all the others. The minimization of risk in this case leads to $p_{i_{\min}} = 1$, where i_{\min} is the least risky asset, in the sense that $A_{i_{\min}}^{\mu} = \min\{A_i^{\mu}\}$.*

One should now distinguish the cases $\mu < 2$ and $\mu > 2$. Despite the fact that the asymptotic power-law behaviour is stable under addition for all values of μ, the tail is progressively 'eaten up' by the centre of the distribution for $\mu > 2$, since the CLT applies. Only when $\mu < 2$ does this tail remain untouched. We thus again recover the arguments of Section 2.3.5, already encountered when we discussed the time dependence of the VaR. One should therefore distinguish two cases: if σ_p^2 is the variance of the portfolio **p** (which is finite if $\mu > 2$), the distribution of the changes δS of the value of the portfolio **p** is approximately Gaussian if $|\delta S| \leq \sqrt{\sigma_p^2 T \log(M)}$, and becomes a power-law with a tail amplitude given by A_p^{μ} beyond this point. The question is thus whether the loss level Λ that one wishes to control is smaller or larger than this crossover value:

- If $\Lambda \ll \sqrt{\sigma_p^2 T \log(M)}$, the minimization of the VaR becomes equivalent to the minimization of the variance, and one recovers the Markowitz procedure explained in the previous paragraph in the case of Gaussian assets.
- If on the contrary $\Lambda \gg \sqrt{\sigma_p^2 T \log(M)}$, then the formulae established in the present section are valid *even when $\mu > 2$.*

Note that the growth of $\sqrt{\log(M)}$ with M is so slow that the Gaussian CLT is not of great help in the present case. The optimal portfolios in terms of the VaR are not those with the minimal variance, and vice versa.

12.1.3 'Exponential' assets

Suppose now that the distribution of price variations is a symmetric exponential around a zero mean value ($m_i = 0$):

$$P(\eta_i) = \frac{\alpha_i}{2} \exp[-\alpha_i |\eta_i|], \tag{12.21}$$

where α_i^{-1} gives the order of magnitude of the fluctuations of X_i (more precisely, $\sqrt{2}/\alpha_i$ is the RMS of the fluctuations.). The variations of the full portfolio **p**, defined as $\delta S = \sum_{i=1}^{M} p_i \eta_i$, are distributed according to:

$$P(\delta S) = \frac{1}{2\pi} \int \frac{\exp(iz\delta S)}{\prod_{i=1}^{M} \left[1 + \left(z p_i \alpha_i^{-1}\right)^2\right]} \, dz, \tag{12.22}$$

where we have used Eq. (2.17) and the fact that the Fourier transform of the exponential distribution Eq. (12.21) is given by:

$$\hat{P}(z) = \frac{1}{1 + (z\alpha^{-1})^2}. \tag{12.23}$$

Now, using the method of residues, it is simple to establish the following expression for $P(\delta S)$ (valid whenever the α_i/p_i are all different):

$$P(\delta S) = \frac{1}{2} \sum_{i=1}^{M} \frac{\alpha_i}{p_i} \frac{1}{\prod_{j \neq i}(1 - [(p_j\alpha_i)/(p_i\alpha_j)]^2)} \exp\left[-\frac{\alpha_i}{p_i}|\delta S|\right]. \tag{12.24}$$

The probability for extreme losses is thus equal to:

$$\mathcal{P}(\delta S < -\Lambda) \underset{\Lambda \to \infty}{\simeq} \frac{1}{2 \prod_{j \neq i^*}(1 - [(p_j\alpha^*)/\alpha_j]^2)} \exp[-\alpha^*\Lambda], \tag{12.25}$$

where α^* is equal to the smallest of all ratios α_i/p_i, and i^* the corresponding value of i. The order of magnitude of the extreme losses is therefore given by $1/\alpha^*$. This is then the quantity to be minimized in a value-at-risk context. This amounts to choosing the p_i's such that $\min_i\{\alpha_i/p_i\}$ is as large as possible.

This minimization problem can be solved using the following trick. One can write formally that:

$$\frac{1}{\alpha^*} = \max\left\{\frac{p_i}{\alpha_i}\right\} = \lim_{\mu \to \infty}\left(\sum_{i=1}^{M} \frac{p_i^\mu}{\alpha_i^\mu}\right)^{\frac{1}{\mu}}. \tag{12.26}$$

This equality is true because in the limit $\mu \to \infty$, the largest term of the sum dominates over all the others. (The choice of the notation μ is on purpose: see below.) For μ large but fixed, one can perform the minimization with respect to the p_i's, using a Lagrange multiplier to enforce normalization. One finds that:

$$p_i^{*\mu-1} \propto \alpha_i^\mu. \tag{12.27}$$

In the limit $\mu \to \infty$, and imposing $\sum_{i=1}^{M} p_i = 1$, one finally obtains:

$$p_i^* = \frac{\alpha_i}{\sum_{j=1}^{M} \alpha_j}, \tag{12.28}$$

which leads to $\alpha^* = \sum_{i=1}^{M} \alpha_i$. In this case, however, all α_i/p_i^* are equal to α^* and the result Eq. (12.24) must be slightly altered. However, the asymptotic exponential fall-off, governed by α^*, is still true (up to polynomial corrections). One again finds that if all the α_i's are comparable, the potential losses, measured through $1/\alpha^*$, are divided by a factor M.

Note that the optimal weights are such that $p_i^* \propto \alpha_i$ if one tries to minimize the probability of extreme losses, whereas one would have found $p_i^* \propto \alpha_i^2$ if the goal was to minimize the variance of the portfolio, corresponding to $\mu = 2$ (cf. Eq. (12.9)). In other words, this is

an explicit example where one can see that minimizing the variance actually increases the value-at-risk, and vice versa.

Formally, as we have noticed in Section 1.8, the exponential distribution corresponds to the limit $\mu \to \infty$ of a power-law distribution: an exponential decay is indeed more rapid than any power-law. Technically, we have indeed established in this section that the minimization of extreme risks in the exponential case is identical to the one obtained in the previous section in the limit $\mu \to \infty$ (see Eq. (12.19)).

12.1.4 General case: optimal portfolio and VaR (*)

In all of the cases treated above, the optimal portfolio is found to be independent of the chosen loss level Λ. For example, in the case of assets with power-law tails, the minimization of the loss probability amounts to minimizing the tail amplitude A_p^μ, independently of Λ. This property is however not true in general, and the optimal portfolio does indeed depend on the risk level Λ, or, equivalently, on the temporal horizon over which risk must be 'tamed'. Let us for example consider the case where all assets are power-law distributed, but with a tail index μ_i that depends on the asset X_i. The probability that the portfolio **p** experiences a loss greater than Λ is given, for large values of Λ, by:[†]

$$\mathcal{P}(\delta S < -\Lambda) = \sum_{i=1}^{M} p_i^{\mu_i} \frac{A_i^{\mu_i}}{\Lambda^{\mu_i}}. \tag{12.29}$$

Looking for the set of p_i's which minimizes the above expression (without constraint on the average return) then leads to:

$$p_i^* = \frac{1}{Z} \left(\frac{\Lambda^{\mu_i}}{\mu_i A_i^\mu} \right)^{\frac{1}{\mu_i-1}} \qquad Z = \sum_{i=1}^{M} \left(\frac{\Lambda^{\mu_i}}{\mu_i A_i^\mu} \right)^{\frac{1}{\mu_i-1}}. \tag{12.30}$$

This example shows that in the general case, the weights p_i^* *explicitly depend on the risk level Λ.* If all the μ_i's are equal, Λ^{μ_i} factors out and disappears from the p_i's.

Another interesting case is that of weakly non-Gaussian assets, such that the first correction to the Gaussian distribution (proportional to the kurtosis κ_i of the asset X_i) is enough to describe faithfully the non-Gaussian effects. The variance of the full portfolio is given by $\sigma_p^2 = \sum_{i=1}^{M} p_i^2 \sigma_i^2$ while the kurtosis is equal to: $\kappa_p = \sum_{i=1}^{M} p_i^4 \sigma_i^4 \kappa_i / \sigma_p^4$. The probability that the portfolio plummets by an amount larger than Λ is therefore given by:

$$\mathcal{P}(\delta S < -\Lambda) \simeq \mathcal{P}_{G>}\left(\frac{\Lambda}{\sqrt{\sigma_p^2 T}} \right) + \frac{\kappa_p}{4!} h\left(\frac{\Lambda}{\sqrt{\sigma_p^2 T}} \right), \tag{12.31}$$

where $\mathcal{P}_{G>}$ is related to the error function (cf. Section 1.6.3) and

$$h(u) = \frac{\exp(-u^2/2)}{\sqrt{2\pi}}(u^3 - 3u). \tag{12.32}$$

[†] The following expression is valid only when the *subleading corrections* to Eq. (12.14) can safely be neglected.

To first order in kurtosis, one thus finds that the optimal weights p_i^ (without fixing the average return) are given by:*

$$p_i^* = \frac{\zeta'}{2\sigma_i^2} - \frac{\kappa_i \zeta'^3}{\sigma_i^6} \tilde{h} \left(\frac{\Lambda}{\sqrt{\sigma_p^2 T}} \right), \tag{12.33}$$

where \tilde{h} is another function, positive for large arguments, and ζ' is fixed by the condition $\sum_{i=1}^M p_i = 1$. Hence, the optimal weights do depend on the risk level Λ, via the kurtosis of the distribution. Furthermore, as could be expected, the minimization of extreme risks leads to a reduction of the weights of the assets with a large kurtosis κ_i.

12.2 Portfolios of correlated assets

The aim of the previous section was to introduce, in a somewhat simplified context, the most important ideas underlying portfolio optimization, in a Gaussian world (where the variance is minimized) or in a non-Gaussian world (where the quantity of interest is the value-at-risk). In reality, the fluctuations of the different assets are often strongly correlated (or anti-correlated). For example, an increase of short-term interest rates often leads to a drop in share prices. All the stocks of the New York Stock Exchange behave, to a certain extent, similarly. These correlations of course modify completely the composition of the optimal portfolios, and actually make diversification more difficult. In a sense, the number of effectively independent assets is decreased from the true number of assets M.

12.2.1 Correlated Gaussian fluctuations

Let us first consider the case where all the fluctuations η_i of the assets X_i are Gaussian, but with arbitrary correlations (see Section 9.1.2). These correlations are described in terms of a (symmetric) correlation matrix C_{ij}, defined as:

$$C_{ij} = \langle \eta_i \eta_j \rangle - m_i m_j. \tag{12.34}$$

As mentioned in Section 9.1.2, an important property of correlated Gaussian variables is that they can be decomposed into a weighted sum of *independent* Gaussian variables e_a, of mean zero and variance equal to σ_a^2 that are the eigenvalues of \mathbf{C}:

$$\eta_i = m_i + \sum_{a=1}^M v_{ia} e_a \qquad \langle e_a e_b \rangle = \delta_{a,b} \sigma_a^2. \tag{12.35}$$

The $\{e_a\}$ are usually referred to as the **explicative factors** (or principal components) for the asset fluctuations. They sometimes have a simple economic interpretation.

The fluctuations δS of the global portfolio \mathbf{p} are then also Gaussian (since δS is a weighted sum of the Gaussian variables e_a), of mean $m_p = \sum_{i=1}^M p_i (m_i - m_0) + m_0$ and variance:

$$\sigma_p^2 = \sum_{i,j=1}^M p_i p_j C_{ij}. \tag{12.36}$$

The minimization of σ_p^2 for a fixed value of the average return m_p (and with the possibility of including the risk-free asset X_0) leads to an equation generalizing Eq. (12.5):

$$2 \sum_{j=1}^{M} C_{ij} p_j^* = \zeta (m_i - m_0), \tag{12.37}$$

which can be inverted as:

$$p_i^* = \frac{\zeta}{2} \sum_{j=1}^{M} C_{ij}^{-1} (m_j - m_0). \tag{12.38}$$

This is Markowitz's classical result (Markowitz (1959), Elton & Gruber (1995)) .

In the case where the risk-free asset is excluded, the minimum variance portfolio is given by:

$$p_i^* = \frac{1}{Z} \sum_{j=1}^{M} C_{ij}^{-1} \qquad Z = \sum_{i,j=1}^{M} C_{ij}^{-1}. \tag{12.39}$$

Actually, the decomposition, Eq. (12.35), shows that, provided one shifts to the basis where all assets are independent (through a linear combination of the original assets), all the results obtained above in the case where the correlations are absent (such as the existence of an efficient border, etc.) are still valid when correlations are present.

In the more general case of non-Gaussian assets of finite variance, the total variance of the portfolio is still given by: $\sum_{i,j=1}^{M} p_i p_j C_{ij}$, where C_{ij} is the correlation matrix. If the variance is an adequate measure of risk, the composition of the optimal portfolio is still given by Eqs. (12.38) and (12.39). The average return of the optimal portfolio is given by:

$$m_p - m_0 = \frac{\zeta}{2} \sum_{i,j=1}^{M} C_{ij}^{-1} (m_i - m_0)(m_j - m_0), \tag{12.40}$$

whereas its volatility is given by:

$$\sigma_p^2 = \frac{\zeta^2}{4} \sum_{i,j=1}^{M} C_{ij}^{-1} (m_i - m_0)(m_j - m_0). \tag{12.41}$$

Therefore, the optimal portfolios are in this case on a line $\sigma_p^2 = (\zeta/2m_p - m_0)$. Comparing with the definition given in Chapter 10, we see that the choice of ζ fixes the risk, measured as the typical drawdown Δ_0 since $\Delta_0 = \sigma_p^2/2(m_p - m_0) = \zeta/4$. Note that the optimal portfolios all have the same (maximum) Sharpe ratio $\mathcal{S} \equiv (m_p - m_0)/\sigma_p$.

Let us however again emphasize that, as discussed in Chapter 9, the empirical determination of the correlation matrix C_{ij} is difficult, in particular when one is concerned with the small eigenvalues of this matrix and their corresponding eigenvectors. Since the above formulae involve the inverse of **C**, the noise on the small eigenvalues leads to large numerical

errors in \mathbf{C}^{-1}. The ideas developed in Section 9.1.3 can actually be used in practice to re-duce the real risk of optimized portfolios. Since the eigenstates corresponding to the 'noise band' are not expected to contain real information, one should not distinguish the different eigenvalues and eigenvectors in this sector. This amounts to replacing the restriction of the empirical correlation matrix to the noise band subspace by the identity matrix with a coefficient such that the trace of the matrix is conserved (i.e. suppressing the measurement broadening due to a finite observation time). This 'cleaned' correlation matrix, where the noise has been (at least partially) removed, is then used to construct an optimal portfolio. We have implemented this idea in practice as follows. Using the same data sets as in Chapter 9, the total available period of time has been divided into two equal sub-periods. We determine the correlation matrix using the first sub-period, 'clean' it, and construct the family of op-timal portfolios and the corresponding efficient frontiers. Here we assume that the investor has perfect predictions on the future average returns m_i, i.e. we take for m_i the observed return on the next sub-period. The results are shown in Figure 12.3: one sees very clearly that using the empirical correlation matrix leads to a dramatic underestimation of the real risk, by over-investing in artificially low-risk eigenvectors. The risk of the optimized portfolio obtained using a cleaned correlation matrix is more reliable, although the real risk is always larger than the predicted one. This comes from the fact that any amount of uncertainty in the correlation matrix produces, through the very optimization procedure, a bias towards

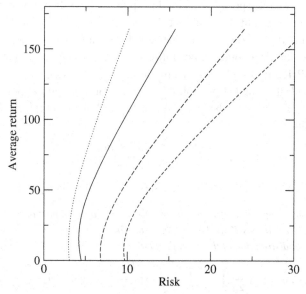

Fig. 12.3. Efficient frontiers from Markowitz optimization, in the return versus volatility plane. The leftmost dotted curve corresponds to the classical Markowitz case using the empirical correlation matrix. The rightmost short-dashed curve is the realization of the same portfolio in the second time period (the risk is underestimated by a factor of three!). The central curves (plain and long-dashed) represent the case of a cleaned correlation matrix. The realized risk is now only a factor of 1.5 larger than the predicted risk.

low-risk portfolios. This can be checked by permuting the two sub-periods: one then finds nearly identical efficient frontiers. (This is expected, since for large correlation matrices these frontiers should be self-averaging.) In other words, even if the cleaned correlation matrices are more stable in time than the empirical correlation matrices, they are not perfectly reflecting future correlations. This might be due to a combination of remaining noise and of a genuine time dependence in the structure of the meaningful correlations.

A somewhat related idea to reduce the noise in the correlation matrix is to artificially increase the amplitude of the diagonal part of \mathbf{C} compared to the off diagonal part, the logic being that off-diagonal cross correlations are not as well determined as (diagonal) volatilities. Therefore, the portfolio optimization formula will involve $\mathbf{C} + \zeta''\mathbf{1}$ instead of \mathbf{C}, where ζ'' is a parameter accounting for the uncertainty in the determination of \mathbf{C}.[†] Interestingly, the final formula for p_i^* is *identical* to that obtained when imposing a fixed effective number of assets Y_2, see Eq. (12.11). Imposing a minimal degree of diversification is therefore, not surprisingly, tantamount to distrusting the information contained in the correlation matrix.

The CAPM and its limitations

Within the above framework, all optimal portfolios are proportional to one another, that is, they only differ through the choice of the factor ζ. Since the problem is linear, this means that the linear superposition of optimal portfolios is still optimal. If all the agents on the market choose their portfolio using this optimization scheme (with the same values for the average return and the correlation coefficients – clearly quite an absurd hypothesis), then the 'market portfolio' (i.e. the one obtained by taking all assets in proportion of their market capitalization) is an optimal portfolio. This remark is at the origin of the 'CAPM' (**Capital Asset Pricing Model**), which aims at relating the average return of an asset with its covariance with the 'market portfolio'. Actually, for any optimal portfolio \mathbf{p}, one can express $m_p - m_0$ in terms of the p^*, and use Eq. (12.38) to eliminate ζ, to obtain the following equality:

$$m_i - m_0 = \beta_i[m_p - m_0] \qquad \beta_i \equiv \frac{\langle(\eta_i - m_i)(\delta S - m_p)\rangle}{\langle(\delta S - m_p)^2\rangle}. \tag{12.42}$$

The covariance coefficient β_i is often called the 'β' of asset i when \mathbf{p} is the market portfolio.

This relation is however not true for other definitions of optimal portfolios. Let us define the generalized kurtosis K_{ijkl} that measures the first correction to Gaussian statistics, from the joint distribution of the asset fluctuations:

$$P(\eta_1, \eta_2, \ldots, \eta_M) = \left(\frac{1}{2\pi}\right)^M \int \int \cdots \int \exp\left[-i\sum_j z_j(\eta_j - m_j)\right.$$

$$\left. -\frac{1}{2}\sum_{ij} z_i C_{ij} z_j + \frac{1}{4!}\sum_{ijkl} K_{ijkl} z_i z_j z_k z_l + \cdots\right]\prod_{j=1}^M dz_j. \tag{12.43}$$

[†] The argument can be formalized within the context of Bayesian estimation theory, where the true correlation matrix is a priori assumed to be distributed around the identity matrix.

If one tries to minimize the probability that the loss is greater than a certain Λ, a generalization of the calculation presented above (cf. Eq. (12.33)) leads to:

$$p_i^* = \frac{\zeta'}{2} \sum_{j=1}^{M} C_{ij}^{-1}(m_j - m_0) - \zeta'^3 \tilde{h} \left(\frac{\Lambda}{\sqrt{\sigma_p^2 T}} \right)$$

$$\times \sum_{j,k,l=1}^{M} \sum_{j',k',l'=1}^{M} K_{ijkl} C_{jj'}^{-1} C_{kk'}^{-1} C_{ll'}^{-1}(m_{j'} - m_0)(m_{k'} - m_0)(m_{l'} - m_0), \qquad (12.44)$$

where \tilde{h} is a certain positive function. The important point here is that the level of risk Λ appears explicitly in the choice of the optimal portfolio. If the different operators choose different levels of risk, their optimal portfolios are no longer related by a proportionality factor, and the above CAPM relation does not hold.

12.2.2 Optimal portfolios with non-linear constraints (*)

In some cases, the constraints on the portfolio composition are non-linear. For example, on futures markets, margin calls require that a certain amount of money is left as a deposit, whether the position is long ($p_i > 0$) or short ($p_i < 0$). One can then impose a *leverage constraint*, such that $\sum_{i=1}^{M} |p_i| = f$, where f is the fraction of wealth invested as a deposit. This constraint leads to a much more complex problem, similar to the one encountered in hard optimization problems, where an exponentially large (in M) number of quasi-degenerate solutions can be found.[†] Let us briefly sketch why this is the case. The solution can again be obtained using two Lagrange multipliers:

$$\frac{\partial}{\partial p_i} \left[\sum_{j,k=1}^{M} p_j C_{jk} p_k - \zeta \sum_{j=1}^{M} p_j m_j - \zeta' \sum_{j=1}^{M} |p_j| \right] \Bigg|_{p_i = p_i^*} = 0. \qquad (12.45)$$

Defining S_i to be the sign of p_i^*, one gets:

$$p_i^* = \zeta \sum_{j=1}^{M} C_{ij}^{-1} m_j + \zeta' \sum_{j=1}^{M} C_{ij}^{-1} S_j, \qquad (12.46)$$

where \mathbf{C}^{-1} is the matrix inverse of \mathbf{C}. Taking the sign of this last equation leads to equations for the S_i's which are identical to those defining the locally stable configurations in a spin-glass (see Mézard et al. (1987)):

$$S_i = \text{sign} \left[H_i + \zeta' \sum_{j=1}^{M} C_{ij}^{-1} S_j \right], \qquad (12.47)$$

where $H_i \equiv \zeta \sum_j^M C_{ij}^{-1} m_j$ and $\zeta' C_{ij}^{-1}$ are the analogue of the 'local magnetic field' and of the spin interaction matrix, respectively. Once the S_i's are known, the p_i^* are determined by Eq. (12.46), and ζ and ζ' are fixed, as usual, so as to satisfy the constraints. Let us consider, for the sake of simplicity, the case where one is interested in the least risky portfolio,

[†] See Galluccio et al. (1998).

which corresponds to $\zeta = 0$.[†] In this case, one sees that only $|\zeta'|$ is fixed by the constraint. Furthermore, the minimum risk is given by $\mathcal{R} = \zeta'^2 \sum_{j,k}^{M} S_j C_{jk}^{-1} S_k$, which is the analogue of the energy of the spin glass. It is now well known that if \mathbf{C} is a generic random matrix, the so called "TAP" equation (12.47), defining the metastable states of a spin glass at zero temperature, generically has an exponentially large (in M) number of solutions (see Mézard et al. (1987)). Note that, as stated above, the multiplicity of solutions is a direct consequence of the nonlinear constraint.

The above calculation counts all local optima, disregarding the associated value of the residual risk \mathcal{R}. One can also obtain the number of solutions for a given \mathcal{R}^*. In analogy with Parisi & Potters (1995), one expects this number to grow as $\exp M f(\mathcal{R})$, where $f(\mathcal{R})$ has a certain parabolic shape, which goes to zero for a certain 'minimal' risk \mathcal{R}^*. But for any small interval around \mathcal{R}^*, there will already be an exponentially large number of solutions. The most interesting feature, with particular reference to applications in economics, finance or social sciences, is that all these quasi-degenerate portfolios can be very far from each other: in other words, it can be rational to follow two totally different strategies! Furthermore, these solutions are usually 'chaotic' in the sense that a small change of the matrix \mathbf{C}, or the addition of an extra asset, completely shuffles the order of the solutions (in terms of their risk). Some solutions might even disappear, while others appear.

12.2.3 'Power-law' fluctuations – linear model (*)

The minimization of large risks also requires, as in the Gaussian case detailed above, the knowledge of the correlations between large events, and therefore an adapted measure of these correlations. Now, in the extreme case $\mu < 2$, the covariance (as the variance), is infinite. A suitable generalization of the covariance is therefore necessary. Even in the case $\mu > 2$, where this covariance is *a priori* finite, the value of this covariance is a mix of the correlations between large negative moves, large positive moves, and all the 'central' (i.e. not so large) events. Again, the definition of a 'tail covariance', directly sensitive to the large negative events, is needed, and was introduced and discussed in Section 11.1.5. In this section, we show how this tool can be used to obtain minimal Value-at-Risk portfolios when the tail of the distributions are power-laws.

Let us again assume that the correlations between the η_i's are well described by a linear model:

$$\eta_i = m_i + \sum_{a=1}^{M} v_{ia} e_a, \tag{12.48}$$

where the e_a are *independent* power-law random variables, the distribution of which is:

$$P(e_a) \underset{e_a \to \pm\infty}{\simeq} \frac{\mu A_a^{\mu}}{|e_a|^{1+\mu}}. \tag{12.49}$$

[†] The following results still hold for small enough ζ. For large ζ, the S_i will obviously be polarized by the strong 'local field' and have a well determined sign.

The fluctuations of the portfolio \mathbf{p} can then be written as:

$$\delta S \equiv \sum_{i=1}^{M} p_i \eta_i = \sum_{a=1}^{M} \left(\sum_{i=1}^{M} p_i v_{ia} \right) e_a. \tag{12.50}$$

Due to our assumption that the e_a are symmetric power-law variables, δS is also a symmetric power-law variable with a tail amplitude given by:

$$A_p^\mu = \sum_{a=1}^{M} \left| \sum_{i=1}^{M} p_i v_{ia} \right|^\mu A_a^\mu. \tag{12.51}$$

In the simple case where one tries to minimize the tail amplitude A_p^μ without any constraint on the average return, one then finds:[†]

$$\sum_{a=1}^{M} v_{ia} A_a^\mu V_a^* = \frac{\zeta'}{\mu}, \tag{12.52}$$

where the vector $V_a^* = \mathrm{sign}(\sum_{j=1}^{M} v_{ja} p_j^*) |\sum_{j=1}^{M} v_{ja} p_j^*|^{\mu-1}$. Using the results of Section 11.1.5, one sees that once the tail covariance matrix \mathbf{C}_t is known, the optimal weights p_i^* are thus determined as follows:

(i) From Eq. (11.20), one sees that the diagonalization of \mathbf{C}_t gives access to a rotation matrix O, from which one can construct the matrix $v = \mathrm{sign}(O)|O|^{2/\mu}$ (understood for each element of the matrix), and the diagonal matrix A.
(ii) The matrix vA is then inverted, and applied to the vector $(\zeta'/\mu)\vec{1}$. This gives the vector $\vec{V}^* = \zeta'/\mu (vA)^{-1}\vec{1}$.
(iii) From the vector \vec{V}^* one can then obtain the weights p_i^* by applying the matrix inverse of v^\dagger to the vector $\mathrm{sign}(V^*)|V^*|^{1/(\mu-1)}$;
(iv) The Lagrange multiplier ζ' is then determined such as $\sum_{i=1}^{M} p_i^* = 1$.

Rather symbolically, one thus has:

$$\mathbf{p} = (v^\dagger)^{-1} \left(\frac{\zeta'}{\mu} (vA)^{-1} \vec{1} \right)^{\frac{1}{\mu-1}}. \tag{12.53}$$

In the case $\mu = 2$, one recovers the previous prescription, since in that case: $vA = vD = Cv$, $(vA)^{-1} = v^{-1}C^{-1}$, and $v^\dagger = v^{-1}$, from which one gets:

$$\mathbf{p} = \frac{\zeta'}{2} vv^{-1}C^{-1}\vec{1} = \frac{\zeta'}{2} C^{-1}\vec{1}, \tag{12.54}$$

which coincides with Eq. (12.39).

The problem of the minimization of the value-at-risk in a strongly non-Gaussian world, where distributions behave asymptotically as power-laws, and in the presence of correlations

[†] If one wants to impose a non-zero average return, one should replace ζ'/μ by $\zeta'/\mu + \zeta m_i/\mu$, where ζ is determined such as to give to m_p the required value.

between tail events, can thus be solved using a procedure rather similar to the one proposed by Markowitz for Gaussian assets.

12.2.4 'Power-law' fluctuations – Student model (*)

As discussed in Section 9.2.5, another natural model of correlations that induces power-law tails is the Student multivariate model, where a stochastic (fat-tailed) volatility factor affects a set of Gaussian correlated returns. The multivariate distribution of returns is given by Eq. (9.24). In this case, it is not hard to show that the minimum Value-at-Risk portfolio is given by the Markowitz formula, Eq. (12.39). This would actually be true for any multi-variate distribution of returns that can be written as a function of the rotationally invariant combination $\sum_{i,j} \eta_i (C^{-1})_{ij} \eta_j$. In this case, any measure of risk turns out to be an increasing function of the variable $\sigma_p^2 = \sum_{ij} p_i C_{ij} p_j$. Therefore, one should, as in the Markowitz case, minimize σ_p^2.

The only difference with the Gaussian case is that the matrix \mathbf{C} appearing in Eq. (12.39) should be determined from the empirical data using the maximum likelihood formula (9.29) giving $\tilde{\mathbf{C}}$ rather than using the empirical correlation matrix.

12.3 Optimized trading

In this section, we will discuss a slightly different problem of portfolio optimization: how to dynamically optimize, as a function of time, the number of shares of a given stock that one holds in order to minimize the risk for a fixed level of return. We shall actually encounter a similar problem in the next chapter on options when the question of the optimal hedging strategy will be addressed.

We will suppose that the trader holds a certain number of shares $\phi_n(g_n)$, where ϕ depends both on the (discrete) time $t_n = n\tau$, and on the present value of the realized gains g_n. For the time being, we assume that the interest rates are negligible and that the total gains at the end of the investment period are given by:

$$g_N = \sum_{k=0}^{N-1} \phi_k(g_k) \delta x_k, \tag{12.55}$$

where $\delta x_k = x_{k+1} - x_k$ is the change of price between time k and time $k+1$. (See Sections 13.1.3 and 13.2.2 for a more detailed discussion of Eq. (12.55).) Let us define the target gain as \mathcal{G} and \mathcal{R} as the error of the final wealth:

$$\mathcal{R}^2 = \langle (g_N - \mathcal{G})^2 \rangle. \tag{12.56}$$

The question is then to find the optimal trading strategy $\phi_k^*(g_k)$, such that the risk \mathcal{R} is minimized, for a given value of \mathcal{G}.

The method to determine $\phi_k^*(g_k)$ is to work backwards in time, starting from $k = N - 1$. The optimal strategy for the last step is easy to determine, since one has to minimize:

$$\mathcal{R}^2 = \langle (g_{N-1} + \phi_{N-1} \delta x_{N-1} - \mathcal{G})^2 \rangle. \tag{12.57}$$

We will assume that the δx_k are random iid increments, of mean $m\tau$ and variance $D\tau$. Taking into account the fact that δx_{N-1} is independent of the value of g_{N-1}, the above expression reads:

$$\mathcal{R}^2 = (D\tau + m^2\tau^2)\phi_{N-1}^2 + 2(g_{N-1} - \mathcal{G})m\tau + (g_{N-1} - \mathcal{G})^2. \tag{12.58}$$

Taking the derivative of this expression with respect to ϕ_{N-1} and setting the result to zero immediately leads to:

$$\phi_{N-1}^* = \frac{m}{D + m^2\tau}(\mathcal{G} - g_{N-1}). \tag{12.59}$$

The determination of ϕ_k^* for $k < N - 1$ proceeds similarly, in a recursive way, by using the already known optimal strategy for all $\ell > k$. The result is, perhaps surprisingly, that the above result holds for all k:

$$\phi_k^*(g_k) = \frac{m}{D + m^2\tau}(\mathcal{G} - g_k). \tag{12.60}$$

The strategy is thus to invest more when one is far from the target, and reduce the investement when the gains close in the target. Let us discuss more precisely the above result by taking the continuous time limit $\tau \to 0$ and assuming that the resulting price process is a continuous time Brownian motion $X(t)$. Then the current value of the gains evolves according to:

$$dg = \phi^* dX = \frac{m}{D}(\mathcal{G} - g)(m\,dt + \sqrt{D}\,dW), \tag{12.61}$$

where $W(t)$ is a Brownian motion of drift zero and unit volatility. Since the price increment is posterior to the determination of the optimal strategy ϕ^*, the above stochastic differential equation must be interpreted in the Ito sense (see Section 3.2.1). One recognizes the equation defining a geometric Brownian walk for $\mathcal{G} - g$. With the initial condition $g(t = 0) = 0$, its solution reads after time T:

$$\frac{\mathcal{G} - g}{\mathcal{G}} = \exp\left[-\frac{3m^2}{2D}T + \frac{m}{\sqrt{D}}W(T)\right]. \tag{12.62}$$

Because $W(T) \sim \sqrt{T}$, the above equation shows that for large times $T \to \infty$, the gain g tends almost surely towards the target \mathcal{G}. The typical time for convergence is $T \sim D/m^2$, which is precisely the 'security time' \hat{T} discussed above.

However, the distribution of gains is found to be log-normal, which means that the 'negative' tail, which governs the cases where the gains turn out to be much less than the target, are much fatter than Gaussian. In particular, the qth moment of the gain difference reads:

$$\left\langle \left(\frac{\mathcal{G} - g}{\mathcal{G}}\right)^q \right\rangle = \exp\left[\frac{(q^2 - 3q)m^2}{2D}T\right]. \tag{12.63}$$

In particular, the average gain and the variance converge asymptotically to zero as $e^{-m^2 T/D}$. However, the fourth moment of $\mathcal{G} - g$, because of the fat tails of the log-normal, actually diverge as $e^{2m^2 T/D}$!

Coming back to the general form of the optimal strategy, it is interesting to compare Eq. (12.60) with the naive strategy which consists in investing a constant value $\phi_0 = \mathcal{G}/mT$, where T is the investment horizon. Clearly, the average gain of this strategy is indeed \mathcal{G}. However, the relative variance of the result is D/m^2T, which is much larger (for large T) than the exponential result found above. On the other hand, the initial optimal investment, at $T = 0$, is (for $D \gg m^2\tau$) given by $m\mathcal{G}/D$, which represents a much larger amount than the naive strategy ϕ_0 whenever $T \gg \hat{T}$, since $D\phi_0/m\mathcal{G} = \hat{T}/T$. This would appear risky to bet so heavily on the asset, but remember that our assumption is that the drift m is perfectly known. Unfortunately, the average return is never very precisely known. A way to take into account the drift uncertainty is to replace in the above formulae D by $D + (\Delta m)^2 T$, where Δm is the root mean square uncertainty. For large T, this becomes the major source of risk. Another important point is that the above strategy allows the possibility of tremendous losses for intermediate times, because these will on average be compensated by the positive trend. In practice, however, these large losses cannot be accepted.

12.4 Value-at-risk for general non-linear portfolios (*)

A very important issue for the control of risk of a complex portfolio, which involves many non-linear assets such as options, is to be able to estimate its value-at-risk reliably. This is a difficult problem, since both the non-Gaussian nature of the fluctuations of the underlying assets and the non-linearities of the price of the derivatives must be dealt with. A solution, which is very costly in terms of computation time and not very precise, is the use of Monte-Carlo simulations. We shall show in this section that in the case where the fluctuations of the 'explicative variables' are strong (a more precise statement will be made below), an approximate formula can be obtained for the value-at-risk of a general non-linear portfolio. Interestingly, this method can be seen as a correctly probabilized 'scenario' calculation (or 'stress-testing'), much used in the financial industry.

12.4.1 Outline of the method: identifying worst cases

Let us assume that the variations of the value of the portfolio can be written as a function $\delta f(e_1, e_2, \ldots, e_M)$ of a set of M independent random variables $e_a, a = 1, \ldots, M$, such that $\langle e_a \rangle = 0$ and $\langle e_a e_b \rangle = \delta_{a,b}\sigma_a^2$. The sensitivity of the portfolio to these 'explicative variables' can be measured as the derivatives of the value of the portfolio with respect to the e_a. We shall therefore introduce the Δ's and Γ's as:

$$\Delta_a = \frac{\partial f}{\partial e_a} \qquad \Gamma_{a,b} = \frac{\partial^2 f}{\partial e_a \partial e_b}. \tag{12.64}$$

We are interested in the probability for a large fluctuation δf^* of the portfolio. We will surmise that this is due to a particularly large fluctuation of one explicative factor, say $a = 1$, that we will call the dominant factor. This is not always true, and depends on the statistics of the fluctuations of the e_a. A condition for this assumption to be true will be

discussed below, and requires in particular that the tail of the dominant factor should not decrease faster than an exponential. Fortunately, this is a good assumption in financial markets.

The aim is to compute the value-at-risk of a certain portfolio, i.e. the value δf^* such that the probability that the variation of f exceeds δf^* is equal to a certain probability p: $\mathcal{P}_>(\delta f^*) = p$. Our assumption about the existence of a dominant factor means that these events correspond to a market configuration where the fluctuation δe_1 is large, whereas all other factors are relatively small. Therefore, the large variations of the portfolio can be approximated as:

$$\delta f(e_1, e_2, \ldots, e_M) = \delta f(e_1) + \sum_{a=2}^{M} \Delta_a e_a + \frac{1}{2} \sum_{a,b=2}^{M} \Gamma_{a,b} e_a e_b, \tag{12.65}$$

where $\delta f(e_1)$ is a shorthand notation for $\delta f(e_1, 0, \ldots, 0)$. Now, we use the fact that:

$$\mathcal{P}_>(\delta f^*) = \int P(e_1, e_2, \ldots, e_M)\Theta[\delta f(e_1, e_2, \ldots, e_M) - \delta f^*] \prod_{a=1}^{M} de_a, \tag{12.66}$$

where $\Theta(x > 0) = 1$ and $\Theta(x < 0) = 0$. Expanding the Θ function to second order leads to:

$$\Theta(\delta f(e_1) - \delta f^*) + \left[\sum_{a=2}^{M} \Delta_a e_a + \frac{1}{2} \sum_{a,b=2}^{M} \Gamma_{a,b} e_a e_b \right] \delta(\delta f(e_1) - \delta f^*)$$

$$+ \frac{1}{2} \sum_{a,b=2}^{M} \Delta_a \Delta_b e_a e_b \delta'(\delta f(e_1) - \delta f^*), \tag{12.67}$$

where δ' is the derivative of the δ-function with respect to δf. In order to proceed with the integration over the variables e_a in Eq. (12.66), one should furthermore note the following identity:

$$\delta(\delta f(e_1) - \delta f^*) = \frac{1}{\Delta_1^*} \delta(e_1 - e_1^*), \tag{12.68}$$

where e_1^* is such that $\delta f(e_1^*) = \delta f^*$, and Δ_1^* is computed for $e_1 = e_1^*$, $e_{a>1} = 0$. Inserting the above expansion of the Θ function into Eq. (12.66) and performing the integration over the e_a then leads to:

$$\mathcal{P}_>(\delta f^*) = \mathcal{P}_>(e_1^*) + \sum_{a=2}^{M} \frac{\Gamma_{a,a}^* \sigma_a^2}{2\Delta_1^*} P(e_1^*) - \sum_{a=2}^{M} \frac{\Delta_a^{*2} \sigma_a^2}{2\Delta_1^{*2}} \left(P'(e_1^*) + \frac{\Gamma_{1,1}^*}{\Delta_1^*} P(e_1^*) \right), \tag{12.69}$$

where $P(e_1)$ is the probability distribution of the first factor, defined as:

$$P(e_1) = \int P(e_1, e_2, \ldots, e_M) \prod_{a=2}^{M} de_a. \tag{12.70}$$

In order to find the value-at-risk δf^*, one should thus solve Eq. (12.69) for e_1^* with $\mathcal{P}_>(\delta f^*) = p$, and then compute $\delta f(e_1^*, 0, \ldots, 0)$. Note that the equation is not trivial since the Greeks must be estimated at the solution point e_1^*.

Let us discuss the general result, Eq. (12.69), in the simple case of a linear portfolio of assets, such that no convexity is present: the Δ_a's are constant and the $\Gamma_{a,a}$'s are all zero. The equation then takes the following simpler form:

$$P_>(e_1^*) - \sum_{a=2}^{M} \frac{\Delta_a^2 \sigma_a^2}{2\Delta_1^2} P'(e_1^*) = p. \tag{12.71}$$

Naively, one could have thought that in the dominant factor approximation, the value of e_1^* would be the value-at-risk value of e_1 for the probability p, defined as:

$$P_>(e_{1,\text{VaR}}) = p. \tag{12.72}$$

However, the above equation shows that there is a correction term proportional to $P'(e_1^*)$. Since the latter quantity is negative, one sees that e_1^* is actually larger than $e_{1,\text{VaR}}$, and therefore $\delta f^* > \delta f(e_{1,\text{VaR}})$. This reflects the effect of all other factors, which tend to increase the value-at-risk of the portfolio.

The result obtained above relies on a second-order expansion; when are higher-order corrections negligible? It is easy to see that higher-order terms involve higher-order derivatives of $P(e_1)$. A condition for these terms to be negligible in the limit $p \to 0$, or $e_1^* \to \infty$, is that the successive derivatives of $P(e_1)$ become smaller and smaller. This is true provided that $P(e_1)$ decays more slowly than exponentially, for example as a power-law. On the contrary, when $P(e_1)$ decays faster than exponentially (for example in the Gaussian case), then the expansion proposed above completely loses its meaning, since higher and higher corrections become dominant when $p \to 0$. This is expected: in a Gaussian world, a large event results from the accidental superposition of many small events, whereas in a power-law world, large events are associated to one single large fluctuation which dominates over all the others. The case where $P(e_1)$ decays as an exponential is interesting, since it is often a good approximation for the tail of the fluctuations of financial assets. Taking $P(e_1) \approx \alpha_1 \exp -\alpha_1 e_1$, one finds that e_1^* is the solution of:

$$e^{-\alpha_1 e_1^*} \left[1 - \sum_{a=2}^{M} \frac{\Delta_a^2 \alpha_1^2 \sigma_a^2}{2\Delta_1^2} \right] = p. \tag{12.73}$$

Since one has $\sigma_1^2 \propto \alpha_1^{-2}$, the correction term is small provided that the variance of the portfolio generated by the dominant factor is much larger than the sum of the variance of all other factors.

Coming back to Eq. (12.69), one expects that if the dominant factor is correctly identified, and if the distribution is such that the above expansion makes sense, an approximate solution is given by $e_1^* = e_{1,\text{VaR}} + \epsilon$, with:

$$\epsilon \simeq \sum_{a=2}^{M} \frac{\Gamma_{a,a} \sigma_a^2}{2\Delta_1} - \sum_{a=2}^{M} \frac{\Delta_a^2 \sigma_a^2}{2\Delta_1^2} \left(\frac{P'(e_{1,\text{VaR}})}{P(e_{1,\text{VaR}})} + \frac{\Gamma_{1,1}}{\Delta_1} \right), \tag{12.74}$$

where now all the 'Greeks' Δ, Γ are estimated at $e_{1,\text{VaR}}$.

In some cases, it appears that a 'one-factor' approximation is not enough to reproduce the correct VaR value. This can be traced back to the fact that there are actually other different

Table 12.1. Comparison between the value-at-risk of the linear and quadratic portfolios, computed either using precise Monte-Carlo simulations, or the different approximation schemes discussed in the text, for various levels of the probability p.

	$p = 1\%$		$p = 0.5\%$		$p = 0.1\%$	
	\mathcal{L}	\mathcal{Q}	\mathcal{L}	\mathcal{Q}	\mathcal{L}	\mathcal{Q}
MC	2.93	13.3	3.53	18.7	5.30	40.6
Th. 0	2.65	9.7	3.26	13.9	5.07	30.8
Th. 1	2.83	10.9	3.42	15.1	5.20	32.2
Th. 2	2.93	12.1	3.52	17.2	5.30	38.6

dangerous market configurations which contribute to the VaR. The above formalism can however easily be adapted to the case where two (or more) dangerous configurations need to be considered. The general equations read:

$$\mathcal{P}_{>a} = \mathcal{P}_{>}(e_a^*) + \sum_{b \neq a}^{M} \frac{\Gamma_{b,b}^* \sigma_b^2}{2\Delta_a^*} P(e_a^*) - \sum_{b \neq a}^{M} \frac{\Delta_2^{*2} \sigma_b^*}{2\Delta_a^{*2}} \left(P'(e_a^*) + \frac{\Gamma_{a,a}^*}{\Delta_a^*} P(e_a^*) \right), \qquad (12.75)$$

where $a = 1, \ldots, K$ are the K different dangerous factors. The e_a^* and therefore δf^*, are determined by the following K conditions:

$$\delta f^*(e_1^*) = \delta f^*(e_2^*) = \cdots = \delta f^*(e_K^*) \qquad \mathcal{P}_{>l} + \mathcal{P}_{>2} + \cdots + \mathcal{P}_{>K} = p. \qquad (12.76)$$

12.4.2 Numerical test of the method

One can test numerically the above idea by calculating the VaR of a simple portfolios which depend on four independent factors e_1, \ldots, e_4, that we chose to be independent Student variables of unit variance with a tail exponent $\mu = 4$. We have considered two different portfolios, 'linear' (L), such that its variations are given by $d\mathcal{L} = e_1 + \frac{1}{2}e_2 + \frac{1}{5}e_3 + \frac{1}{20}e_4$, and 'quadratic' (Q), such that its variations are given by $d\mathcal{Q} \equiv d\mathcal{L} + (d\mathcal{L})^2$. We have determined the 99%, 99.5% and 99.9% VaR of these portfolios both using a Monte-Carlo (MC) scheme and the above approach. The results are summarized in Table 12.1, where we give the simple estimate based on the VaR of the most dangerous factor $e_{1,VaR}$, (called Th. 0) and the estimate which includes the contribution of the other factors e_1^* (called Th. 1, see Eq. (12.71)). For both L and Q, one sees that the latter estimate represents a significant improvement over the naive $e_{1,VaR}$ estimate. One the other hand, our calculation still underestimates the MC result, in particular for the Q portfolio. This can be traced back to the fact that there are actually other different dangerous market configurations which contribute to the VaR for this particular choice of parameters.

We have therefore computed the VaR for the different portfolios in the $K = 2$ approximation. The results are reported in Table 12.1 under the name 'Th. 2'; one sees that the agreement with the MC simulations is improved, and is actually excellent in the linear case, where both the configurations e_1 positive and large and e_2 positive and large contribute.

In the quadratic case, the two most dangerous configurations are e_1 positive and large and e_1 negative and large. In order to get perfect agreement with the Monte-Carlo result, one should extend the calculation to $K = 3$, taking into account the configuration where e_2 positive and large.

12.5 Summary

- In a Gaussian world, all measures of risk are equivalent. Minimizing the variance or the probability of large losses (or the value-at-risk) lead to the same result, which is the family of optimal portfolios obtained by Markowitz. In the general case, however, the VaR minimization leads to portfolios which are different from those of Markowitz. These portfolios depend on the level of risk Λ (or on the time horizon T), to which the operator is sensitive. An important result is that minimizing the variance can actually increase the VaR.

- One can extend Markowitz's portfolio construction in several ways. One way is to control the diversification by introducing the concept of effective number of assets. Another is to consider the case when the constraints applied to the portfolio are non linear. In this case, perhaps surprisingly, an exponential number of solutions, that can be quite far from each other, appear. Finally, one can minimize the VaR for a portfolio made of power-law assets using the idea of tail covariance. One can also look for optimal temporal trading strategies that minimize the variance of the investment.

- Finally, one can take advantage of the strongly non-Gaussian nature of the fluctuations of financial assets to simplify the analytic calculation of the Value-at-Risk of complicated non linear portfolios. Interestingly, the calculation allows one to visualize directly the 'dangerous' market configurations which correspond to extreme risks. In this sense, this method corresponds to a correctly probabilized 'scenario' calculation.

- FURTHER READING

H. Markowitz, *Portfolio Selection: Efficient Diversification of Investments*, Wiley, New York, 1959.

M. Mézard, G. Parisi, M. A. Virasoro, *Spin Glasses and Beyond*, World Scientific, 1987.

T. E. Copeland, J. F. Weston, *Financial Theory and Corporate Policy*, Addison-Wesley, Reading, MA, 1983.

E. Elton, M. Gruber, *Modern Portfolio Theory and Investment Analysis*, Wiley, New York, 1995.

S. Rachev, S. Mittnik, *Stable Paretian Models in Finance*, Wiley 2000.

- SPECIALIZED PAPERS

V. Bawa, E. Elton, M. Gruber, *Simple rules for optimal portfolio selection in stable Paretian markets*, J. Finance, **39**, 1041 (1979).

L. Belkacem, J. Lévy-Vehel, Ch. Walter, *CAPM, Risk and portfolio selection in α-stable markets*, Fractals, **8**, 99 (2000).

J.-P. Bouchaud, D. Sornette, Ch. Walter, J.-P. Aguilar, *Taming large events: portfolio selection for strongly fluctuating assets*, Int. Jour. Theo. Appl. Fin., **1**, 25 (1998).

E. Fama, *Portfolio analysis in a stable Paretian market*, Management Science, **11**, 404–419 (1965).

S. Galluccio, J.-P. Bouchaud, M. Potters, *Rational decisions, random matrices and spin glasses*, Physica A, **259**, 449 (1998).

L. Gardiol, R. Gibson, P.-A. Barès, R. Cont, S. Guger, *A large deviation approach to portfolio management*, Int. Jour. Theo. Appl. Fin., **3**, 617 (2000).

I. Kondor, *Spin glasses in the trading book*, Int. Jour. Theo. Appl. Fin., **3**, 537 (2000).

J.-F. Muzy, D. Sornette, J. Delour, A. Arneodo, *Multifractal returns and hierarchical portfolio theory*, Quantitative Finance, **1**, 131 (2001).

G. Parisi and M. Potters, *Mean-field equations for spin models with orthogonal interaction matrices*, J. Phys., **A 28**, 5267 (1995).

D. Sornette, D. P. Simonetti, J. V. Andersen, ϕ^q*-field theory for portfolio optimisation: fat tails and non linear correlations*, Phys. Rep., **335**, 19 (2000).

13

Futures and options: fundamental concepts

Les personnes non averties sont sujettes à se laisser induire en erreur.[†]
(Lord Raglan, 'Le tabou de l'inceste', quoted by Boris Vian in *L'automne à Pékin*.)

13.1 Introduction

13.1.1 Aim of the chapter

The aim of this chapter is to introduce the general theory of derivative pricing in a simple and intuitive, but rather unconventional, way. The usual presentation, which can be found in all the available books on the subject,[‡] relies on particular models where it is possible to construct *riskless hedging strategies*, which replicate exactly the corresponding derivative product.[§] Since the risk is strictly zero, there is no ambiguity in the price of the derivative: it is equal to the cost of the hedging strategy. In the general case, however, these 'perfect' strategies do not exist. Not surprisingly for the layman, zero risk is the exception rather than the rule. Correspondingly, a suitable theory must include risk as an essential feature, which one would like to *minimize*. The following chapters thus aim at developing simple methods to obtain optimal strategies, residual risks, and prices of derivative products, which take into account in an adequate way the peculiar statistical nature of financial markets, that have been described in Chapters 6, 7, 8.

13.1.2 Strategies in uncertain conditions

A *derivative* product is an asset the value of which depends on the price history of another asset, the 'underlying'. The best known examples, on which the following chapters will focus in detail, are futures and options. For example, one could sell a contract (called a 'digital' option) which pays $1 whenever the price of – say – gold exceeds some given value at a time T in the future, and zero otherwise. This is a derivative. Obviously, one can

[†] Unwarned people may easily be fooled.
[‡] See e.g. Hull (1997), Wilmott (1998), Baxter & Rennie (1996) and Chapter 16.
[§] A hedging strategy is a trading strategy allowing one to reduce, and sometimes eliminate, the risk.

invent a great number of such contracts, with different rules – some of them will be detailed in Chapter 17.

Now, the obvious question that the seller of such a derivative asks himself is: how much money will I make, or lose, when the contract has expired?

Since the answer is by definition uncertain, a more precise question is: what is the probability distribution $P(\Delta W)$ of the wealth balance ΔW for the seller of the contract?

The more heavily skewed towards the positive side, the better for the seller of the contract. The interesting point is that the seller can craft, to some extent, the distribution $P(\Delta W)$, such as to reach a shape which is optimal given his risk constraints. For example, if he raises the price of the contract (and is still able to sell it), the distribution will obviously shift to the right, but without changing shape (see Fig. 13.1). Less trivially, the seller can follow certain trading strategies that also narrow the distribution, or reduce the weight of the left tail (Fig. 13.1). The optimal strategy will depend on the shape one wishes to achieve, as we shall illustrate in the following.

From a general point of view, the shape of $P(\Delta W)$ can be characterized by its different moments. The average of $P(\Delta W)$ is obviously the expected gain (or loss) from selling the

Fig. 13.1. *Chiseling distributions*: Generic form of the distribution of the wealth balance $P(\Delta W)$ associated to a certain financial operation, such as writing an option. By increasing the price of the option, the seller can shift $P(\Delta W)$ to the right. By *hedging* the option, the seller can narrow $P(\Delta W)$, and sometimes even affect its skewness.

contract. The variance of $P(\Delta W)$ is the simplest measure of the uncertainty of the final wealth of the seller, and is therefore often taken as a measure of risk. If this definition of risk is deemed suitable, the hedging strategy should be such that the variance of the wealth balance is minimum. On the other hand, one is often worried about tail events, in particular the left tail. In this case, measures that include the third cumulant (skewness) or the fourth cumulant (kurtosis) of the wealth balance could be more adapted, or criteria based on value-at-risk or expected shortfall (see Section 10.3.3). The main idea to remember is that in general, the problem of derivative pricing and hedging has no unique solution, because the 'optimal shape' of the distribution of ΔW will very much depend on one's perception of risk. The perception of risk of the buyer and the seller of the contract are in general different: this in fact makes the trade possible.

In a sense, finance is the art of chiseling probability distributions. Only in very special cases will the solution be unique: when a perfect strategy exists, such that $P(\Delta W)$ can be made infinitely sharp around a certain value, in which case the risk is zero according to all possible definitions. In this case, the only possible price for the contract is such that the average of $P(\Delta W)$ is also zero, otherwise a riskless profit (called an arbitrage opportunity) would follow, either for the seller if the contract is overpriced, or for the buyer if the contract is underpriced. These special zero risk situations will appear below, in the context of futures contracts, or in the case of options within the Black-Scholes model.

13.1.3 Trading strategies and efficient markets

In the previous chapters, we have insisted on the fact that if the detailed prediction of future market moves is probably impossible, its statistical description is a reasonable and useful idea, at least as a first approximation. This approach only relies on a certain degree of stability (in time) in the way markets behave and the prices evolve. A reasonable hypothesis is that the statistical 'texture' of the markets (i.e. the shape of the probability distributions) is stable over time, but that the *parameters* which fix the amplitude of the fluctuations can be time dependent – see Chapter 7 and Section 13.3.4. Let us thus assume that one can determine (using a statistical analysis of past time series) the probability density $P(x, t | x_0, t_0)$, which gives the probability that the price of the asset X is equal to x (to within dx) at time t, knowing that at a previous time t_0, the price was equal to x_0. As in previous chapters, we shall denote as $\langle \mathcal{O} \rangle$ the average (over the 'historical' probability distribution) of a certain observable \mathcal{O}:

$$\langle \mathcal{O}(x, t) \rangle \equiv \int P(x, t | x_0, 0) \mathcal{O}(x, t) \, dx. \tag{13.1}$$

As we have shown in Section 6.2.2, the price fluctuations are somewhat correlated for small time intervals (a few minutes), but become rapidly uncorrelated (but not necessarily independent!) on longer time scales. In the following, we shall choose as our elementary time scale τ an interval a few times larger than the correlation time – say $\tau = 30$ minute on liquid markets. We shall thus assume that the correlations of price increments on two

different intervals of size τ are negligible.[†] When correlations are small, the information on future movements based on the study of past fluctuations is weak. In this case, no systematic trading strategy can be more profitable (on the long run) than holding the market index – of course, one can temporarily 'beat' the market through sheer luck. This property corresponds to the *efficient market hypothesis*.[‡]

It is interesting to translate this property into more formal terms. Let us suppose that at time $t_n = n\tau$, an investor has a portfolio containing, in particular, a quantity $\phi_n(x_n)$ of the asset X, quantity which can depend on the price of the asset $x_n = x(t_n)$ at time t_n (this strategy could actually depend on the price of the asset for all previous times: $\phi_n(x_n, x_{n-1}, x_{n-2}, \ldots)$). Between t_n and t_{n+1}, the price of X varies by δx_n. This generates a profit (or a loss) for the investor equal to $\phi_n(x_n)\delta x_n$. Note that the change of wealth is *not* equal to $\delta(\phi x) = \phi\delta x + x\delta\phi$, since the second term only corresponds to converting some stocks in cash, or vice versa, but not to a real change of wealth. The wealth difference between time $t = 0$ and time $t_N = T = N\tau$, due to the trading of asset X is, for zero interest rates, equal to:

$$\Delta W_X = \sum_{n=0}^{N-1} \phi_n(x_n)\delta x_n. \tag{13.2}$$

Now, we assume that price returns are random variables following a multiplicative model:

$$\frac{\delta x_k}{x_k} \equiv \eta_k = m + \epsilon_k\sigma_k, \tag{13.3}$$

where ϵ_k are iid random variables of zero mean and unit variance, and σ_k corresponds to the local volatility of the fluctuations, which can be correlated in time and also, possibly, with the random variables ϵ_j with $j < k$ (see Chapter 8). The point however here is that the ϵ_k are *unpredictable* from the knowledge of any past information. Since the trading strategy $\phi_n(x_n)$ can only be chosen *before* the price actually changes, the absence of correlations means that the average value of ΔW_X (with respect to the historical probability distribution) reads:

$$\langle\Delta W_X\rangle = \sum_{n=0}^{N-1} \langle\phi_n(x_n)\delta x_n\rangle \equiv m\tau \sum_{n=0}^{N-1} \langle x_n\phi_n(x_n)\rangle, \tag{13.4}$$

where m is the average return of the asset X, defined in (13.3).

The above equation (13.4) thus means that the average value of the profit is fixed by the average return of the asset, weighted by the level of investment across the considered period.

[†] The presence of small correlations on larger time scales is however difficult to exclude, in particular on the scale of several years, which might reflect economic cycles for example. Note also that the 'volatility' fluctuations exhibit long-range correlations – cf. Section 7.2.2.

[‡] The existence of successful systematic hedge funds, which have consistently produced returns higher than the average for several years, suggests that some sort of 'hidden' correlations do exist, at least on certain markets. But if they exist these correlations must be small and therefore are not relevant for our concern, at least in a first approximation.

Trading in the presence of temporal correlations

It is interesting to investigate the case where correlations are not zero. For simplicity, we shall assume an additive model where the fluctuations δx_n are stationary Gaussian variables of zero mean ($m_1 = 0$). The correlation function is then given by $\langle \delta x_n \delta x_k \rangle \equiv C_{nk}$. The C_{nk}^{-1}'s are the elements of the matrix inverse of \mathbf{C}. If one knows the sequence of past increments $\delta x_0, \ldots, \delta x_{n-1}$, the distribution of the next δx_n conditioned to such an observation is simply given by:

$$P(\delta x_n) = \mathcal{N} \exp -\frac{C_{nn}^{-1}}{2}[\delta x_n - m_n]^2, \tag{13.5}$$

$$m_n \equiv -\frac{\sum_{i=0}^{n-1} C_{in}^{-1} \delta x_i}{C_{nn}^{-1}}, \tag{13.6}$$

where \mathcal{N} is a normalization factor, and m_n the mean of δx_n conditioned to the past, which is non-zero precisely because some correlations are present.[†] A simple strategy which exploits these correlations is to choose:

$$\phi_n(x_n, x_{n-1}, \ldots) = \text{sign}(m_n), \tag{13.7}$$

which means that one buys (resp. sells) one stock if the expected next increment is positive (resp. negative). The average profit is then obviously given by $\langle |m_n| \rangle > 0$. We further assume that the correlations are short ranged, such that only C_{nn}^{-1} and C_{nn-1}^{-1} are non-zero. The unit time interval is then the above correlation time τ. If this strategy is used during the time $T = N\tau$, the average profit is given by:

$$\langle \Delta W_X \rangle = Na\langle |\delta x| \rangle \text{ where } a \equiv \left| \frac{C_{nn-1}^{-1}}{C_{nn}^{-1}} \right|, \tag{13.8}$$

(a does not depend on n for a stationary process). With typical values, $\tau = 30$ min, $\langle |\delta x| \rangle \approx 10^{-3} x_0$, and a of about 0.1 (cf. Section 6.2.2), the average profit would reach 50% annual![‡] Hence, the presence of correlations (even rather weak) allows in principle one to make rather substantial gains. However, one should remember that some transaction costs are always present in some form or other (see Section 5.3.6). Since our assumption is that the correlation time is equal to τ, it means that the frequency with which our trader has to 'flip' his strategy (i.e. $\phi \to -\phi$) is τ^{-1}. One should thus subtract from Eq. (13.8) a term on the order of $-N\nu x_0$, where ν represents the fractional cost of a transaction. A very small value of the order of 10^{-4} is thus enough to cancel completely the profit resulting from the observed correlations on the markets. Interestingly enough, the 'basis point' (10^{-4}) is precisely the order of magnitude of the friction faced by traders with the most direct access to the markets. More generally, transaction costs allow the existence of correlations, and thus of 'theoretical' inefficiencies of the markets. The strength of the 'allowed' correlations on a certain time T is on the order of the transaction costs ν divided by the volatility of the asset on the scale of T.

[†] The notation m_n has already been used in Chapter 1 with a different meaning. Note also that a general formula exists for the distribution of δx_{n+k} for all $k \geq 0$, and can be found in books on optimal filter theory, see references.

[‡] A strategy that allows one to generate consistent abnormal returns with zero or minimal risk is called an **arbitrage opportunity**.

At this stage, it would appear that devising a 'theory' for trading is useless: in the absence of correlations, speculating on stock markets is tantamount to playing the roulette (the zero playing the role of transaction costs!). In fact, as will be discussed in the present chapter, the existence of riskless assets such as bonds, which yield a known return, and the emergence of more complex financial products such as futures contracts or options, demand an adapted theory for pricing and hedging. This theory turns out to be predictive, even in the complete absence of temporal correlations.

13.2 Futures and forwards

13.2.1 Setting the stage

Before turning to the rather complex case of options, we shall first focus on the very simple case of forward contracts, which allows us to define a certain number of notions (such as arbitrage and hedging) and notations. A forward, or futures contract F amounts to buying or selling today an asset X (the 'underlying') with payment and delivery at a date $T = N\tau$ in the future (see Section 5.2.6). What is the price \mathcal{F} of this contract, knowing that it must be paid at the date of expiry?[†] The naive answer that first comes to mind is a 'fair game' condition: the price must be adjusted such that, *on average*, the two parties involved in the contract fall even on the day of expiry. For example, taking the point of view of the writer of the contract (who sells the forward), the wealth balance associated with the forward reads:

$$\Delta W_F = \mathcal{F} - x(T). \tag{13.9}$$

This actually assumes that the writer has not considered the possibility of simultaneously trading the underlying stock X to reduce his risk, which of course he should do, see below.

Under this assumption, the fair game condition amounts to set $\langle \Delta W_F \rangle = 0$, which gives for the forward price:

$$\mathcal{F}_B = \langle x(T) \rangle \equiv \int x P(x, T | x_0, 0) \, dx, \tag{13.10}$$

if the price of X is x_0 at time $t = 0$. This price, that can we shall call the 'Bachelier' price, is not satisfactory since the seller takes the risk that the price at expiry $x(T)$ ends up far above $\langle x(T) \rangle$, which could prompt him to increase his price above \mathcal{F}_B.

Actually, the Bachelier price \mathcal{F}_B is not related to the market price for a simple reason: the seller of the contract can suppress his risk completely if he buys now the underlying asset X at price x_0 and waits for the expiry date. However, this strategy is costly: the amount of cash frozen during that period does not yield the riskless interest rate. The cost of the strategy is thus $x_0 e^{rT}$, where r is the interest rate per unit time. From the view point of the buyer, it would certainly be absurd to pay more than $x_0 e^{rT}$, which is the cost of borrowing

[†] Note that if the contract was to be paid now, its price would be exactly equal to that of the underlying asset (barring the risk of delivery default and neglecting cost and benefits of ownership such as dividends).

the cash needed to pay the asset right away. The only viable price for the forward contract is thus $\mathcal{F} = x_0 e^{rT} \neq \mathcal{F}_B$, and is, in this particular case, completely unrelated to the potential moves of the asset X!

An elementary argument thus allows one to know the price of the forward contract and to follow a perfect hedging strategy: buy a quantity $\phi = 1$ of the underlying asset during the whole life of the contract. The aim of next paragraph is to establish this rather trivial result, which has not required any maths, in a much more sophisticated way. The importance of this procedure is that one needs to learn how to write down a proper wealth balance in order to price more complex derivative products such as options, on which we shall focus in the next sections.

13.2.2 Global financial balance

Let us write a general financial balance which takes into account the trading strategy of the the underlying asset X. The difficulty lies in the fact that the amount $\phi_n x_n$ which is invested in the asset X rather than in bonds is effectively costly: one 'misses' the risk-free interest rate. Furthermore, this loss cumulates in time. It is not *a priori* obvious how to write down the correct balance. Suppose that only two assets have to be considered: the risky asset X, and a bond B. The whole capital W_n at time $t_n = n\tau$ is shared between these two assets:

$$W_n \equiv \phi_n x_n + B_n. \tag{13.11}$$

The time evolution of W_n is due both to the change of price of the asset X, and to the fact that the bond yields a known return through the interest rate r:

$$W_{n+1} - W_n = \phi_n(x_{n+1} - x_n) + B_n\rho, \qquad \rho = r\tau. \tag{13.12}$$

On the other hand, the amount of capital in bonds evolves both mechanically, through the effect of the interest rate $(+B_n\rho)$, but also because the quantity of stock does change in time $(\phi_n \to \phi_{n+1})$, and causes a flow of money from bonds to stock or vice versa. Hence:

$$B_{n+1} - B_n = B_n\rho - x_{n+1}(\phi_{n+1} - \phi_n). \tag{13.13}$$

Note that Eq. (13.11) is obviously compatible with the following two equations. The solution of the second equation reads:

$$B_n = (1 + \rho)^n B_0 - \sum_{k=1}^{n} x_k(\phi_k - \phi_{k-1})(1 + \rho)^{n-k}. \tag{13.14}$$

Plugging this result in Eq. (13.11), and renaming $k - 1 \to k$ in the second part of the sum, one finally ends up with the following central result for W_n:

$$W_n = W_0(1 + \rho)^n + \sum_{k=0}^{n-1} \psi_k^n (x_{k+1} - x_k - \rho x_k), \tag{13.15}$$

with $\psi_k^n \equiv \phi_k(1 + \rho)^{n-k-1}$.

This last expression has an intuitive meaning: the gain or loss incurred between time k and $k + 1$ must include the cost of the hedging strategy $-\rho x_k$; furthermore, this gain or loss must be forwarded up to time n through interest rate effects, hence the extra factor $(1 + \rho)^{n-k-1}$. Another useful way to write and understand Eq. (13.15) is to introduce the discounted prices $\tilde{x}_k \equiv x_k(1 + \rho)^{-k}$. One then has:

$$W_n = (1 + \rho)^n \left(W_0 + \sum_{k=0}^{n-1} \phi_k(\tilde{x}_{k+1} - \tilde{x}_k) \right). \tag{13.16}$$

The effect of interest rates can be thought of as an erosion of the value of the money itself. The discounted prices \tilde{x}_k are therefore the 'true' prices, and the difference $\tilde{x}_{k+1} - \tilde{x}_k$ the 'true' change of wealth. The overall factor $(1 + \rho)^n$ then converts this true wealth into the current value of the money.

The global balance associated with the forward contract contains two further terms: the price of the forward \mathcal{F} that one wishes to determine, and the price of the underlying asset at the delivery date. Hence, finally:

$$W_N = \mathcal{F} - x_N + (1 + \rho)^N \left(W_0 + \sum_{k=0}^{N-1} \phi_k(\tilde{x}_{k+1} - \tilde{x}_k) \right). \tag{13.17}$$

Since by identity $\tilde{x}_N = \sum_{k=0}^{N-1}(\tilde{x}_{k+1} - \tilde{x}_k) + x_0$, this last expression can also be written as:

$$W_N = \mathcal{F} + (1 + \rho)^N \left(W_0 - x_0 + \sum_{k=0}^{N-1}(\phi_k - 1)(\tilde{x}_{k+1} - \tilde{x}_k) \right). \tag{13.18}$$

13.2.3 Riskless hedge

In this last formula, all the randomness, the uncertainty on the future evolution of the prices, only appears in the last term. But if one chooses ϕ_k to be identically equal to one, then the global balance is completely independent of the evolution of the stock price. The final result is *not* random, and reads:

$$W_N = \mathcal{F} + (1 + \rho)^N (W_0 - x_0). \tag{13.19}$$

Now, the wealth of the writer of the forward contract at time $T = N\tau$ cannot be, with certitude, greater (or less) than the one he would have had if the contract had not been signed, i.e. $W_0(1 + \rho)^N$. If this was the case, one of the two parties would be losing money in a totally predictable fashion (since Eq. (13.19) does not contain any unknown term).

Since any reasonable participant is reluctant to give away his money with no hope of return, one concludes that the forward price is indeed given by:

$$\mathcal{F} = x_0(1 + \rho)^N \approx x_0 e^{rT} \neq \mathcal{F}_B, \tag{13.20}$$

which does not rely on any statistical information on the price of X!

Dividends

In the case of a stock that pays a constant dividend rate $\delta = d\tau$ per interval of time τ, the global wealth balance is obviously changed into:

$$W_N = \mathcal{F} - x_N + W_0(1 + \rho)^N + \sum_{k=0}^{N-1} \psi_k^N (x_{k+1} - x_k + (\delta - \rho)x_k). \tag{13.21}$$

It is easy to redo the above calculations in this case. One finds that the riskless strategy is now to hold:

$$\phi_k \equiv \frac{(1 + \rho - \delta)^{N-k-1}}{(1 + \rho)^{N-k-1}} \tag{13.22}$$

stocks. The wealth balance breaks even if the price of the forward is set to:

$$\mathcal{F} = x_0(1 + \rho - \delta)^N \approx x_0 e^{(r-d)T}, \tag{13.23}$$

which again could have been obtained using a simple no arbitrage argument of the type presented below.

Variable interest rates

In reality, the interest rate is not constant in time but rather also varies randomly. More precisely, as explained in Chapter 19, at any instant of time the whole interest rate curve for different maturities is known, but evolves with time. The generalization of the global balance, as given by formula Eq. (13.17), depends on the maturity of the bonds included in the portfolio, and thus on the whole interest rate curve. Assuming that only short-term bonds are included, yielding the (time-dependent) 'spot' rate ρ_k, one has:

$$W_N = \mathcal{F} - x_N + W_0 \prod_{k=0}^{N-1}(1 + \rho_k)$$

$$+ \sum_{k=0}^{N-1} \psi_k^N (x_{k+1} - x_k - \rho_k x_k), \tag{13.24}$$

with: $\psi_k^N \equiv \phi_k \prod_{\ell=k+1}^{N-1}(1 + \rho_\ell)$. It is again quite obvious that holding a quantity $\phi_k \equiv 1$ of the underlying asset leads to zero risk in the sense that the fluctuations of X disappear. However, the above strategy is not immune to interest rate risk. The interesting complexity of interest rate problems (such as the pricing and hedging of interest rate derivatives) comes from the fact that one may choose bonds of arbitrary maturity to construct the

hedging strategy. In the present case, one may take a bond of maturity equal to that of the forward. In this case, risk disappears entirely, and the price of the forward reads:

$$\mathcal{F} = \frac{x_0}{B(0, T)}, \tag{13.25}$$

where $B(0, T)$ stands for the value, at time 0, of the bond maturing at time $T = N\tau$.

13.2.4 Conclusion: global balance and arbitrage

From the simple example of forward contracts, one should bear in mind the following points, which are the key concepts underlying the derivative pricing theory as presented in this book. After writing down the *complete* financial balance, taking into account the trading of all assets used to cover the risk, it is quite natural (at least from the view point of the writer of the contract) to determine the trading strategy for all the hedging assets so as to minimize the risk associated to the contract (see the discussion in Section 13.1.2). After doing so, a reference price is obtained by demanding that the global balance is zero on average, corresponding to a fair price from the point of view of both parties. In certain cases (such as the forward contracts described above), the minimum risk is zero and the true market price cannot differ from the fair price, or else arbitrage would be possible. On the example of forward contracts, the price, Eq. (13.20), indeed corresponds to the *absence of arbitrage opportunities* (AAO), that is, of riskless profit. Suppose for example that the price of the forward is *below* $\mathcal{F} = x_0(1 + \rho)^N$. One can then sell the underlying asset now at price x_0 and simultaneously buy the forward at a price $\mathcal{F}' < \mathcal{F}$, that must be paid for on the delivery date. The cash x_0 is used to buy bonds with a yield rate ρ. On the expiry date, the forward contract is used to buy back the stock and close the position. The wealth of the trader is then $x_0(1 + \rho)^N - \mathcal{F}'$, which is positive under our assumption – and furthermore fully determined at time zero: there is profit, but no risk. Similarly, a price $\mathcal{F}' > \mathcal{F}$ would also lead to an opportunity of arbitrage. More generally, if the hedging strategy is perfect, this means that the final wealth is known in advance. Thus, increasing the price as compared to the fair price leads to a riskless profit for the seller of the contract, and vice versa. This AAO principle is at the heart of most derivative pricing theories currently used. Unfortunately, this principle cannot be used in the general case, where the minimal risk is non-zero, or when transaction costs are present (and absorb the potential profit, see the discussion in Section 13.1.3 above). When the risk is non-zero, there exists a fundamental ambiguity in the price, since one should expect that a risk premium is added to the fair price (for example, as a bid-ask spread). This risk premium depends both on the risk-averseness of the market maker, but also on the liquidity of the derivative market: if the price asked by one market maker is too high, less greedy market makers will make the deal. However, this mechanism does not operate for 'over-the-counter' operations (OTC, that is between two individual parties, as opposed to through an organized market). We shall come back to this important discussion in Section 15.4.1 below.

> Let us emphasize that the proper accounting of all financial elements in the wealth balance is crucial to obtain the correct fair price.

For example, we have seen above that if one forgets the term corresponding to the trading of the underlying stock, one ends up with the intuitive, but wrong, Bachelier price, Eq. (13.10).

13.3 Options: definition and valuation

13.3.1 Setting the stage

As mentioned in Section 5.2.6, a buy option (or 'call' option) is an insurance policy, protecting the owner against the potential increase of the price of a given asset, which he will need to buy in the future. The call option provides to its owner the certainty of not paying the asset more than a certain price. Symmetrically, a 'put' option protects against drawdowns, by insuring to the owner a minimum value for his stock.

More precisely, in the case of a so-called 'European' option,[†] the contract is such that at a given date in the future (the 'expiry date' or 'maturity') $t = T$, the owner of the option will not pay the asset more than x_s (the 'exercise price', or 'strike' price). Knowing that the price of the underlying asset is x_0 now (i.e. at $t = 0$), what is the price (or 'premium') C of the call? What is the optimal hedging strategy that the writer of the option should follow in order to minimize his risk?

The very first scientific theory of option pricing dates back to Bachelier in 1900. His proposal was, following a fair game argument, that the option price should equal the average value of the pay-off of the contract at expiry. Bachelier calculated this average by assuming the price increments δx_n to be independent Gaussian random variables, which leads to the formula, Eq. (13.43) below. However, Bachelier did not discuss the possibility of hedging, and therefore did not include in his wealth balance the term corresponding to the trading strategy that we have discussed in the previous section. As we have seen, this is precisely the term responsible for the difference between the forward 'Bachelier price' \mathcal{F}_B (cf. Eq. (13.10)) and the true price, Eq. (13.20). The problem of the optimal trading strategy must thus, in principle, be solved before one can fix the price of the option.[‡] This is the problem that was solved by Black and Scholes in 1973, when they showed that for a continuous-time Gaussian process, there exists a perfect strategy, in the sense that the risk associated to writing an option is *strictly zero*, as is the case for forward contracts. The determination of this perfect hedging strategy allows one to fix completely the price of the option using an AAO argument. Unfortunately, as repeatedly discussed below, this ideal

[†] Many other types of options exist: 'American', 'Asian', 'Lookback', 'Digitals', etc. (Nelken (1996)). We will discuss some of them in Chapter 17.

[‡] In practice, however, the influence of the trading strategy on the price (but not on the risk!) is quite small, see below.

strategy only exists in a continuous-time, Gaussian world,[†] that is, if the market correlation time was infinitely short, and if no discontinuities in the market price were allowed – both assumptions rather remote from reality. The hedging strategy cannot, in general, be perfect. However, an *optimal* hedging strategy can always be found, for which the risk is minimal (cf. Section 14.2). This optimal strategy thus allows one to calculate the fair price of the option and the associated residual risk. One should nevertheless bear in mind that, as empha-sized in Sections 13.2.4 and 14.6, there is no such thing as a unique option price whenever the risk is non-zero.

Let us now discuss these ideas more precisely. Following the method and notations introduced in the previous section, we write the global wealth balance for the writer of an option between time $t = 0$ and $t = T$ as:

$$W_N = [W_0 + \mathcal{C}](1 + \rho)^N - \max(x_N - x_s, 0)$$

$$+ \sum_{k=0}^{N-1} \psi_k^N (x_{k+1} - x_k - \rho x_k), \tag{13.26}$$

which reflects the fact that:

- The premium \mathcal{C} is paid immediately (i.e. at time $t = 0$).
- The writer of the option incurs a *loss* $x_N - x_s$ only if the option is exercised ($x_N > x_s$).
- The hedging strategy requires that the writer convert a certain amount of bonds into the underlying asset, as was discussed before Eq. (13.17).

A crucial difference with forward contracts comes from the *non-linear* nature of the pay-off, which is equal, in the case of a European option, to $\mathcal{Y}(x_N) \equiv \max(x_N - x_s, 0)$. This must be contrasted with the forward pay-off, which is linear (and equal to x_N). It is ultimately the non-linearity of the pay-off which, combined with the non-Gaussian nature of the fluctuations, leads to a non-zero residual risk.

An equation for the call price \mathcal{C} is obtained by requiring that the excess return due to writing the option, $\Delta W = W_N - W_0(1 + \rho)^N$, is zero on average:

$$(1 + \rho)^N \mathcal{C} = \left[\langle \max(x_N - x_s, 0) \rangle - \sum_{k=0}^{N-1} \langle \psi_k^N (x_{k+1} - x_k - \rho x_k) \rangle \right]. \tag{13.27}$$

This price therefore depends, in principle, on the strategy $\psi_k^N = \phi_k^N (1 + \rho)^{N-k-1}$. This price corresponds to the fair price, to which a risk premium will in general be added (for example in the form of a bid-ask spread).

In the rather common case where the underlying asset is not a stock but a forward on the stock, the hedging strategy is less costly since only a small fraction f of the value of the stock is required as a deposit. In the case where $f = 0$, the wealth balance appears to

[†] A property also shared by a discrete binomial evolution of prices, where at each time step, the price increment δx can only take two values, see Chapter 16. However, the risk is non-zero as soon as the number of possible price changes exceeds two.

take a simpler form, since the interest rate is not lost while trading the underlying forward contract:

$$\mathcal{C} = (1 + \rho)^{-N} \left[\langle \max(\mathcal{F}_N - x_s, 0) \rangle - \sum_{k=0}^{N-1} \langle \phi_k (\mathcal{F}_{k+1} - \mathcal{F}_k) \rangle \right]. \tag{13.28}$$

However, one can check that if one expresses the price of the forward in terms of the underlying stock, Eqs. (13.27) and (13.28) are actually identical. (Note in particular that $\mathcal{F}_N = x_N$.)

13.3.2 Orders of magnitude

Let us suppose that the maturity of the option is such that non-Gaussian 'tail' effects can be neglected (this might in some cases correspond to quite long time scales), so that the distribution of the terminal price $x(T)$ can be approximated by a Gaussian of mean mT and variance $DT = \sigma^2 x_0^2 T$.[†] If the average trend is small compared to the RMS \sqrt{DT}, a direct calculation for 'at-the-money' options gives:[‡]

$$\langle \max(x(T) - x_s, 0) \rangle = \int_{x_s}^{\infty} \frac{x - x_s}{\sqrt{2\pi DT}} \exp \left(-\frac{(x - x_0 - mT)^2}{2DT} \right) dx$$

$$\simeq \sqrt{\frac{DT}{2\pi}} + \frac{mT}{2} + O \left(\sqrt{\frac{m^4 T^3}{D}} \right). \tag{13.29}$$

Taking $T = 100$ days, a daily volatility of $\sigma = 1\%$, an annual return of $m = 5\%$, and a stock such that $x_0 = 100$ points, one gets:

$$\sqrt{\frac{DT}{2\pi}} \approx 4 \text{ points} \qquad \frac{mT}{2} \approx 0.67 \text{ points}. \tag{13.30}$$

In fact, the effect of a non-zero average return of the stock is much less than the above estimation would suggest. The reason is that one must also take into account the last term of the balance equation, Eq. (13.27), which comes from the hedging strategy. This term corrects the price by an amount $-\langle \phi \rangle mT$. Now, as we shall see later, the optimal strategy for an at-the-money option is to hold, on average $\langle \phi \rangle = \frac{1}{2}$ stock per option. Hence, this term precisely compensates the increase (equal to $mT/2$) of the average pay-off, $\langle \max(x(T) - x_s, 0) \rangle$. This strange compensation is actually *exact* in the Black–Scholes model (where the price of the option turns out to be completely independent of the value of m) but only approximate in more general cases. However, it is a very good first approximation to neglect the dependence of the option price on the average return of the stock: we shall come back to this important point in detail in Section 15.1.

The interest rate appears in two different places in the balance equation, Eq. (13.27): in front of the call premium \mathcal{C}, and in the cost of the trading strategy. For $\rho = r\tau$ corresponding

[†] We again neglect the difference between Gaussian and log-normal statistics in the following order of magnitude estimate.

[‡] A call option is called **at-the-money** if the strike price is equal to the current price of the underlying stock ($x_s = x_0$), **out-of-the-money** if $x_s > x_0$ and **in-the-money** if $x_s < x_0$.

to 5% per year, the factor $(1 + \rho)^N$ corrects the option price by a very small amount, which, for $T = 100$ days, is equal to 0.06 points. However, the cost of the trading strategy is not negligible. Its order of magnitude is $\sim \langle \phi \rangle x_0 r T$; in the above numerical example, it corresponds to a price increase of $\frac{2}{3}$ of a point (16% of the option price), see Section 17.1.

13.3.3 Quantitative analysis – option price

We shall assume in the following that the price increments can be written as in Eq. (13.3):

$$x_{k+1} - x_k = \rho x_k + \delta x_k, \tag{13.31}$$

with $\delta x_k = \sigma_k \epsilon_k x_k$. The above order of magnitude estimate suggests that the influence of a non-zero mean value of δx_k leads to small corrections in comparison with the potential fluctuations of the stock price. We shall thus temporarily set $\langle \epsilon_k \rangle = 0$ to simplify the following discussion, and come back to the corrections brought about by a non-zero value of the average return in Chapter 15.

As already discussed above, the hedging strategy ψ_k must obviously be determined *before* the next random change of price δx_k. Since ϵ_k is assumed to be independent of all past information, one has:

$$\langle \psi_k \delta x_k \rangle = \langle \psi_k \sigma_k x_k \rangle \langle \epsilon_k \rangle = 0 \qquad (\langle \epsilon_k \rangle = 0). \tag{13.32}$$

In this case, the hedging strategy disappears from the price of the option, which is given by the following 'Bachelier'-like formula, generalized to the case where the increments are not necessarily Gaussian:

$$\begin{aligned}
\mathcal{C} &= (1 + \rho)^{-N} \langle \max(x_N - x_s, 0) \rangle \\
&\equiv (1 + \rho)^{-N} \int_{x_s}^{\infty} (x - x_s) P(x, N | x_0, 0) \, dx.
\end{aligned} \tag{13.33}$$

In order to use Eq. (13.33) concretely, one needs to specify a model for price increments.

The Black-Scholes model

We first recover the classical model of Black and Scholes, by assuming that the *relative* returns $\eta_k = \delta x_k / x_k$ are iid random variables with a constant volatility $\sigma_k = \sigma_1$. If one knows (from empirical observation) the distribution $P_1(\eta_k)$ of returns over the elementary time scale τ, one can easily reconstruct (using the independence of the returns) the distribution $P(x, N | x_0, 0)$ needed to compute \mathcal{C}. After changing variables to $x \to x_0(1 + \rho)^N e^y$, the formula, Eq. (13.33), is transformed into:

$$\mathcal{C} = x_0 \int_{y_s}^{\infty} (e^y - e^{y_s}) P_N(y) \, dy, \tag{13.34}$$

where $y_s \equiv \log(x_s / x_0[1 + \rho]^N)$ and $P_N(y) \equiv P(y, N | 0, 0)$. Note that y_s involves the ratio of the strike price to the *forward* price of the underlying asset, $x_0[1 + \rho]^N$ – see the discussion in Section 17.1.

Setting $x_k = x_0(1 + \rho)^k e^{y_k}$, the evolution of the y_k's is given by:

$$y_{k+1} - y_k \simeq \frac{\eta_k}{1 + \rho} - \frac{\eta_k^2}{2} \qquad y_0 = 0, \tag{13.35}$$

where third-order terms $(\eta^3, \eta^2\rho, \ldots)$ have been neglected. The distribution of the quantity $y_N = \sum_{k=0}^{N-1} [(\eta_k/(1 + \rho)) - (\eta_k^2/2)]$ is then obtained, in Fourier space, as:

$$\hat{P}_N(z) = [\tilde{P}_1(z)]^N, \tag{13.36}$$

where we have defined, in the right-hand side, a modified Fourier transform:

$$\tilde{P}_1(z) = \int P_1(\eta) \exp\left[iz\left(\frac{\eta}{1 + \rho} - \frac{\eta^2}{2}\right)\right] d\eta. \tag{13.37}$$

The Black and Scholes limit

We can now examine the Black–Scholes limit, where $P_1(\eta)$ is a Gaussian of zero mean and RMS equal to $\sigma_1 = \sigma\sqrt{\tau}$. Using the above Eqs. (13.36) and (13.37) one finds, for N large:[†]

$$P_N(y) = \frac{1}{\sqrt{2\pi N\sigma_1^2}} \exp\left(-\frac{\left(y + N\sigma_1^2/2\right)^2}{2N\sigma_1^2}\right). \tag{13.38}$$

The Black–Scholes model also corresponds to the limit where the elementary time interval τ goes to zero, in which case one has $N = T/\tau \to \infty$. This limit is of course not realistic since any transaction takes at least a few seconds and, more significantly, that some correlations are seen to persist over at least several minutes. Notwithstanding, if one takes the mathematical limit where $\tau \to 0$, with $N = T/\tau \to \infty$ but keeping the product $N\sigma_1^2 = T\sigma^2$ finite, one constructs the continuous-time log-normal process, or 'geometrical Brownian motion'. In this limit, using the above form for $P_N(y)$ and $(1 + \rho)^N \to e^{rT}$, one obtains the celebrated Black–Scholes formula:

$$\boxed{\begin{aligned} \mathcal{C}_{\mathrm{BS}}(x_0, x_s, T) &= x_0 \int_{y_s}^{\infty} \frac{(e^y - e^{y_s})}{\sqrt{2\pi\sigma^2 T}} \exp\left(-\frac{(y + \sigma^2 T/2)^2}{2\sigma^2 T}\right) dy \\ &= x_0 \mathcal{P}_{\mathrm{G>}}\left(\frac{y_-}{\sigma\sqrt{T}}\right) - x_s e^{-rT} \mathcal{P}_{\mathrm{G>}}\left(\frac{y_+}{\sigma\sqrt{T}}\right), \end{aligned}} \tag{13.39}$$

where $y_s = \log(x_s/x_0) - rT$, $y_{\pm} = \log(x_s/x_0) - rT \pm \sigma^2 T/2$ and $\mathcal{P}_{\mathrm{G>}}(u)$ is the cumulative normal distribution defined by Eq. (2.36). The way $\mathcal{C}_{\mathrm{BS}}$ varies with the four parameters x_0, x_s, T and σ is quite intuitive, and discussed at length in all the books which deal with options. The derivatives of the price with respect to these parameters are now called the

[†] In fact, the variance of $P_N(y)$ is equal to $N\sigma_1^2/(1 + \rho)^2$, but we neglect this small correction which disappears in the limit $\tau \to 0$.

Greeks, because they are denoted by Greek letters (see Section 16.1.6 below). For example, the larger the price of the stock x_0, the more probable it is to reach the strike price at expiry, and the more expensive the option. Hence the so-called '**Delta**' of the option ($\Delta = \partial \mathcal{C}/\partial x_0$) is positive. As we shall show below, this quantity is actually directly related to the optimal hedging strategy. The variation of the hedging strategy Δ with the price of the underlying is noted Γ (**Gamma**) and is defined as: $\Gamma = \partial \Delta/\partial x_0$. Similarly, the dependence in maturity T and volatility σ (which in fact only appear through the combination $\sigma\sqrt{T}$ if rT is very small) leads to the definition of two further 'Greeks': **Theta** $\Theta = -\partial \mathcal{C}/\partial T = \partial \mathcal{C}/\partial t < 0$ and '**Vega**' $\mathcal{V} = \partial \mathcal{C}/\partial \sigma > 0$. In this limit, $\mathcal{V} = -2T/\sigma\Theta$. The signs reflect the fact that if the call premium is an increasing function of the volatility on the scale of the maturity of the option, $\sigma\sqrt{T}$.

Bachelier's Gaussian limit

Suppose now that the price process is additive rather than multiplicative. This means that the increments δx_k themselves, rather than the relative increments, should be considered as independent random variables. (In other words, we consider as case where $\sigma_k x_k$ in Eq. (13.3) is a constant.) Using the change of variable $x_k = (1 + \rho)^k \tilde{x}_k$ in Eq. (13.31), one finds that x_N can be written as:

$$x_N = x_0(1 + \rho)^N + \sum_{k=0}^{N-1} \delta x_k (1 + \rho)^{N-k-1}. \tag{13.40}$$

When N is large, the difference $x_N - x_0(1 + \rho)^N$ becomes, according to the CLT, a Gaussian variable of mean zero (if the ϵ_k have zero mean) and of variance equal to:

$$c_2(T) = D\tau \sum_{\ell=0}^{N-1} (1 + \rho)^{2\ell} \simeq DT[1 + \rho(N - 1) + O(\rho^2 N^2)] \tag{13.41}$$

where $D\tau$ is the variance of the individual increments, related to the volatility through: $D \equiv \sigma^2 x_0^2$. The price, Eq. (13.33), then takes the following form:

$$\mathcal{C}_G(x_0, x_s, T) = e^{-rT} \int_{x_s}^{\infty} \frac{x - x_s}{\sqrt{2\pi c_2(T)}} \exp\left(-\frac{(x - x_0 e^{rT})^2}{2c_2(T)}\right) dx. \tag{13.42}$$

The price formula written down by Bachelier corresponds to the limit of *short maturity* options, where all interest rate effects can be neglected. In this limit, the above equation simplifies to:

$$\mathcal{C}_G(x_0, x_s, T) \simeq \int_{x_s}^{\infty} \frac{x - x_s}{\sqrt{2\pi DT}} \exp\left(-\frac{(x - x_0)^2}{2DT}\right) dx. \tag{13.43}$$

This equation can also be derived directly from the Black–Scholes price, Eq. (13.39), in the small maturity limit, where relative price variations are small: $x_N/x_0 - 1 \ll 1$, allowing one to write $y = \log(x/x_0) \simeq (x - x_0)/x_0$. As emphasized in Section 1.7, this is the limit where the Gaussian and log-normal distributions become very similar.

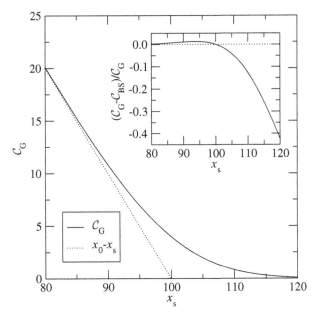

Fig. 13.2. Price of a call of maturity $T = 100$ days as a function of the strike price x_s, in the Bachelier model. The underlying stock is worth 100, and the daily volatility is 1%. The interest rate is taken to be zero. Inset: relative difference between the the log-normal, Black–Scholes price C_{BS} and the Gaussian, Bachelier price (C_G), for the same values of the parameters.

The dependence of $C_G(x_0, x_s, T)$ as a function of x_s is shown in Figure 13.2, where the numerical value of the relative difference between the Black–Scholes and Bachelier price is also plotted.

In a more general additive model, when N is finite, one can reconstruct the full distribution $P(x, N|x_0, 0)$ from the elementary distribution $P_1(\delta x)$ using the convolution rule (slightly modified to take into account the extra factors $(1 + \rho)^{N-k-1}$ in Eq. (13.40)). When N is large, the Gaussian approximation $P(x, N|x_0, 0) = P_G(x, N|x_0(1 + \rho)^N, 0)$ becomes accurate, according to the CLT discussed in Chapter 2.

As far as interest rate effects are concerned, it is interesting to note that both formulae, Eqs. (13.39) and (13.42), can be written as e^{-rT} times a function of $x_0 e^{rT}$, that is of the forward price \mathcal{F}_0. This is quite natural, since an option on a stock and on a forward must have the same price, since they have the same pay-off at expiry. As will be discussed in Section 17.1, this is a rather general property, not related to any particular model for price fluctuations.

13.3.4 Real option prices, volatility smile and 'implied' kurtosis

Stationary distributions and the smile curve

We shall now come back to the case where the distribution of the price increments δx_k is arbitrary, for example a truncated Levy distribution, or a Student distribution. For simplicity, we set the interest rate ρ to zero (a non-zero interest rate can readily be taken into account

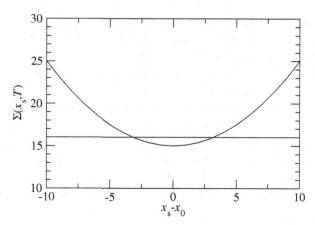

Fig. 13.3. 'Implied volatility' Σ as used by the market to account for non-Gaussian effects. The larger the difference between x_s and x_0, the larger the value of Σ: the curve has the shape of a smile. The function plotted here is a simple parabola, which is obtained theoretically by including the effect of a non-zero kurtosis κ_1, but a zero skewness, of the elementary price increments. Note that for at-the-money options ($x_s = x_0$), the implied volatility is smaller than the true volatility.

by discounting the price of the call on the forward contract), and assume that the maturity is such that an *additive* model for price changes is suitable. The price difference is then the sum of $N = T/\tau$ iid random variables, to which the discussion of the CLT presented in Chapter 2 applies directly. For large N, the terminal price distribution $P(x, N|x_0, 0)$ becomes Gaussian, while for N finite, 'fat tail' effects are important and the deviations from the CLT are noticeable. In particular, short maturity options or out-of-the-money options, or options on very 'jagged' assets, are not adequately priced by the Black–Scholes formula.

In practice, the market corrects for this difference empirically, by introducing in the Black–Scholes formula an *ad hoc* 'implied' volatility Σ, different from the 'true', historical volatility of the underlying asset. Furthermore, the value of the implied volatility needed to price options of different strike prices x_s and/or maturities T properly is not constant, but rather depends both on T and x_s. This defines a 'volatility surface' $\Sigma(x_s, T)$. It is usually observed that the larger the difference between x_s and x_0, the larger the implied volatility: this is the so-called 'smile effect' (Fig. 13.3). On the other hand, the longer the maturity T, the better the Gaussian approximation; the smile thus tends to flatten out with maturity.

It is important to understand how traders on option markets use the Black–Scholes formula. Rather than viewing it as a predictive theory for option prices given an observed (historical) volatility, the Black–Scholes formula allows the trader to translate back and forth between market prices (driven by supply and demand) and an abstract parameter called the implied volatility. In itself this transformation (between prices and volatilities) does not add any new information. However, this approach is useful in practice, since the implied volatility varies much less than option prices themselves as a function of maturity and moneyness. The Black–Scholes formula is thus viewed as a zero-th order approximation that takes into account the gross features of option prices, the traders then correcting for

other effects by adjusting the implied volatility. This view is quite common in experimental sciences where an *a priori* constant parameter of an approximate theory is made into a varying effective parameter to describe more subtle effects.

However, the use of the Black–Scholes formula (which assumes a constant volatility) with an *ad hoc* variable volatility is not very satisfactory from a theoretical point of view. Furthermore, this practice requires the manipulation of a whole volatility surface $\Sigma(x_s, T)$, which can deform in time – this is not particularly convenient.

A simple cumulant expansion allows one to understand the origin and the shape of the volatility smile.[†] Let us assume that the maturity $T = N\tau$ is sufficiently large so that only the skewness and kurtosis of $P_1(\delta x)$ must be taken into account to measure the difference with a Gaussian distribution. As shown below, the formula Eq. (13.34) reads, when the skewness is zero:

$$\Delta\mathcal{C}_\kappa(x_0, x_s, T) = \frac{\kappa_1 \tau}{24T}\sqrt{\frac{DT}{2\pi}}\exp\left[-\frac{(x_s - x_0)^2}{2DT}\right]\left(\frac{(x_s - x_0)^2}{DT} - 1\right), \tag{13.44}$$

where $D \equiv \sigma^2 x_0^2$ and $\Delta\mathcal{C}_\kappa = \mathcal{C}_\kappa - \mathcal{C}_{\kappa=0}$.

One can indeed transform the formula

$$\mathcal{C} = \int_{x_s}^{\infty} (x' - x_s) P(x', T|x_0, 0)\,dx', \tag{13.45}$$

through an integration by parts

$$\mathcal{C} = \int_{x_s}^{\infty} \mathcal{P}_>(x', T|x_0, 0)\,dx'. \tag{13.46}$$

After changing variables $x' \to x' - x_0$, and using Eq. (2.37) of Section 2.3.3, one gets:

$$\mathcal{C} = \mathcal{C}_G + \frac{\sqrt{DT}}{\sqrt{N}}\int_{u_s}^{\infty}\frac{1}{\sqrt{2\pi}}Q_1(u)e^{-u^2/2}\,du$$
$$+\frac{\sqrt{DT}}{N}\int_{u_s}^{\infty}\frac{1}{\sqrt{2\pi}}Q_2(u)e^{-u^2/2}\,du + \cdots \tag{13.47}$$

where $u_s \equiv (x_s - x_0)/\sqrt{DT}$. Now, noticing that:

$$Q_1(u)e^{-u^2/2} = \frac{\lambda_3}{6}\frac{d^2}{du^2}e^{-u^2/2}, \tag{13.48}$$

and

$$Q_2(u)e^{-u^2/2} = -\frac{\lambda_4}{24}\frac{d^3}{du^3}e^{-u^2/2} - \frac{\lambda_3^2}{72}\frac{d^5}{du^5}e^{-u^2/2}, \tag{13.49}$$

[†] The idea of a cumulant expansion for option prices was pioneered by Jarrow & Rudd (1982). The present results have been obtained in Potters et al. (1997) in the context of an additive model, and very similar results for the multiplicative model can be found in Backus et al. (1997). More recently, similar results have also been recovered in a very different language in Fouque et al. (2000), in the framework of a stochastic volatility model with short range correlations. As discussed in Section 7.1.4, a stochastic volatility gives rise to a non zero kurtosis decaying as $1/N$, which is indeed the small expansion parameter of Fouque et al. (2001).

the integrations over u are readily performed, yielding:

$$C = C_G + \sqrt{DT}\frac{e^{-u_s^2/2}}{\sqrt{2\pi}}\left(-\frac{\lambda_3}{6\sqrt{N}}u_s + \frac{\lambda_4}{24N}(u_s^2 - 1)\right.$$
$$\left. + \frac{\lambda_3^2}{72N}(u_s^4 - 6u_s^2 + 3) + \cdots\right), \tag{13.50}$$

which indeed coincides with Eq. (13.44) for $\lambda_3 = 0$. In the practical case $\lambda_3^2 \ll \lambda_4$, and we will neglect the last term in this equation.

Note that we refer here to *additive* (rather than multiplicative) price increments: the Black–Scholes volatility smile is often asymmetrical merely because of the use of a skewed log-normal distribution to describe a nearly symmetrical process (see the extensive discussion in Chapter 8).

On the other hand, a simple calculation shows that the variation of $C_{\kappa=0}(x_0, x_s, T)$ (as given by Eq. (13.43)) when the volatility changes by a small quantity $\delta D = 2\sigma x_0^2 \delta\sigma$ is given by:

$$\delta C_{\kappa=0}(x_0, x_s, T) = \delta\sigma\, x_0\sqrt{\frac{T}{2\pi}}\exp\left[-\frac{(x_s - x_0)^2}{2DT}\right]. \tag{13.51}$$

The effect of a non-zero skewness and kurtosis can thus be reproduced (to first order) by a Gaussian pricing formula, but at the expense of using an *effective* volatility $\Sigma(x_s, T) = \sigma + \delta\sigma$ given by:

$$\Sigma(x_s, T) = \sigma\left[1 + \frac{\varsigma(T)}{6}\mathcal{M} + \frac{\kappa(T)}{24}(\mathcal{M}^2 - 1)\right], \tag{13.52}$$

where $\mathcal{M} = (x_s - x_0)/\sqrt{DT}$ is the rescaled moneyness, $\varsigma(T)$ and $\kappa(T)$ are the skewness and kurtosis of the price increments on the scale of the maturity of the option.[†] This very simple formula, plotted in Figure 13.3, allows one to understand intuitively the shape and amplitude of the smile. For example, for $\kappa(T) = 1$ (typical of one month options), the kurtosis correction to the volatility is on the order of $\delta\sigma/\sigma \approx 33\%$ for strongly out-of-the-money options such that $\mathcal{M} = 3$. Note that the effect of the kurtosis is to *reduce* the implied at-the-money volatility in comparison with its true value. The effect of the skewness is to make the smile asymmetric around the strike price x_s: for a negative skewness, the smile remains a parabola, but centered around a shifted strike price $x_s \to x_s - 2\varsigma(T)\sigma\sqrt{T}/\kappa(T)$. When the skew is negative and large, the smile looks very much, around the money, like a straight line of negative slope. This shape is indeed observed for option on stock indices, for which the historical skew is indeed large (see Chapter 8).

[†] A similar formula, obtained in the context of the multiplicative (log-normal) Brownian motion was obtained in Backus et al. (1997), in the limit where the volatility σ is small. In this limit, one indeed expect the multiplicative and additive models to coincide.

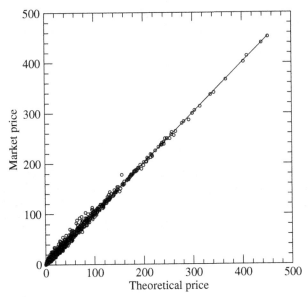

Fig. 13.4. Comparison between theoretical and observed option prices. Each point corresponds to an option on the Bund (traded on the LIFFE in London), with different strike prices and maturities. The x coordinate of each point is the theoretical price of the option (determined using the empirical terminal price distribution), while the y coordinate is the market price of the option. If the theory is good, one should observe a cloud of points concentrated around the line $y = x$. The line corresponds to a linear regression, and gives $y = 0.998x + 0.02$ (in basis points units).

Figure 13.4 shows some 'experimental' data, concerning options on the Bund (futures) contract – for which a weakly non-Gaussian model is satisfactory. The associated option market is, furthermore, very liquid; this tends to reduce the width of the bid-ask spread, and thus, probably to decrease the difference between the market price and a theoretical 'fair' price. This is not the case for OTC options, when the overhead charged by the financial institution writing the option can in some case be very large, cf. Section 14.1 below. In Figure 13.4, each point corresponds to an option of a certain maturity and strike price. The coordinates of these points are the theoretical price, Eq. (13.34), along the x axis (calculated using the historical distribution of the Bund), and the observed market price along the y axis. If the theory is good, one should observe a cloud of points concentrated around the line $y = x$. Figure 13.4 includes points from the first half of 1995, which was rather 'calm', in the sense that the volatility remained roughly constant during that period (see Fig. 13.5). Correspondingly, the assumption that the process is stationary is reasonable.

The agreement between theoretical and observed prices is however much less convincing if one uses data from 1994, when the volatility of interest rate markets has been high. The following subsection aims at describing these discrepancies in greater detail.

Non-stationarity and 'implied' kurtosis

As discussed in details in Chapter 7, a more precise analysis reveals that the *scale* of the fluctuations of the underlying asset (here the Bund contract) itself varies noticeably around

Fig. 13.5. Time dependence of the volatility σ, obtained from the analysis of the high frequency (intra-day) fluctuations of the underlying asset (here the Bund), or from the implied volatility of at-the-money options. More precisely, the historical determination of σ comes from the daily average of the absolute value of the 5-min price increments. The two curves are then smoothed over 5 days. These two determinations are strongly correlated, showing that the option price mostly reflects the instantaneous volatility of the underlying asset itself.

its mean value; these 'scale fluctuations' are furthermore correlated with the option prices. More precisely, one postulates that the distribution of price increments δx has a constant *shape*, but a width σ_k which is time dependent (cf. Chapter 7 and Eq. (13.3)).

Figure 13.5 shows the evolution of the volatility σ (filtered over 5 days in the past) as a function of time and compares it to the implied volatility $\Sigma(x_s = x_0)$ extracted from at-the-money options during the same period. One thus observes that the option prices are rather well tracked by adjusting the factor σ_k through a short-term estimate of the volatility of the underlying asset. This means that the option prices primarily reflects the short term average volatility of the underlying asset itself, rather than an 'anticipated' average volatility on the life time of the option. More precisely, the short term average volatility is found empirically to be on average a better predictor of future realized volatilities than the implied volatility. Of course, in some market conditions, the volatility is expected to rise or fall because of – say – political events, an effect that cannot be captured by the analysis of the past time series.

It is interesting to notice that the mere existence of volatility fluctuations leads to a non-zero kurtosis κ_N of the asset fluctuations (see Section 7.1). This kurtosis has an anomalous time dependence (cf. Eq. (7.15)), in agreement with the direct observation of the volatility fluctuations. On the other hand, an 'implied kurtosis' can be extracted from the market price of options, using Eq. (13.52) as a fit to the empirical smile, see Figure 13.6. Remarkably

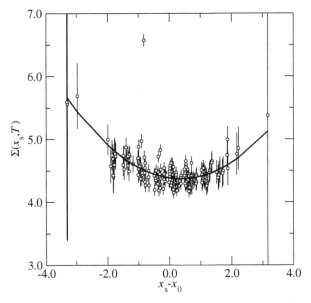

Fig. 13.6. Implied Black–Scholes volatility and fitted parabolic smile. The circles correspond to all quoted prices on 26th April, 1995, on options of 1-month maturity. The error bars correspond to an error on the price of ± 1 basis point. The curvature of the parabola allows one to extract the implied kurtosis $\kappa(T)$ using Eq. (13.52).

enough, this implied kurtosis is in close correspondence with the historical kurtosis – note that Figure 13.7 does not require any further adjustable parameter.

As a conclusion of this section, it seems that market operators have empirically corrected the Black–Scholes formula to account for two distinct, but related effects:

- The presence of market jumps, implying fat tailed distributions ($\kappa > 0$) of short-term price increments. This effect is responsible for the volatility smile (and also, as we shall discuss next, for the fact that options are risky).
- The fact that the volatility is not constant, but fluctuates in time, leading to an anomalously slow decay of the kurtosis (slower than $1/N$) and, correspondingly, to a non-trivial deformation of the smile with maturity.

It is interesting to note that, through trial and errors, the market as a whole has evolved to allow for such non-trivial statistical features – at least on most actively traded markets. This might be called 'market efficiency'; but contrarily to stock markets where it is difficult to judge whether the stock price is or is not the 'true' price (which might be an empty concept), option markets offer a remarkable testing ground for this idea (the fact that options have a finite life time allows one to judge ex post if their price is fair on average). Option markets

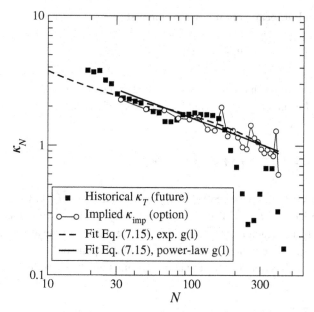

Fig. 13.7. Plot (in log–log coordinates) of the average implied kurtosis κ_{imp} (determined by fitting the implied volatility for a fixed maturity by a parabola) and of the empirical kurtosis κ_N (determined directly from the historical movements of the Bund contract), as a function of the reduced time scale $N = T/\tau$, $\tau = 30$ min. All transactions of options on the Bund future from 1993 to 1995 were analysed along with 5 minute tick data of the Bund future for the same period. We show for comparison a fit with $\kappa_N \approx N^{-0.42}$ (dark line), as suggested by the results of Section 7.1.4. A fit with an exponentially decaying volatility correlation function is however also acceptable (dashed line).

offer a nice example of adaptation of a population (the option traders) to a complex and hostile environment, which has taken place in only a few decades!

13.3.5 The case of an infinite kurtosis

Since there is some evidence that the kurtosis might not always be finite for financial time series, the above cumulant expansion might be problematic. It is interesting to consider a particular case where the kurtosis of the terminal distribution is infinite and where an explicit formula for the price of options can be written down. It turns out that when the distribution of price changes obeys a Student distribution with $\mu = 3$ (i.e. with the value of the tail exponent in agreement with recent studies, see Chapter 6), a very simple formula for option prices can be obtained.

Suppose that the price change $\Delta x = x_T - x_0$ between now $t = 0$ and maturity $t = T$ is distributed according to:

$$P(\Delta x) = \frac{2}{\pi} \frac{(DT)^{3/2}}{(DT + \Delta x^2)^2},$$

(13.53)

where DT is the variance of the price difference on the time scale T.

The price of a European option, for zero interest rates, is given by:

$$\mathcal{C}(x_0, x_s, T) = \int_{x_s}^{\infty} (x - x_s) P(x - x_0) \, dx. \tag{13.54}$$

It turns out that the above integral can be computed exactly and leads to a rather simple result (see Eq. 1.40). Introducing the rescaled moneyness of the option $\mathcal{M} = (x_s - x_0)/\sqrt{DT}$, the option price reads:

$$\mathcal{C}(x_0, x_s, T) = \frac{\sqrt{DT}}{2\pi} \left(2 - \pi\mathcal{M} + 2\mathcal{M} \arctan \mathcal{M}\right). \tag{13.55}$$

At the money ($\mathcal{M} = 0$), the price of the option is equal to \sqrt{DT}/π, which is smaller than the Gaussian price $\sqrt{DT/2\pi}$. The 'implied volatility' is therefore reduced, compared to the true volatility, by a factor $\sqrt{2/\pi} = 0.798$. This reduction of the at-the-money volatility was already noted above in the context of a cumulant expansion.

For deep out-of-the-money calls, the above formula can be expanded as:

$$\mathcal{C}(x_0, x_s, T) \simeq \frac{\sqrt{DT}}{3\pi\mathcal{M}^2} + \cdots; \tag{13.56}$$

where the slow \mathcal{M}^{-2} decay reflects that of the Student distribution above. Finally, for deep in-the-money calls, one finds:

$$\mathcal{C}(x_0, x_s, T) \simeq (x_0 - x_s) + \frac{\sqrt{DT}}{\pi} + \cdots \tag{13.57}$$

Another way to present the results is to draw the implied volatility $\Sigma(\mathcal{M})$ such that a Gaussian formula with such an implied volatility exactly reproduces Eq. (13.55) above. The shape of the smile in this case is shown in Fig. (13.8). Note that for large moneyness, the smile becomes approximately linear.

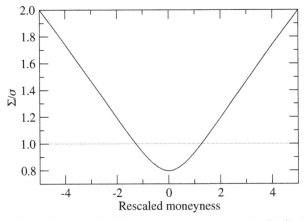

Fig. 13.8. Shape of the smile for an infinite kurtosis case, here a Student distribution with $\mu = 3$.

An explicit formula for the price of the option can actually be obtained for all integer values of μ (see e.g. Eq. (1.41)).

13.4 Summary

- The problem of derivative pricing and hedging consist in controlling and 'chiseling' the distribution of profits and losses associated to the writing of a contract. Both the average value and the shape of this distribution are affected by the price of the contract and the hedging strategy. The variance of this distribution can be taken as a simple indicator of risk, although other indicators, based on tail events, can also be considered.

- The fair price of a contract is such that the average profit, including the cost of the hedging strategy, is zero. When the risk can be made zero, the fair price is such that there are no arbitrage opportunities. When the risk is non zero, the price of the contract may reflect a risk premium, which can be large in an asymmetric market where buyers and sellers of the contract have different perceptions of risk.

- In the case of futures contracts, a perfect hedging strategy exists. This uniquely fixes the price of futures as $x_0 e^{rT}$, when x_0 is the current price of the underlying and T the maturity.

- In the case of options, the fair-price is, when the mean return of the underlying is equal to the risk free rate, the average of the final pay-off of the option over the distribution of price histories. For a Gaussian process, one finds the Bachelier formula (for the additive case) and the Black-Scholes formula (for the multiplicative case). In the case of non-Gaussian markets, corrections to these formula appear. These can be encoded in a modified ('implied') value of the volatility used in the Black-Scholes formula, which becomes both strike and maturity dependent.

- Part of the information contained in this implied volatility surface used by market participants can be explained by an adequate statistical model of the underlying asset fluctuations. In particular, in weakly non-Gaussian markets, important parameters are the time-dependent skewness and kurtosis, see Eq. (13.52), that give respectively the skewness and curvature of the smile. The anomalous maturity dependence of the kurtosis encodes the fact that the volatility itself has long range correlations.

- FURTHER READING

L. Bachelier, *Théorie de la Spéculation*, Gabay, Paris, 1995.

M. Baxter, A. Rennie, *Financial Calculus, An Introduction to Derivative Pricing*, Cambridge University Press, Cambridge, 1996.

J. P. Fouque, G. Papanicolaou, R. Sircar, *Derivatives in Financial Markets with Stochastic Volatility*, Cambridge University Press, 2000.

J. C. Hull, *Futures, Options and Other Derivatives*, Prentice Hall, Upper Saddle River, NJ, 1997.

M. Musiela, M. Rutkowski, *Martingale Methods in Financial Modelling*, Springer, New York, 1997.

D. Lamberton, B. Lapeyre, *Introduction au calcul stochastique appliqué à la finance*, Ellipses, Paris, 1991.

M. Musiela, M. Rutkowski, *Martingale Methods in Financial Modelling*, Springer, New York, 1997.

I. Nelken (ed.), *The Handbook of Exotic Options*, Irwin, Chicago, IL, 1996.

P. Wilmott, J. N. Dewynne, S. D. Howison, *Option Pricing: Mathematical Models and Computation*, Oxford Financial Press, Oxford, 1993.

P. Wilmott, *Derivatives, The theory and practice of financial engineering*, John Wiley, 1998.

P. Wilmott, *Paul Wilmott Introduces Quantitative Finance*, John Wiley, 2001.

• SPECIALIZED PAPERS: SOME CLASSICS

F. Black, M. Scholes, *The pricing of options and corporate liabilities*, Journal of Political Economy, **3**, 637 (1973).

J. C. Cox, S. A. Ross, M. Rubinstein, *Option pricing: a simplified approach*, Journal of Financial Economics, **7**, 229 (1979).

R. C. Merton, *Theory of rational option pricing*, Bell Journal of Economics and Management Science, **4**, 141 (1973).

• SPECIALIZED PAPERS: ALTERNATIVE MODELS

L. Borland, *Option pricing formulas based on a non-Gaussian stock price model*, Physical Review Letters, **89**, 09870 (2002); Quantitative Finance, **2**, 415 (2002).

P. Carr, H. Geman, D. Madan, M. Yor, *The fine structure of asset returns: an empirical investigation*, forthcoming, Journal of Business (2000).

E. Derman, I. Kani, *Stochastic implied trees: arbitrage pricing with stochastic term and strike structure of volatility*, Int. Jour. Theo. Appl. Fin., **1**, 61 (1998).

E. Eberlein, U. Keller, *Option pricing with hyperbolic distributions*, Bernoulli, **1**, 281 (1995).

J.-P. Fouque, G. Papanicolaou, K. R. Sircar, *Mean-reverting stochastic volatility*, Int. Jour. Theo. Appl. Fin., **3**, 101 (2000).

S. L. Heston, *A closed-form solution for options with stochastic volatility with applications to bond and currency options*, Review of Financial Studies, **6**, 327 (1993).

J. K. Hoogland, C. D. D. Newmann, *Local scale invariance and contingent claim pricing I*, Int. Jour. Theo. Appl. Fin., **4**, 1 (2001).

J. K. Hoogland, C. D. D. Newmann, *Local scale invariance and contingent claim pricing II*, Int. Jour. Theo. Appl. Fin., **4**, 23 (2001).

J. C. Hull, A. White, *The pricing of options on assets with stochastic volatility*, Journal of Finance, **XLII**, 281 (1987).

J. C. Hull, A. White, *An analysis of the bias in option pricing caused by stochastic volatility*, Advances in Futures and Option Research, **3**, 29 (1988).

R. W. Lee, *Implied and local volatilities under stochastic volatility*, Int. Jour. Theo. Appl. Fin., **4**, 45 (2001).

A. Matacz, *Financial Modelling and Option Theory with the Truncated Levy Process*, Int. Jour. Theo. Appl. Fin., **3**, 703 (2000).

• SPECIALIZED PAPERS: CUMULANT EXPANSION AND SMILES

D. Backus, S. Foresi, K. Lai, L. Wu, *Accounting for biases in Black-Scholes*, Working paper of NYU Stern School of Business, 1997.

C. J. Corrado, T. Su, *Skewness and kurtosis in S&P500 index returns implied by option prices*, Journal of Financial Research, **XIX**, 175 (1996).

R. Cont, J. da Fonseca, *Dynamics of implied volatility surfaces*, Quantitative Finance, **2**, 45 (2002), and references therein.

J. C. Jackwerth, M. Rubinstein, *Recovering probability distributions from option prices*, Journal of Finance, **51**, 1611.

R. Jarrow, A. Rudd, *Approximate option valuation for arbitrary stochastic processes*, Journal of Financial Economics, **10**, 347 (1982).

B. Pochart, J.-P. Bouchaud, *The skewed multifractal random walk with applications to option smiles*, Quantitative Finance, **2**, 303 (2002).

M. Potters, R. Cont, J.-P. Bouchaud, *Financial markets as adaptive ecosystems*, Europhysics Letters, **41**, 239 (1998).

• SOME RECENT READS

R. Cont, P. Tankov, *Financial Modelling with Jump Processes*, Chapman & Hall / CRC Press, 2003.

J. Gatheral, *The Volatility Surface: A Practitioner's Guide*, Wiley Finance, 2006.

14

Options: hedging and residual risk

La nécessité des affaires nous oblige souvent à nous determiner avant que nous ayons
eu le loisir de les examiner soigneusement.[†]

(René Descartes, *Méditations*.)

14.1 Introduction

14.2 Residual risk

In the previous chapter, we have chosen a model of price increments such that the average
cost of the hedging strategy (i.e. the term $\langle \psi_k \delta x_k \rangle$) could be neglected, which is justified if
the excess return m is small, or for short maturity options. Beside the fact that the correction
to the option price induced by a non-zero return is important to assess (this will be done in
the next chapter), the determination of an 'optimal' hedging strategy and the corresponding
minimal residual risk is crucial for the following reason. Suppose that an adequate measure
of the risk taken by the writer of the option is given by the variance of the global wealth
balance associated to the operation, i.e.:[‡]

$$\mathcal{R}^2 = \langle \Delta W^2 \rangle. \tag{14.1}$$

As we shall find below, there is a special strategy ϕ^* such that the above quantity reaches a
minimum value. Within the hypotheses of the Black–Scholes model, this minimum risk is
even, rather surprisingly, *strictly zero*. Under less restrictive and more realistic hypotheses,
however, the residual risk $\mathcal{R}^* \equiv \sqrt{\langle \Delta W^2 [\phi^*] \rangle}$ actually amounts to a substantial fraction
of the option price itself. It is thus rather natural for the writer of the option to try to
reduce further the loss probability by overpricing the option, adding to the 'fair price' a risk
premium proportional to \mathcal{R}^*.

 Therefore, a market maker on option markets will offer to buy and to sell options at
slightly different prices (the 'bid-ask' spread), centred around the fair price \mathcal{C}. The am-
plitude of the spread is presumably governed by the residual risk, and is thus $\pm \lambda \mathcal{R}^*$,
where λ is a certain numerical factor, which measures the price of risk. The search
for minimum risk corresponds to the need of keeping the bid-ask spread as small as

[†] Some matters force us to take decisions before we have had time to examine them carefully.
[‡] The case where a better measure of risk is the loss probability or the value-at-risk will be discussed in Section 14.4.

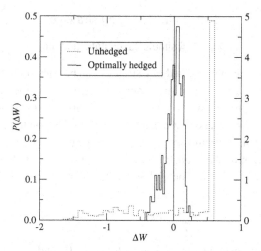

Fig. 14.1. Histogram of the wealth balance ΔW associated to writing of a certain at-the-money option of maturity 60 trading days. The dotted line corresponds to the case where the option is not hedged ($\phi \equiv 0$). The RMS of the distribution is 1.04, to be compared with the price of the option $C = 0.6$. The 'peak' at 0.6 thus corresponds to the cases where the option is not exercised. The full line corresponds to the optimal hedge ($\phi = \phi^*$), recalculated every half-hour. The RMS \mathcal{R}^* is clearly smaller ($= 0.28$), but non-zero. Note that the distribution is skewed towards $\Delta W < 0$.

possible, because several competing market makers are present. One therefore expects the coefficient λ to be smaller on liquid markets, and the price closer to the fair price. On the contrary, the writer of an OTC option usually claims a rather high risk premium λ.

Let us illustrate this idea by Figure 14.1, generated using real market data (these figures should be compared to the schematized graph, Figure 13.1). We show the histogram of the global wealth balance ΔW, corresponding to an at-the-money option, of maturity equal to 3 months, in the case of a bare position ($\phi \equiv 0$), and in the case where one follows the optimal strategy ($\phi = \phi^*$) prescribed below. The fair price of the option is such that $\langle \Delta W \rangle = 0$ (vertical thick line). It is clear that without hedging, option writing is a very risky operation. The optimal hedge substantially reduces the risk, though the residual risk remains quite high. Increasing the price of the option by an amount $\lambda \mathcal{R}^*$ corresponds to a shift of the vertical thick line to the left of Figure 14.1, thereby reducing the weight of unfavourable events.

Another way of expressing this idea is to use \mathcal{R}^* as a scale to measure the difference between the market price of an option \mathcal{C}_M and its theoretical price:

$$\lambda = \frac{\mathcal{C}_M - \mathcal{C}}{\mathcal{R}^*}.$$

(14.2)

An option with a large value of λ is an expensive option, which includes a large risk premium.

14.2 Optimal hedging strategies

14.2.1 A simple case: static hedging

Let us now discuss how an optimal hedging strategy can be constructed, by focusing first on the simple case where the amount of the underlying asset held in the portfolio is fixed once and for all when the option is written, i.e. at $t = 0$. This extreme case would correspond to very high transaction costs, so that changing one's position on the market is very costly. Another situation is that of illiquid assets, such as hedge funds for example, which are often only liquid once a month. We shall furthermore assume, for simplicity, that interest rate effects are negligible, i.e. $\rho = 0$. The global wealth balance, Eq. (13.28), then reads:

$$\Delta W = \mathcal{C} - \max(x_N - x_s, 0) + \phi \sum_{k=0}^{N-1} \delta x_k. \tag{14.3}$$

In the case where the average return is zero ($\langle \delta x_k \rangle = 0$) and where the increments are uncorrelated (i.e. $\langle \delta x_k \delta x_l \rangle = D\tau \delta_{k,l}$), the variance of the final wealth, $\mathcal{R}^2 = \langle \Delta W^2 \rangle - \langle \Delta W \rangle^2$, reads:

$$\mathcal{R}^2 = ND\tau\phi^2 - 2\phi\langle(x_N - x_0)\max(x_N - x_s, 0)\rangle + \mathcal{R}_0^2, \tag{14.4}$$

where \mathcal{R}_0^2 is the intrinsic risk, associated to the 'bare', unhedged option ($\phi = 0$):

$$\mathcal{R}_0^2 = \langle \max(x_N - x_s, 0)^2 \rangle - \langle \max(x_N - x_s, 0) \rangle^2. \tag{14.5}$$

The dependence of the risk on ϕ is shown in Figure 14.2, and indeed reveals the existence of an optimal value $\phi = \phi^*$ for which \mathcal{R} takes a minimum value. Taking the derivative of Eq. (14.4) with respect to ϕ,

$$\left.\frac{d\mathcal{R}}{d\phi}\right|_{\phi=\phi^*} = 0, \tag{14.6}$$

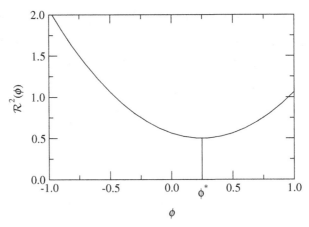

Fig. 14.2. Dependence of the residual risk \mathcal{R} as a function of the strategy ϕ, in the simple case where this strategy does not vary with time. Note that \mathcal{R} is minimum for a well-defined value of ϕ.

one finds:

$$\phi^* = \frac{1}{D\tau N} \int_{x_s}^{\infty} (x - x_s)(x - x_0) P(x, N | x_0, 0) \, dx, \tag{14.7}$$

thereby fixing the optimal hedging strategy within this simplified framework. Interestingly, if $P(x, N | x_0, 0)$ is Gaussian (which becomes a better approximation as $N = T/\tau$ increases), one can write:

$$\frac{1}{DT} \int_{x_s}^{\infty} \frac{1}{\sqrt{2\pi DT}} (x - x_s)(x - x_0) \exp\left[-\frac{(x - x_0)^2}{2DT}\right] dx =$$

$$-\int_{x_s}^{\infty} \frac{1}{\sqrt{2\pi DT}} (x - x_s) \frac{\partial}{\partial x} \exp\left[-\frac{(x - x_0)^2}{2DT}\right] dx, \tag{14.8}$$

giving, after an integration by parts: $\phi^* \equiv \mathcal{P}$, or else the probability, calculated from $t = 0$, that the option is exercised at maturity: $\mathcal{P} \equiv \int_{x_s}^{\infty} P(x, N | x_0, 0) \, dx$.

Hence, buying a certain fraction of the underlying allows one to reduce the risk associated to the option. As we have formulated it, risk minimization appears as a *variational problem*. The result therefore depends on the family of 'trial' strategies ϕ. A natural idea is thus to generalize the above procedure to the case where the strategy ϕ is allowed to vary both with time, and with the price of the underlying asset. In other words, one can certainly do better than holding a certain fixed quantity of the underlying asset, by adequately readjusting this quantity in the course of time.

14.2.2 The general case and 'Δ' hedging

If one writes again the complete wealth balance, Eq. (13.26), as:

$$\Delta W = \mathcal{C}(1 + \rho)^N - \max(x_N - x_s, 0) + \sum_{k=0}^{N-1} \psi_k^N(x_k) \delta x_k, \tag{14.9}$$

the calculation of $\langle \Delta W^2 \rangle$ involves, as in the simple case treated above, three types of terms: quadratic in ψ, linear in ψ, and independent of ψ. The last family of terms can thus be forgotten in the minimization procedure. The first two terms of \mathcal{R}^2 are:

$$\sum_{k=0}^{N-1} \langle (\psi_k^N)^2 \rangle \langle \delta x_k^2 \rangle_k - 2 \sum_{k=0}^{N-1} \langle \psi_k^N \delta x_k \max(x_N - x_s, 0) \rangle_k. \tag{14.10}$$

The subscript k indicates that the average is taken conditional to all information available at time k. We have used $\langle \delta x_k \rangle_k = 0$ and assumed the increments to be of finite variance and uncorrelated (but not necessarily stationary nor independent):[†]

$$\langle \delta x_k \delta x_l \rangle_k = \langle \delta x_k^2 \rangle_k \delta_{k,l}. \tag{14.11}$$

We introduce again the distribution $P(x, k | x_0, 0)$ for arbitrary intermediate times k. The dynamic strategy ψ_k^N may now depend on the value of the price x_k. One can express the

[†] The case of correlated Gaussian increments is treated in Section 15.3.

terms appearing in Eq. (14.10) in the following form:

$$\langle (\psi_k^N)^2 \rangle \langle \delta x_k^2 \rangle_k = \int \left[\psi_k^N(x) \right]^2 P(x, k | x_0, 0) \langle \delta x_k^2 \rangle_k \, dx, \tag{14.12}$$

and

$$\langle \psi_k^N \delta x_k \max(x_N - x_s, 0) \rangle_k = \int \psi_k^N(x) P(x, k | x_0, 0) \, dx$$
$$\times \int_{x_s}^{+\infty} \langle \delta x_k \rangle_{(x,k) \to (x',N)} (x' - x_s) P(x', N | x, k) \, dx', \tag{14.13}$$

where the notation $\langle \delta x_k \rangle_{(x,k) \to (x',N)}$ means that the average of δx_k is restricted to those trajectories which start from point x at time k and end at point x' at time N, conditional to all information available at time k. Without the restriction on the end points, the average would of course be zero since we have assumed $\langle \delta x_k \rangle$ to be zero.

The *functions* $\psi_k^N(x)$, for $k = 1, 2, \ldots, N$, must be chosen so that the risk \mathcal{R} is as small as possible. Technically, one must 'functionally' minimize Eq. (14.10). We shall not try to justify this procedure mathematically; one can simply imagine that the integrals over x are discrete sums (which is actually true, since prices are expressed in cents and therefore are not continuous variables). The function $\psi_k^N(x)$ is thus determined by the discrete set of values of $\psi_k^N(i)$. One can then take usual derivatives of Eq. (14.10) with respect to these $\psi_k^N(i)$. The continuous limit, where the points i become infinitely close to one another, allows one to define the functional derivative $\partial / \partial \psi_k^N(x)$, but it is useful to keep in mind its discrete interpretation. Using Eqs. (14.12) and (14.13), one thus finds the following fundamental equation:

$$\frac{\partial \mathcal{R}}{\partial \psi_k^N(x)} = 2\psi_k^N(x) P(x, k | x_0, 0) \langle \delta x_k^2 \rangle_k$$
$$- 2P(x, k | x_0, 0) \int_{x_s}^{+\infty} \langle \delta x_k \rangle_{(x,k) \to (x',N)} (x' - x_s) P(x', N | x, k) \, dx'. \tag{14.14}$$

Setting this functional derivative to zero then provides a general and rather explicit expression for the optimal strategy $\psi_k^{N*}(x)$, where the *only assumption* is that the increments δx_k are uncorrelated (cf. Eq. (14.11)):[†]

$$\psi_k^{N*}(x) = \frac{1}{\langle \delta x_k^2 \rangle_k} \int_{x_s}^{+\infty} \langle \delta x_k \rangle_{(x,k) \to (x',N)} (x' - x_s) P(x', N | x, k) \, dx'. \tag{14.15}$$

This formula can be simplified in the case where the increments δx_k are identically distributed (of variance $\langle \delta x_k^2 \rangle_k \equiv D\tau$) and when the interest rate is zero. As shown in

[†] This is true within the variational space considered here, where ϕ only depends on x and t. If some volatility correlations are present, one could in principle consider strategies that also depend on the past values of the volatility.

Appendix D, one then has:

$$\langle \delta x_k \rangle_{(x,k)\to(x',N)} = \frac{x' - x}{N - k}. \tag{14.16}$$

The intuitive interpretation of this result is that all the $N - k$ increments δx_l along the trajectory $(x, k) \to (x', N)$ contribute on average equally to the overall price change $x' - x$. The optimal strategy then finally reads, for $\rho = 0$:

$$\phi_k^{N*}(x) = \int_{x_s}^{+\infty} \frac{x' - x}{D\tau(N - k)}(x' - x_s)P(x', N|x, k)\,\mathrm{d}x'. \tag{14.17}$$

One can show that the above expression varies monotonically from $\phi_k^{N*}(-\infty) = 0$ to $\phi_k^{N*}(+\infty) = 1$; this result is valid without any further assumption on $P(x', N|x, k)$.

In order to show this, note that within an additive model, $P(x', N|x, k)$ only depends on $x' - x$, and we will write $P(x', N|x, k) = f_{N-k}(x - x')$. Introducing the moneyness $M = x_s - x$ of the option, one finds, after setting $u = x' - x$:

$$\phi_k^{N*}(x) = \int_M^{+\infty} \frac{u}{D\tau(N - k)}(u - M)f_{N-k}(u)\,\mathrm{d}u. \tag{14.18}$$

Note that since $f_{N-k}(u)$ is a probability density that we assume to be of finite variance, it must decay faster than $1/u^3$. Using this, and the fact that: $\int u f_{N-k}(u)\mathrm{d}u = 0$ and $\int u^2 f_{N-k}(u)\mathrm{d}u = D\tau(N - k)$, one sees directly that $\phi_k^{N}(-\infty) = 0$ and $\phi_k^{N*}(+\infty) = 1$. Furthermore, taking once the derivative of $\phi_k^{N*}(x)$ with respect to x gives:*

$$\frac{\partial \phi_k^{N*}(x)}{\partial x} = \int_M^{+\infty} \frac{u}{D\tau(N - k)}f_{N-k}(u)\,\mathrm{d}u, \tag{14.19}$$

which is seen to be zero both for $x \to 0$ and to $x \to \infty$. Taking one further derivative with respect to x leads to:

$$\frac{\partial^2 \phi_k^{N*}(x)}{\partial x^2} = \frac{x_s - x}{D\tau(N - k)}f_{N-k}(x_s - x). \tag{14.20}$$

This last quantity has a unique zero for $x = x_s$, is positive for $x > x_s$ and negative for $x < x_s$. Therefore $\partial \phi_k^{N}(x)/\partial x$ must be everywhere non negative.*

If $P(x', N|x, k)$ is well approximated by a Gaussian, the following identity then holds true:

$$\frac{x' - x}{D\tau(N - k)}P_G(x', N|x, k) \equiv \frac{\partial P_G(x', N|x, k)}{\partial x}. \tag{14.21}$$

The comparison between Eqs. (14.17) and (13.33) then leads to the so-called 'Delta' hedging of Black and Scholes:

$$\phi_k^{N*}(x = x_k) = \frac{\partial \mathcal{C}_{BS}[x, x_s, N - k]}{\partial x}\bigg|_{x=x_k}$$
$$\equiv \Delta(x_k, N - k). \tag{14.22}$$

One can actually show that this result is true even if the interest rate is non-zero (see Appendix D), as well as if the relative returns (rather than the absolute returns) are independent Gaussian variables, as assumed in the Black–Scholes model (see Section 14.2.3 below.)

Equation (14.22) has a very simple (perhaps sometimes misleading) interpretation. If between time k and $k + 1$ the price of the underlying asset varies by a small amount dx_k, then to first order in dx_k, the variation of the option price is given by $\Delta[x, x_s, N - k] dx_k$ (using the very definition of Δ as the derivative of the price). This variation is therefore exactly compensated by the gain or loss of the strategy, i.e. $\phi_k^{N*}(x = x_k) dx_k$. In other words, $\phi_k^{N*} = \Delta(x_k, N - k)$ appears to be a perfect hedge, which ensures that the portfolio of the writer of the option does not vary *at all* with time (cf. Chapter 16, when we will discuss the differential approach of Black and Scholes). However, as we discuss now, the relation between the optimal hedge and Δ is not general, and does not hold for non-Gaussian statistics.

Cumulant corrections to Δ hedging

More generally, using Fourier transforms, one can express Eq. (14.17) as a cumulant expansion. Using the definition of the cumulants of P, one has:

$$(x' - x)P(x', N|x, 0) \equiv \frac{1}{2\pi} \int \hat{P}_N(z) \frac{\partial}{\partial(-iz)} e^{-iz(x'-x)} dz$$

$$= -\sum_{n=2}^{\infty} \frac{(-)^n c_{n,N}}{(n-1)!} \frac{\partial^{n-1}}{\partial x'^{m-1}} P(x', N|x, 0), \qquad (14.23)$$

where $\log \hat{P}_N(z) = \sum_{n=2}^{\infty}(iz)^n c_{n,N}/n!$. Assuming that the increments are independent, one can use the additivity property of the cumulants discussed in Chapter 2, i.e.: $c_{n,N} \equiv Nc_{n,1}$, where the $c_{n,1}$ are the cumulants at the elementary time scale τ. One finds the following general formula for the optimal strategy:[†]

$$\phi^{N*}(x) = \frac{1}{D\tau} \sum_{n=2}^{\infty} \frac{c_{n,1}}{(n-1)!} \frac{\partial^{n-1} C[x, x_s, N]}{\partial x^{n-1}}, \qquad (14.24)$$

where $c_{2,1} \equiv D\tau$, $c_{4,1} \equiv \kappa_1[c_{2,1}]^2$, etc. In the case of a Gaussian distribution, $c_{n,1} \equiv 0$ for all $n \geq 3$, and one thus recovers the previous 'Delta' hedging. If the distribution of the elementary increments δx_k is symmetrical, $c_{3,1} = 0$. The first correction is then of order $c_{4,1}/c_{2,1} \times (1/\sqrt{D\tau N})^2 \equiv \kappa/N$, where we have used the fact that $C[x, x_s, N]$ typically varies with x on the scale $\sqrt{D\tau N}$, which allows us to estimate the order of magnitude of its derivatives with respect to x. Equation (14.24) shows that in general, the optimal strategy is not simply given by the derivative of the price of the option with respect to the price of the underlying asset.

It is interesting to compute the difference between ϕ^ and the Black–Scholes strategy as used by the traders, ϕ_M^*, which takes into account the implied volatility of the option $\Sigma(x_s, T = N\tau)$:[‡]*

$$\phi_M^*(x, \Sigma) \equiv \left. \frac{\partial C[x, x_s, T]}{\partial x} \right|_{\sigma=\Sigma}. \qquad (14.25)$$

[†] This formula can actually be generalized to the case where the volatility is time dependent, see Appendix D and Section 5.2.3.

[‡] Note that the market does not compute the total derivative of the option price with respect to the underlying price, which would also include the derivative of the implied volatility.

If one chooses the implied volatility Σ in such a way that the 'correct' price (cf. Eq. (13.52) above) is reproduced, one can show that:

$$\phi^*(x) = \phi_M^*(x, \Sigma) + \frac{\kappa_T}{12} \frac{\mathcal{M}}{\sqrt{2\pi}} \exp\left[-\frac{\mathcal{M}^2}{2}\right], \tag{14.26}$$

where $\mathcal{M} = (x_s - x_0)/\sqrt{DT}$ is the rescaled moneyness, κ_T is the kurtosis on scale T and higher-order cumulants are neglected. The Black–Scholes strategy, even calculated with the implied volatility, is thus not the optimal strategy when the kurtosis is non-zero. However, the difference between the two is zero both at the money and far from the money. It reaches a maximum when $\mathcal{M} = 1$, and is equal to:

$$\delta\phi^* \approx 0.02\kappa_T \qquad T = N\tau. \tag{14.27}$$

For $T = 20$ days, and a typical kurtosis of $\kappa_T = 1$, one finds that the first correction to the optimal hedge is on the order of 0.02, see also Figure 14.3. As will be discussed below, the use of such an approximate hedging induces a rather small increase of the residual risk.

For the infinite kurtosis case treated in Section 13.3.5, the optimal hedge is given by:

$$\phi^*(x_0, x_s, T) = \frac{1}{2} - \frac{1}{\pi}\arctan\mathcal{M}, \tag{14.28}$$

that has the expected limits $\phi^(-\infty) = 1$, $\phi^*(0) = 1/2$, and $\phi^*(+\infty) = 0$.*

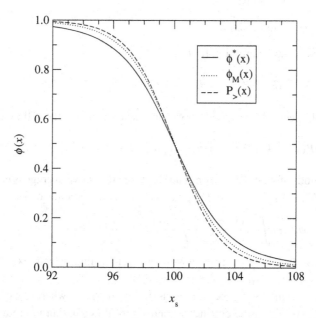

Fig. 14.3. Three hedging strategies for an option on an asset of terminal value x_N distributed according to a TLD, of kurtosis $\kappa = 5$, as a function of the strike price x_s. ϕ^* denotes the optimal strategy, calculated from Eq. (14.17), ϕ_M is the Black–Scholes strategy computed with the appropriate implied volatility, such as to reproduce the correct option price, and $\mathcal{P}_>$ is the probability that the option is exercised at expiry. These three strategies are actually quite close to each other (they coincide at-the-money, and deep in-the-money and out-the-money). Note however that, the variations of ϕ^* with the price are slower (smaller Γ).

14.2.3 Global hedging vs. instantaneous hedging

It is important to stress that the above procedure, where the global risk associated to the option (calculated as the variance of the final wealth balance) is minimized is *equivalent* to the minimization of an 'instantaneous' hedging error – at least when the increments are independent. The latter consists in minimizing the difference of value of a position which is short one option and ϕ long in the underlying stock, between two consecutive times $k\tau$ and $(k+1)\tau$. This is closer to the concern of many option traders, since the value of their option books is calculated every day: the risk is estimated in a 'marked to market' way. If the price increments are iid, we show now that the optimal strategy is indeed identical to the 'global' one discussed above.

The change of wealth between times k and $k+1$ is, in the absence of interest rates, given by:

$$\delta W_k = \mathcal{C}_k - \mathcal{C}_{k+1} + \phi_k(x_k)\delta x_k. \tag{14.29}$$

The global wealth balance is simply given by the sum over k from 0 to $N-1$ of all these δW_k. If the price increments are uncorrelated, so are the δW_k's. Therefore, the variance of the total wealth balance is the sum of the instantaneous variances $\langle \delta W_k^2 \rangle$. Minimizing the global risk therefore requires the instantaneous risk to be minimum.

Let us now calculate the part of $\langle \delta W_k^2 \rangle$ which depends on ϕ_k. Using the fact that the increments are iid, one finds:

$$\phi_k^2(x_k)\langle \delta x_k^2 \rangle - 2\phi_k(x_k)\langle \mathcal{C}_{k+1}\delta x_k \rangle. \tag{14.30}$$

Using now the explicit expression for \mathcal{C}_{k+1}:

$$\mathcal{C}_{k+1} = \int_{x_s}^{\infty} (x' - x_s)P(x', N|x_{k+1}, k+1)\,dx', \tag{14.31}$$

one sees that the second term in the above expression contains the following average:

$$\int (x_{k+1} - x_k)P(x_{k+1}, k+1|x_k, k)P(x', N|x_{k+1}, k+1)\,dx_{k+1}. \tag{14.32}$$

Using the methods of Appendix D, one can show that the above average is equal to $(x' - x_k/N - k)P(x', N|x_k, k)$. Therefore, the part of the risk that depends on ϕ_k is given by:

$$\phi_k^2(x_k)\langle \delta x_k^2 \rangle - 2\phi_k(x_k) \int_{x_s}^{\infty} \frac{(x' - x_s)(x' - x_k)}{N - k} P(x', N|x_k, k)\,dx'. \tag{14.33}$$

Taking the derivative of this expression with respect to ϕ_k finally leads exactly to the optimal strategy, Eq. (14.17), above.

If the price increments can be written as $\delta x_k = f_k(x_k)\epsilon_k$, where $f_k(x_k)$ is an arbitrary function of x_k (possibly time dependent) and ϵ_k are iid Gaussian random variables, there is a more direct path from Eq. (14.30) to the Δ-hedge. Using Novikov's formula (see Eq. (3.24)), one finds that:

$$\langle \mathcal{C}_{k+1}\delta x_k \rangle \equiv \left\langle \frac{\partial \mathcal{C}_{k+1}}{\partial x} \right\rangle \langle \delta x_k^2 \rangle. \tag{14.34}$$

But since $\langle \mathcal{C}_{k+1} \rangle = \mathcal{C}_k(x_k)$, the minimum of Eq. (14.30) is given by:

$$\phi_k(x_k) = \frac{\partial \mathcal{C}_k}{\partial x_k},$$
 (14.35)

which is the Δ-hedge. This formula is valid for a quite general class of Gaussian processes, defined above by the function f_k. For example, the multiplicative Black-Scholes case corresponds to $f_k(x_k) = \sigma_1 x_k$.

This 'instantaneous' viewpoint is actually that of Black-Scholes, on which we shall expand in Chapters 16 and 18.

14.3 Residual risk

14.3.1 The Black–Scholes miracle

We shall now compute the *residual risk* \mathcal{R}^* obtained by substituting ϕ by ϕ^* in Eq. (14.10). In the case where the variance of price increments is constant, one finds:

$$\mathcal{R}^{*2} = \mathcal{R}_0^2 - D\tau \sum_{k=0}^{N-1} \int P(x,k|x_0,0)[\phi_k^{N*}(x)]^2 \, dx,$$
 (14.36)

where \mathcal{R}_0 is the unhedged risk.

The Black–Scholes 'miracle' is the following: in the case where P is Gaussian, the two terms in the above equation *exactly compensate* in the limit where $\tau \to 0$, with $N\tau = T$ fixed.

The 'zero-risk' property is true both within an 'additive' Gaussian model and the Black–Scholes multiplicative model, or for more general continuous-time Brownian processes. This property is easily obtained using Ito's stochastic differential calculus, see Chapters 3 and 16, but the latter formalism is only valid in the continuous-time limit ($\tau = 0$).

This 'miraculous' compensation is due to a very peculiar property of the Gaussian distribution, for which:

$$P_G(x_1,T|x_0,0)\delta(x_1 - x_2) - P_G(x_1,T|x_0,0)P_G(x_2,T|x_0,0) =$$
$$D \int_0^T \int P_G(x,t|x_0,0)\frac{\partial P_G(x_1,T|x,t)}{\partial x}\frac{\partial P_G(x_2,T|x,t)}{\partial x} \, dx \, dt.$$
 (14.37)

Integrating this equation with respect to x_1 and x_2 after multiplication by the corresponding payoff functions, the left-hand side gives \mathcal{R}_0^2. As for the right-hand side, one recognizes the limit of $\sum_{k=0}^{N-1} \int P(x,k|x_0,0)[\phi_k^{N}(x)]^2 \, dx$ when $\tau = 0$, where the sum becomes an integral.*

Therefore, in the continuous-time Gaussian case, the Δ-hedge of Black and Scholes allows one to eliminate completely the risk associated with an option, as was the case for

the forward contract. However, contrarily to the forward contract where the elimination of risk is possible whatever the statistical nature of the fluctuations (and follows from the linearity of the pay-off), the result of Black and Scholes is only valid in a very special limiting case. For example, as soon as the elementary time scale τ is finite (which is the case in reality), the residual risk \mathcal{R}^* is non-zero even if the increments are Gaussian. The calculation is easily done in the limit where τ is small compared to T: the residual risk then comes from the difference between a continuous integral $D \int dt' \int P_G(x, t'|x_0, 0)\phi^{*2}(x, t') dx$ (which is equal to \mathcal{R}_0^2) and the corresponding discrete sum appearing in Eq. (14.36). This difference is given by the Euler–McLaurin formula and is equal to:

$$\mathcal{R}^{*2} = \frac{D\tau}{2}\mathcal{P}(1 - \mathcal{P}) - \frac{(D\tau)^{3/2}}{24\sqrt{2\pi}}P(x_s, T|x_0, 0) + \cdots, \tag{14.38}$$

where \mathcal{P} is the probability (at $t = 0$) that the option is exercised at expiry ($t = T$), and we have written explicitly the sub-leading term, of order $\tau^{3/2}$ (and not τ^2).[†] In the limit $\tau \to 0$, one indeed recovers $\mathcal{R}^* = 0$, which also holds if $\mathcal{P} \to 0$ or 1, since the outcome of the option then becomes certain. However, Eq. (14.38) already shows that in reality, the residual risk is *not small*. Take for example an at-the-money option, for which $\mathcal{P} = \frac{1}{2}$. The comparison between the residual risk and the price of the option allows one to define a 'quality' ratio Q for the hedging strategy, which is larger for better hedges:

$$\frac{1}{Q} \equiv \frac{\mathcal{R}^*}{\mathcal{C}} \approx \sqrt{\frac{\pi}{4N}}, \tag{14.39}$$

with $N = T/\tau$. For an option of maturity 1 month, rehedged daily, $N \approx 25$. Assuming Gaussian fluctuations, the quality ratio is then $Q \approx 5$. In other words, the residual risk is one-fifth of the price of the option itself. Even if one rehedges every 30 min in a Gaussian world, the quality ratio is only $Q \approx 20$. If the increments are not Gaussian, then \mathcal{R}^* can never reach zero. This is actually rather intuitive, the presence of unpredictable price 'jumps' jeopardizes the differential strategy of Black–Scholes. The miraculous compensation of the two terms in Eq. (14.36) no longer takes place. Figure 14.4 gives the residual risk as a function of τ for an option on the Bund contract, calculated using the optimal strategy ϕ^*, and assuming independent increments. As expected, \mathcal{R}^* increases with τ, but does *not* tend to zero when τ decreases: the theoretical quality ratio saturates around $Q \approx 7$, reflecting the non zero kurtosis (jumps) of the underlying process. In fact, the real risk is even larger than the theoretical estimate since the model is imperfect. In particular, it neglects the volatility fluctuations. (The importance of these volatility fluctuations for determining the correct price was discussed above in Section 13.3.4.) In other words, the theoretical curve shown in Figure 14.4 neglects what is usually called the 'volatility' risk. A Monte-Carlo simulation of the profit and loss associated to an at-the-money option, hedged using the optimal strategy ϕ^* determined above leads to the histogram shown in Figure 14.1. The empirical variance of this histogram corresponds to $Q \approx 3.6$, substantially smaller than the theoretical estimate that neglects volatility risk.

[†] The above formula is only correct in the additive limit, but can be generalized to any Gaussian model, in particular the log-normal Black–Scholes model: see Eq. (14.40) below.

Fig. 14.4. Residual risk \mathcal{R}^* as a function of the time τ (in days) between adjustments of the hedging strategy ϕ^*, for at-the-money options of maturity equal to 60 trading days. The risk \mathcal{R}^* decreases when the trading frequency increases, but only rather slowly: the risk only falls by 10% when the trading time drops from 1 day to 30 min. Furthermore, \mathcal{R}^* does not tend to zero when $\tau \to 0$, at variance with the Gaussian case (dotted line). Finally, the present model neglects the volatility risk, which leads to an even larger residual risk (marked by the star symbol), corresponding to a Monte-Carlo simulation using real price changes.

Let us also consider the case of a multiplicative Gaussian model in discrete time. A similar Euler–Mc Laurin treatment finally leads to:

$$\mathcal{R}^{*2} = \frac{\sigma^2 \tau}{2} \left(\langle x^2 \rangle|_s - \langle x \rangle|_s^2 \right) + \cdots, \tag{14.40}$$

where $\langle x^q \rangle|_s = \int_{x_s}^{\infty} x^q P(x, T | x_0, 0) \, dx$. One can check that in the limit $\sigma^2 T \ll 1$ this formula becomes identical to the additive formula (14.38). On the other hand, in the limit $T \to \infty$, \mathcal{R}^{*2} grows without limit for any finite τ.

14.3.2 The 'stop-loss' strategy does not work

There is a very simple strategy that leads, at first sight, to a perfect hedge. This strategy is to hold $\phi = 1$ of the underlying as soon as the price x exceeds the strike price x_s, and to sell everything ($\phi = 0$) as soon as the price falls below x_s. For zero interest rates, this 'stop-loss' strategy would obviously lead to zero risk, since either the option is exercised at time T, but the writer of the option has bought the underlying when its value was x_s, or the option is not exercised, but the writer does not possess any stock. If this were true, the price

of the option would actually be zero, since in the global wealth balance, the term related to the hedge perfectly matches the option pay-off!

In fact, this strategy does not work at all. A way to see this is to realize that when the strategy takes place in discrete time, the price of the underlying is never *exactly* at x_s, but slightly above, or slightly below. If the trading time is τ, the difference between x_s and x_k (where k is the time in discrete units) is, for a random walk, of the order of $\sqrt{D\tau}$. The difference between the ideal strategy, where the buy or sell order is always executed precisely at x_s, and the real strategy is thus of the order of $N_\times \sqrt{D\tau}$, where N_\times is the number of times the price crosses the value x_s during the lifetime T of the option. For an at-the-money option, this is equal to the number of times a random walk returns to its starting point in a time T. The result is well known (see Feller (1971)), and is $N_\times \propto \sqrt{T/\tau}$ for $T \gg \tau$. Therefore, the uncertainty in the final result due to the accumulation of the above small errors is found to be of order \sqrt{DT}, independently of τ, and therefore does not vanish in the continuous-time limit $\tau \to 0$. This uncertainty is of the order of the option price itself, even for the case of a Gaussian process. Hence, the 'stop-loss' strategy is not the optimal strategy, and was therefore not found as a solution of the functional minimization of the risk presented above.

14.2.3 Instantaneous residual risk and kurtosis risk

It is also illuminating to consider the risk from an instantaneous point of view, as in Section 14.2.3 above. We write again the local wealth balance as:

$$\delta W_k = C_k - C_{k+1} + \phi_k(x_k)\delta x_k. \tag{14.41}$$

Now, we suppose δx_k small and Taylor expand in powers of δx_k the difference $C_k - C_{k+1}$:

$$C_k - C_{k+1} \simeq -\frac{\partial C_k}{\partial t}\tau - \frac{\partial C_k}{\partial x_k}\delta x_k - \frac{1}{2}\frac{\partial^2 C_k}{\partial x_k^2}(\delta x_k)^2 + \cdots. \tag{14.42}$$

The choice $\phi_k = \partial C/\partial x_k$ obviously removes the first order term in δx_k. The residual risk to lowest order therefore reads:[†]

$$\mathcal{R}_k^2 = \langle \delta W_k^2 \rangle - \langle \delta W_k \rangle^2 = \frac{1}{4}\left(\frac{\partial^2 C_k}{\partial x_k^2}\right)^2 [\langle(\delta x_k)^4\rangle - \langle(\delta x_k)^2\rangle^2]. \tag{14.43}$$

Using the very definition of kurtosis κ_1 on scale τ, this expression can be rewritten as:

$$\mathcal{R}_k^2 = \frac{2+\kappa_1}{4}\Gamma_k^2 D^2 \tau^2, \tag{14.44}$$

where $\Gamma_k = C_k''$ is the Gamma of the option at time k. In the absence of kurtosis, the total residual risk is the sum of $N = T/\tau$ terms, each of order τ^2. The total risk is thus of order τ,

[†] Note that since $\delta W_k - \langle \delta W_k \rangle$ is proportional to δx_k^2 with a negative coefficient (minus the Gamma of the option), the distribution of δW_k is negatively skewed for the option seller, a result already discussed in the caption of Fig. 14.1.

as already found above, Eq. (14.38). On the other hand, if κ_T is finite for finite time lags, the elementary kurtosis κ_1 is inversely proportional to τ. The contribution of the kurtosis to the residual risk remains *finite* when the rehedging time becomes small, as seen in Fig. 14.4. Taking into account that for Gaussian increments $D\tau \langle \Gamma_k^2 \rangle \sim 1/\sqrt{k(2N-k)}$, the total residual risk for an at the money option is, for τ small, of order:

$$\mathcal{R}^{*2} \approx \frac{D\kappa_1 \tau}{8\pi} \sum_{k=1}^{N} \frac{1}{\sqrt{k(2N-k)}} \approx \frac{DT\kappa(T)}{16}. \tag{14.45}$$

> Therefore, tail effects, as measured by the kurtosis, constitute the major source of the residual risk when the trading time τ is small.

This remark is the echo of the discussion in Section 4.2.4, where we showed that a non zero kurtosis becomes the major source of uncertainty for the high frequency volatility: the risk is non zero precisely because the volatility is not perfectly known.

14.2.4 Stochastic volatility models*

Along the same line of thought, it is instructive to consider the case where the fluctuations are purely Gaussian, but where the volatility itself is randomly varying. In other words, the instantaneous variance is given by $D_k = \overline{D} + \delta D_k$, where δD_k is itself a random variable of variance $\overline{(\delta D)^2}$. If the different δD_k's are uncorrelated, this model leads to a non-zero kurtosis (cf. Eq. (7.15) and Section 7.1) equal to $\kappa_1 = 3\overline{(\delta D)^2}/\overline{D}^2$.

Suppose then that the option trader follows the Black–Scholes strategy corresponding to the *average* volatility \overline{D}. This strategy is close, but not identical to, the optimal strategy which he would follow if he knew all the future values of δD_k (this strategy is given in Appendix D). If $\delta D_k \ll \overline{D}$, one can perform a perturbative calculation of the residual risk associated with the uncertainty on the volatility. This excess risk is certainly of order $(\delta D)^2$ since one is close to the absolute minimum of the risk, which is reached for $\delta D \equiv 0$. The calculation does indeed yield a relative increase in risk whose order of magnitude is given by:

$$\frac{\delta \mathcal{R}^*}{\mathcal{R}^*} \propto \frac{\overline{(\delta D)^2}}{\overline{D}^2}. \tag{14.46}$$

If the observed kurtosis is entirely due to this stochastic volatility effect, one finds $\delta \mathcal{R}^*/\mathcal{R}^* \propto \kappa_1$. One thus recovers the result of the previous section. Again, this volatility risk can represent an appreciable fraction of the real risk, especially for frequently hedged options. Figure 14.4 actually suggests that the fraction of the risk due to the fluctuating volatility is comparable to that induced by the intrinsic kurtosis of the distribution.

14.4 Hedging errors. A variational point of view

> The formulation of the hedging problem as a variational problem has a rather interesting consequence, which is a certain amount of stability against hedging errors.

Suppose indeed that instead of following the optimal strategy $\phi^*(x, t)$, one uses (as the market does) a suboptimal strategy close to ϕ^*, such as the Black–Scholes hedging strategy with the value of the implied volatility, discussed in Section 14.2.2. Denoting the difference between the actual strategy and the optimal one by $\delta\phi(x, t)$, one can show that for small $\delta\phi$, the increase in residual risk is given by:

$$\delta[\mathcal{R}^2] = D\tau \sum_{k=0}^{N-1} \int [\delta\phi(x, t_k)]^2 P(x, k|x_0, 0)\, dx, \tag{14.47}$$

which is *quadratic* in the hedging error $\delta\phi$, as expected since we expand around the optimal hedge. Therefore, the extra risk is rather small. For example, we have estimated in Section 14.2.2 that within a first-order cumulant expansion, $\delta\phi$ is at most equal to $0.02\kappa_N$, where κ_N is the kurtosis corresponding to the terminal distribution. (See also Fig. 14.3.) Therefore, one has:

$$\delta[\mathcal{R}^2] \leq 4\ 10^{-4}\kappa_N^2 DT. \tag{14.48}$$

For at-the-money options $\mathcal{C} \sim \sqrt{DT/2\pi}$. One can reexpress the above result in terms of the quality ratio of the optimal hedge Q, to find:

$$\frac{\delta\mathcal{R}}{\mathcal{R}^*} \leq 1.2\ 10^{-3}\kappa_N^2 Q^2. \tag{14.49}$$

For a typical hedging quality ratio $Q = 4$ and $\kappa_N = 1$, this represents a rather small relative increase equal to 2% at most. In reality, as numerical simulations show, the increase in risk induced by the use of the Black–Scholes Δ-hedge instead of the optimal hedge is indeed only of a few per cent for 1-month maturity options. This difference however increases as the maturity of the options decreases.

14.5 Other measures of risk – hedging and VaR (*)

It is conceptually illuminating to consider a model where the price increments δx_k are Lévy variables of index $\mu < 2$, for which the variance is infinite. Such a model is useful to describe extremely volatile markets, such as emergent countries (like the Mexican Peso), or very speculative assets (cf. Chapter 6). In this case, the variance of the global wealth balance ΔW is meaningless and cannot be used as a reliable measure of the risk associated to the option. Indeed, one expects that the distribution of ΔW behaves, for large negative ΔW, as a power-law of exponent μ.

In order to simplify the discussion, let us come back to the case where the hedging strategy ϕ is independent of time, and suppose interest rate effects negligible. The wealth balance, Eq. (13.28), then clearly shows that the catastrophic losses occur in two complementary cases:

- Either the price of the underlying soars rapidly, far beyond both the strike price x_s and x_0. The option is exercised, and the loss incurred by the writer is then:

$$|\Delta W_+| = x_N(1 - \phi) - x_s + \phi x_0 - C \approx x_N(1 - \phi). \tag{14.50}$$

The hedging strategy ϕ, in this case, limits the loss, and the writer of the option would have been better off holding $\phi = 1$ stock per written option.
- Or, on the contrary, the price plummets far below x_0. The option is then not exercised, but the strategy leads to important losses since:

$$|\Delta W_-| = \phi(x_0 - x_N) - C. \tag{14.51}$$

In this case, the writer of the option should not have held any stock at all ($\phi = 0$).

However, both dangers are *a priori* possible. Which strategy should one follow? Summing the probability of the above events, the probability p that the loss exceeds (in absolute value) a certain level \mathcal{R}_p is given by:

$$p = \mathcal{P}(\Delta W < -\mathcal{R}_p) = \mathcal{P}_> \left(\frac{\mathcal{R}_p + C + (x_s - x_0)}{1 - \phi} \right) + \mathcal{P}_< \left(-\frac{\mathcal{R}_p + C}{\phi} \right), \tag{14.52}$$

where $\mathcal{P}_{>,<}$ are the cumulative distributions of the price difference $x_N - x_0$. One has at this stage two a rather different options for the optimization: either minimizing the probability of loss p for a given threshold \mathcal{R}_p, or minimizing the *Value-at-Risk* \mathcal{R}_p for a given probability of loss p. Let us first investigate the first choice, which is slightly simpler. The equation fixing the optimal ϕ reads:

$$\left(\frac{\phi^*}{1 - \phi^*} \right)^2 = \frac{\mathcal{R}}{\mathcal{R} + x_s - x_0} \frac{P[-\mathcal{R}/\phi^*]}{P[(x_s - x_0 + \mathcal{R})/(1 - \phi^*)]}, \tag{14.53}$$

where P is the probability density corresponding to $\mathcal{P}_{>,<}$, and $\mathcal{R} \equiv \mathcal{R}_p + C$ is the part of the loss due to the trading. This equation is general and does not depend on the precise shape of P. Consider the case where this distribution decreases as a power-law for large arguments:

$$P(\Delta x = x_N - x_0) \underset{\Delta x \to \pm \infty}{\simeq} \frac{\mu A_\pm^\mu}{|\Delta x|^{1+\mu}}. \tag{14.54}$$

For at the money options, such that $\mathcal{M} = x_s - x_0 = 0$, the solution of the optimization equation reads, for large enough \mathcal{R}:

$$\phi^*(\mathcal{M} = 0) = \frac{A_+^\zeta}{A_+^\zeta + A_-^\zeta} \quad \zeta \equiv \frac{\mu}{\mu - 1}, \tag{14.55}$$

for $1 < \mu \leq 2$.[†] For $\mu < 1$, ϕ^* is equal to 0 if $A_- > A_+$ or to 1 in the opposite case. For a general moneyness, the optimal strategy can be expressed in terms of $\phi^*(\mathcal{M} = 0)$ as:

$$\phi^*_{\mathcal{M}} = \frac{\phi^*_{\mathcal{M}=0}}{(1 - \phi^*_{\mathcal{M}=0})F(\mathcal{M}) + \phi^*_{\mathcal{M}=0}} \tag{14.56}$$

with:

$$F(\mathcal{M}) = \left(1 + \frac{\mathcal{M}}{\mathcal{R}}\right)^{\frac{\mu+1}{\mu-1}}. \tag{14.57}$$

Several points are worth emphasizing:

- The hedge ratio ϕ^* becomes independent of moneyness \mathcal{M} when extreme risks are considered, i.e. when $\mathcal{R} \to \infty$. In this limit, $F(\mathcal{M}) = 1$. The result (14.55) could have been obtained directly by noting that the distribution of losses also decays as a power-law for large arguments:

$$P(\Delta W) \underset{\Delta W \to -\infty}{\simeq} \frac{\mu \Delta W_0^\mu}{|\Delta W|^{1+\mu}} \qquad \Delta W_0^\mu \equiv A_+^\mu (1 - \phi)^\mu + A_-^\mu \phi^\mu. \tag{14.58}$$

 The optimal ϕ^* obtained above for $\mathcal{R} \to \infty$ can indeed be seen to minimize the tail amplitude ΔW_0^μ.

 More generally, the optimal VaR hedging strategy varies with the price of the underlying on the scale of the value at risk itself, since only the ratio \mathcal{M}/\mathcal{R} enters the expression of ϕ^*.

- Similarly, when hedging extreme risks, the strategy ϕ^* is time independent, and also corresponds to the optimal instantaneous hedge, where the VaR between times k and $k + 1$ is minimum. The fact that this strategy is time independent is quite interesting from the point of view of transaction costs (see Section 17.1.4).

- Even when the tail amplitude ΔW_0 of the loss probability is minimum, the variance of the final wealth is still infinite for $\mu < 2$. However, $\Delta W_0^* = \Delta W_0(\phi^*)$ sets the order of magnitude of probable losses, for example with 95% confidence. As in the example of the optimal portfolio discussed in Chapter 12, infinite variance does not necessarily mean that the risk cannot be diversified. The option price, fixed for $\mu > 1$ by Eq. (13.33), ought to be corrected by a risk premium proportional to ΔW_0^*. Note also that with such violent fluctuations, the smile becomes a spike!

- It is instructive to note that the histogram of ΔW is completely asymmetrical, since extreme events only contribute to the loss side of the distribution. As for gains, they are limited, since the distribution decreases very fast for positive ΔW.[‡] In this case, the analogy between an option and an insurance contract is most clear, and shows that buying or selling an option are not at all equivalent operations, as they appear to be in a Black–Scholes world. Note that the asymmetry of the histogram of ΔW is visible even in the case of weakly non-Gaussian assets (Fig. 14.1).

[†] Note that the above strategy is still valid for $\mu > 2$, and corresponds to the optimal VaR hedge for power-law-tailed assets, see below.

[‡] For at-the-money options, one can actually show that $\Delta W \leq \mathcal{C}$.

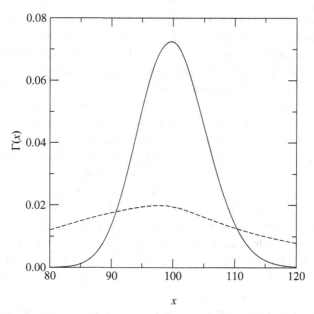

Fig. 14.5. The price dependence of the Γ of an option hedged using the Black–Scholes Δ (plain line), or a hedge that minimizes $\langle \Delta W^4 \rangle$ (dashed line). The strike is 100, and the terminal distribution has a kurtosis $\kappa = 3$.

• Finally, one could have considered the problem where the probability of loss p, rather than the level of loss \mathcal{R}_p, is fixed and one looks for the minimum VaR \mathcal{R}_p^*. In this case, the relation between ϕ^* and \mathcal{R}_p^* is still given by Eq. (14.53). The use of Eq. (14.52) for a given value of p then gives the second relation between ϕ^* and \mathcal{R}_p^*, needed to determine both.

Other measure of risks can also be considered, such as the fourth moment of the wealth balance, $\langle \Delta W^4 \rangle$.[†] As a rule of thumb, the more sensitive to extreme events the risk measure is, the 'flatter' the hedging strategy as a function of the underlying price. In a Black–Scholes language: hedging extreme risks reduces the Gamma, as illustrated in Fig. 14.5

14.6 Conclusion of the chapter: the pitfalls of zero-risk

The traditional approach to derivative pricing is to find an ideal hedging strategy, which perfectly duplicates the derivative contract. Its price is then, using an arbitrage argument, equal to that of the hedging strategy, and the residual risk is zero. This argument appears as such in nearly all the available books on derivatives and on the Black–Scholes model. For example, the last chapter of Hull (1997), called 'Review of Key Concepts', starts by the following sentence: *The pricing of derivatives involves the construction of riskless hedges from traded securities.* Although there is a rather wide consensus on this point of view, we feel that it is unsatisfactory to base a whole theory on exceptional situations: as explained above, both the continuous-time Gaussian model and the binomial model (see Section 16.3)

[†] see Selmi & Bouchaud (2000).

are very special models indeed. We think that it is more appropriate to start from the ingredient which allows the derivative markets to exist in the first place, namely *risk*. In this respect, it is interesting to compare the above quote from Hull to the following disclaimer, found on most Chicago Board Options Exchange documents: *Option trading involves risk*!

The idea that zero risk is the exception rather than the rule is important for a better pedagogy of financial risks in general; an adequate estimate of the residual risk – inherent to the trading of derivatives – has actually become one of the major concerns of risk management. The idea that the risk is zero is inadequate because zero cannot be a good approximation of anything. It furthermore provides a feeling of apparent security which can prove disastrous on some occasions. For example, the Black–Scholes strategy allows one, in principle, to hold an insurance against the fall of one's portfolio without buying a true Put option, but rather by following the associated hedging strategy. This is called a **portfolio insurance**, and was much used in the 1980s, when faith in the Black–Scholes model was at its highest. The idea is simply to sell a certain fraction of the portfolio when the market goes down. This fraction is fixed by the Black–Scholes Δ of the virtual option, where the strike price is the security level below which the investor does not want to plummet. During the 1987 crash, this strategy has been particularly inefficient: not only because crash situations are the most extremely non-Gaussian events that one can imagine (and thus the zero-risk idea is totally absurd), but also because this strategy feeds back onto the market to make it crash further (a drop of the market mechanically triggers further sell orders when 'put' options are Δ-hedged). According to the Brady commission, this mechanism has indeed significantly contributed to enhance the amplitude of the crash (see the discussion in Hull (1997)): around 80 billion dollars of stocks were managed this way!

14.6 Summary

- One can find an optimal hedging strategy, in the sense that the risk (measured by the variance of the change of wealth) is minimum. This strategy can be obtained explicitly, with very few assumptions, and is given by Eqs. (14.17), or (14.24).

- For a general non-linear pay-off (i.e. for any derivative except the futures contract) the residual risk is non-zero, and actually represents an appreciable fraction of the price of the option itself. The exception is the Black–Scholes model where the risk, rather miraculously, disappears. Kurtosis effects (or volatility fluctuations) lead to irreducible residual risks.

- The theory presented here generalizes that of Black and Scholes, and is formulated as a variational theory. Interestingly, this means that small errors in the hedging strategy increases the risk only to second order. Risk is therefore stable against small hedging errors.

- Other measures of risk, more sensitive to tail events, can be considered. The optimal strategies are found the vary less rapidly with the price of the underlying than the standard Black-Scholes Δ-hedge.

14.8 Appendix D: computation of the conditional mean

On many occasions in this chapter, we needed to compute the mean value of the instantaneous increment δx_k, restricted on trajectories starting from x_k and ending at x_N. We assume that the δx_k's are identically distributed, up to a volatility factor σ_k. In other words:

$$P_{1k}(\delta x_k) \equiv \frac{1}{\sigma_k} P_{10}\left(\frac{\delta x_k}{\sigma_k}\right). \tag{14.59}$$

The quantity we wish to compute is then:

$$P(x_N, N|x_k, k)\langle \delta x_k\rangle_{(x_k,k)\to(x_N,N)} =$$
$$\int \delta x_k\, \delta\left(x_N - x_k - \sum_{j=k}^{N-1} \delta x_j\right) \left[\prod_{j=k}^{N-1} P_{10}\left(\frac{\delta x_j}{\sigma_j}\right)\frac{d\delta x_j}{\sigma_j}\right], \tag{14.60}$$

where the δ function insures that the sum of increments is indeed equal to $x_N - x_k$. Using the Fourier representation of the δ function, the right-hand side of this equation reads:

$$\frac{1}{2\pi} \int e^{iz(x_N-x_k)} i\sigma_k \hat{P}'_{10}(\sigma_k z) \left[\prod_{j=k+1}^{N-1} \hat{P}_{10}(z\sigma_j)\right] dz. \tag{14.61}$$

- In the case where all the σ_k's are equal to σ_0, one recognizes:

$$\frac{1}{2\pi} \int e^{iz(x_N-x_k)} \frac{i}{N-k} \frac{\partial}{\partial z} [\hat{P}_{10}(z\sigma_0)]^{N-k} \, dz. \tag{14.62}$$

Integrating by parts and using the fact that:

$$P(x_N, N|x_k, k) \equiv \frac{1}{2\pi} \int e^{iz(x_N-x_k)} [\hat{P}_{10}(z\sigma_0)]^{N-k} \, dz, \tag{14.63}$$

one finally obtains the expected result:

$$\langle \delta x_k\rangle_{(x_k,k)\to(x_N,N)} \equiv \frac{x_N - x_k}{N-k}. \tag{14.64}$$

- In the case where the σ_k's are different from one another, one can write the result as a cumulant expansion, using the cumulants $c_{n,1}$ of the distribution P_{10}. After a simple computation, one finds:

$$P(x_N, N|x_k, k)\langle \delta x_k\rangle_{(x_k,k)\to(x_N,N)} = \sum_{n=2}^{\infty} \frac{(-\sigma_k)^n c_{n,1}}{(n-1)!} \frac{\partial^{n-1}}{\partial x_N^{n-1}} P(x_N, N|x_k, k), \tag{14.65}$$

which allows one to generalize the optimal strategy in the case where the volatility is time dependent. In the Gaussian case, all the c_n for $n \geq 3$ are zero, and the last expression

boils down to:

$$\langle \delta x_k \rangle_{(x_k,k) \to (x_N,N)} = \frac{\sigma_k^2 (x_N - x_k)}{\sum_{j=k}^{N-1} \sigma_j^2}. \tag{14.66}$$

This expression can be used to show that the optimal strategy is indeed the Black–Scholes Δ, for arbitrary Gaussian increments in the presence of a non-zero interest rate. Suppose that

$$x_{k+1} - x_k = \rho x_k + \delta x_k, \tag{14.67}$$

where the δx_k are identically distributed. One can then write x_N as:

$$x_N = x_0 (1 + \rho)^N + \sum_{k=0}^{N-1} \delta x_k (1 + \rho)^{N-k-1}. \tag{14.68}$$

The above formula, Eq. (14.66), then reads:

$$\langle \delta x_k \rangle_{(x_k,k) \to (x_N,N)} = \frac{\gamma_k^2 (x_N - x_k (1 + \rho)^{N-k})}{\sum_{j=k}^{N-1} \gamma_j^2}. \tag{14.69}$$

with $\gamma_k = (1 + \rho)^{N-k-1}$.

- FURTHER READING

M. Baxter, A. Rennie, *Financial Calculus, An Introduction to Derivative Pricing*, Cambridge University Press, Cambridge, 1996.

N. Dunbar, *Inventing Money*, John Wiley, 2000.

W. Feller, *An Introduction to Probability Theory and its Applications*, Wiley, New York, 1971.

J. C. Hull, *Futures, Options and Other Derivatives*, Prentice Hall, Upper Saddle River, NJ, 1997.

D. Lamberton, B. Lapeyre, *Introduction au Calcul Stochastique Appliqué à la Finance*, Ellipses, Paris, 1991.

M. Musiela, M. Rutkowski, *Martingale Methods in Financial Modelling*, Springer, New York, 1997.

N. Taleb, *Dynamical Hedging: Managing Vanilla and Exotic Options*, Wiley, New York, 1998.

P. Wilmott, *Derivatives, The theory and practice of financial engineering*, John Wiley, 1998.

P. Wilmott, *Paul Wilmott Introduces Quantitative Finance*, John Wiley, 2001.

- SPECIALIZED PAPERS

E. Aurell, S. Simdyankin, *Pricing risky options simply*, International Journal of Theoretical and Applied Finance, **1**, 1 (1998).

J.-P. Bouchaud, D. Sornette, *The Black–Scholes option pricing problem in mathematical finance: generalization and extensions for a large class of stochastic processes*, Journal de Physique I, **4**, 863–881 (1994).

S. Esipov, I. Vaysburd, *On the profit and loss distribution of dynamic hedging strategies*, Int. Jour. Theo. Appl. Fin., **2**, 131 (1998).

S. Fedotov and S. Mikhailov, *Option pricing for incomplete markets via stochastic optimization: transaction costs, adaptive control and forecast*, Int. Jour. Theo. Appl. Fin., **4**, 179 (2001).

H. Follmer, D. Sondermann, *Hedging of non redundant contigent claims*, in *Essays in Honor of G. Debreu*, (W. Hildenbrand, A. Mas-Collel Etaus, eds), North-Holland, Amsterdam, 1986.

H. Follmer, P. Leukert, *Efficient hedges: cost versus shortfall risk*, Finance and Stochastics, **4**, 117 (2000).

J.-P. Laurent, H. Pham, *Dynamic programming and mean-variance hedging*, Finance and Stochastics, **3**, 83 (1999).

M. Schäl, *On quadratic cost criteria for option hedging*, The Mathematics of Operations Research, **19**, 121–131 (1994).

M. Schweizer, *Variance–optimal hedging in discrete time*, The Mathematics of Operations Research, **20**, 1–32 (1995).

M. Schweizer, *Risky options simplified*, International Journal of Theoretical and Applied Finance, **2**, 59 (1999).

F. Selmi, J.-P. Bouchaud, *Hedging large risks reduces the transaction costs*, e-print cond-mat/0005148 (2000), Wilmott magazine, March (2003).

15

Options: The role of drift and correlations

Où est la raison de tout cela ? Pourquoi la fumée de cette pipe va-t-elle à droite plutôt qu'à gauche ? Fou! trois fois fou à lier, celui qui calcule ses chances, qui met la raison de son côté !†

(A. de Musset, *Les caprices de Marianne*.)

15.1 The influence of drift on an optimally hedged option

We should now come back to the case where the excess return (or drift) $m_1 \equiv \langle \delta x_k \rangle = m\tau$ is non-zero. This case is very important conceptually: indeed, one of the most striking results of Black and Scholes (besides the zero risk property) is that the price of the option and the hedging strategy are *totally independent* of the value of m. This may sound at first rather strange, since one could think that if m is very large and positive, the price of the underlying asset on average increases fast, thereby increasing the average pay-off of the option. On the contrary, if m is large and negative, the option should be worthless.

This intuitive argument does not take into account the impact of the hedging strategy on the global wealth balance, which is itself proportional to m. In other words, the pay-off $\max(x(N) - x_s, 0)$, averaged with the historical distribution $P_m(x, N|x_0, 0)$, such that:

$$P_m(x, N|x_0, 0) = P_{m=0}(x - Nm_1, N|x_0, 0), \tag{15.1}$$

is obviously strongly dependent on the value of m. However, this dependence is partly compensated when one includes the trading strategy, and even vanishes in the Black–Scholes model.

15.1.1 A perturbative expansion

Let us first present a perturbative calculation, assuming that m is small, or more precisely that $(mT)^2/DT \ll 1$. Typically, for indices, for $T = 100$ days, $mT = 5\% 100/365 = 0.014$ and $\sqrt{DT} \approx 1\%\sqrt{100} \approx 0.1$. The term of order m^2 that we neglect corresponds to a relative error of $(0.14)^2 \approx 0.02$.

† What is the logic of all this? Why is the smoke of this pipe going to the right rather than to the left? Mad, completely mad, the one who calculates his bets, who puts reason on his side!

The average gain (or loss) induced by the hedge is equal to:[†]

$$\langle \Delta W_S \rangle = +m_1 \sum_{k=0}^{N-1} \int P_m(x, k|x_0, 0)\phi_k^{N*}(x)\,dx. \tag{15.2}$$

To order m, one can consistently use the general result Eq. (14.24) for the optimal hedge ϕ^*, established above for $m = 0$, and also replace P_m by $P_{m=0}$:[‡]

$$\langle \Delta W_S \rangle = -m_1 \sum_{k=0}^{N-1} \int P_0(x, k|x_0, 0)$$
$$\times \int_{x_s}^{+\infty} (x' - x_s) \sum_{n=2}^{\infty} \frac{(-)^n c_{n,1}}{D\tau(n-1)!} \frac{\partial^{n-1}}{\partial x'^{n-1}} P_0(x', N|x, k)\,dx'\,dx, \tag{15.3}$$

where P_0 is the unbiased distribution ($m = 0$).

Now, using the Chapman–Kolmogorov equation for conditional probabilities:

$$\int P_0(x', N|x, k)P_0(x, k|x_0, 0)\,dx = P_0(x', N|x_0, 0), \tag{15.4}$$

one easily derives, after an integration by parts, and using the fact that $P_0(x', N|x_0, 0)$ only depends on $x' - x_0$, the following expression:

$$\langle \Delta W_S \rangle = m_1 N \left[\int_{x_s}^{+\infty} P_0(x', N|x_0, 0)\,dx' \right.$$
$$\left. + \sum_{n=3}^{\infty} \frac{c_{n,1}}{D\tau(n-1)!} \frac{\partial^{n-3}}{\partial x_0^{n-3}} P_0(x_s, N|x_0, 0) \right]. \tag{15.5}$$

On the other hand, the increase of the average pay-off, due to a non-zero mean value of the increments, is calculated as:

$$\langle \max(x(N) - x_s, 0)\rangle_m \equiv \int_{x_s}^{\infty} (x' - x_s)P_m(x', N|x_0, 0)\,dx'$$
$$= \int_{x_s - m_1 N}^{\infty} (x' + m_1 N - x_s)P_0(x', N|x_0, 0)\,dx', \tag{15.6}$$

where in the second integral, the shift $x_k \to x_k + m_1 k$ allowed us to remove the bias m, and therefore to substitute P_m by P_0. To first order in m, one then finds:

$$\langle \max(x(N) - x_s, 0)\rangle_m \simeq \langle \max(x(N) - x_s, 0)\rangle_0 + m_1 N \int_{x_s}^{\infty} P_0(x', N|x_0, 0)\,dx'. \tag{15.7}$$

[†] In the following, we shall again stick to an additive model and discard interest rate effects, in order to focus on the main concepts.
[‡] The equation for optimal strategy when $m \neq 0$ is given in Appendix E.

Hence, grouping together Eqs. (15.5) and (15.7), one finally obtains the price of the option in the presence of a non-zero average return as:

$$C_m = C_0 - \frac{mT}{D\tau} \sum_{n=3}^{\infty} \frac{c_{n,1}}{(n-1)!} \frac{\partial^{n-3}}{\partial x_0^{n-3}} P_0(x_s, N|x_0, 0). \tag{15.8}$$

Quite remarkably, *the correction terms are zero in the Gaussian case*, since all the cumulants $c_{n,1}$ are zero for $n \geq 3$. In fact, in the Gaussian case, this property holds to *all* orders in m (cf. Section 16.2), and is the 'second miracle' of the Black-Scholes theory. However, for non-Gaussian fluctuations, one finds that a non-zero return should in principle affect the price of the options. Using again Eq. (14.23), one can rewrite Eq. (15.8) in a simpler form as:

$$C_m = C_0 + mT[\mathcal{P} - \phi^*], \tag{15.9}$$

where \mathcal{P} is the probability that the option is exercised, and ϕ^* the optimal strategy, both calculated at $t = 0$. From Fig. 14.3 one sees that in the presence of 'fat tails', a positive average return makes out-of-the-money options less expensive ($\mathcal{P} < \phi^*$), whereas in-the-money options should be more expensive ($\mathcal{P} > \phi^*$). Again, the Gaussian model (for which $\mathcal{P} = \phi^*$) is misleading:[†] the independence of the option price with respect to the market 'trend' only holds for Gaussian processes, and is no longer valid in the presence of 'jumps'. Note however that the correction is usually numerically quite small: for $x_0 = 100, m = 10\%$ per year, $T = 100$ days, and $|\mathcal{P} - \phi^*| \sim 0.02$, one finds that the price change is of the order of 0.05 points, to be compared to $\mathcal{C} \approx 4$ points (1% correction).

15.1.2 'Risk neutral' probability and martingales

It is interesting to notice that the result, Eq. (15.8), can alternatively be rewritten in a suggestive form as:

$$C_m = \int_{x_s}^{\infty} (x - x_s) Q(x, N|x_0, 0) \, dx, \tag{15.10}$$

with an 'effective probability' (called **risk neutral probability**, or **pricing kernel** in the mathematical literature) Q defined as:

$$Q(x, N|x_0, 0) = P_0(x, N|x_0, 0) - \frac{m_1}{D\tau} \sum_{n=3}^{\infty} \frac{c_{n,N}}{(n-1)!} \frac{\partial^{n-1}}{\partial x_0^{n-1}} P_0(x, N|x_0, 0). \tag{15.11}$$

By inspection, one sees that Q satisfies the following equations:

$$\int Q(x, N|x_0, 0) \, dx \equiv 1, \tag{15.12}$$

$$\int (x - x_0) Q(x, N|x_0, 0) \, dx \equiv 0. \tag{15.13}$$

[†] The fact that the optimal strategy Δ is equal to the probability \mathcal{P} of exercising the option also holds in the Black–Scholes model, up to small σ^2 correction terms.

Note that the second equation means that the evolution of x under the 'probability' Q is unbiased, i.e. $\langle x \rangle|_Q = x_0$. This is the definition of a *martingale* process (see, e.g. Baxter & Rennie (1996), Musiela & Rutkowski (1997)). Equation (15.10), with the very same Q, actually holds for an arbitrary pay-off function $\mathcal{Y}(x_N)$, replacing $\max(x_N - x_s, 0)$ in the above equation. Using Eq. (14.23), Eq. (15.11) can also be written in a more compact way as:

$$Q(x, N|x_0, 0) = P_0(x, N|x_0, 0) - \frac{m_1}{D\tau}(x - x_0)P_0(x, N|x_0, 0)$$

$$+ m_1 N \frac{\partial P_0(x, N|x_0, 0)}{\partial x_0}. \tag{15.14}$$

Note however that Eq. (15.10) is rather formal, since nothing ensures that $Q(x, N|x_0, 0)$ is everywhere positive, except in the Gaussian case where $Q(x, N|x_0, 0) = P_0(x, N|x_0, 0)$. In fact, one can construct cases where $Q(x, N|x_0, 0)$ becomes negative for some values of x; correspondingly, the *fair-price* of some options can be negative. This happens when m is quite large, and for very far out-of-the money options, i.e. in rather unrealistic situations. Nevertheless, a negative price calls for more comments, which we defer until the next section.

15.2 Drift risk and delta-hedged options

15.2.1 Hedging the drift risk

We have shown above that for non-Gaussian processes, the price of an optimally hedged option (in the context of variance minimization) depends on the drift m of the underlying, see Eq. (15.9). This formula could be trusted if the value of the drift was perfectly known. Unfortunately, however, this is never the case. The drift itself is an unknown, random quantity. Eq. (15.9) shows that the optimal hedging of an option leads to an exposure of the writer of the option to the drift of the underlying. Since \mathcal{C}_m is chosen such as to make the wealth balance vanish on average in the presence of a known drift m, any deviation δm from this expected drift will lead to a residual bias given by:

$$\delta W = -(\delta m)T[\mathcal{P} - \phi^*] = -(\delta m)T[\Delta - \phi^*]. \tag{15.15}$$

Since δm is unknown, this induces an extra source of risk, which we may call a drift risk, equal to:

$$\delta_m \mathcal{R}^2 = \sigma_m^2 T^2 [\Delta - \phi^*]^2, \tag{15.16}$$

where σ_m^2 measures the uncertainty on the drift. This extra source of risk adds to the residual risk \mathcal{R}^{*2} computed in the previous chapter.

If one chooses a hedging strategy ϕ intermediate between Δ and ϕ^*, the total extra risk will therefore be given by:

$$\delta \mathcal{R}^2 = \sigma_m^2 T^2 [\Delta - \phi]^2 + DT[\phi^* - \phi]^2, \tag{15.17}$$

where the second term comes from the quadratic error around the optimal hedge ϕ^*. The optimal strategy ϕ^{**} in the presence of drift risk is thus given by:

$$\phi^{**} = \frac{\sigma_m^2 T^2 \Delta + DT\phi^*}{\sigma_m^2 T^2 + DT}. \tag{15.18}$$

This formula shows that if the uncertainty on the drift is zero, one should choose $\phi^{**} = \phi^*$, as expected. On the other hand, when the uncertainty on the drift is strong, the optimal strategy will depart from ϕ^* such as to minimize the drift risk, (15.16), and approach the Black-Scholes Δ-hedge. Note that for long maturity options, the drift risk becomes the dominant factor. As we show below, when the hedge is chosen to be the Δ-hedge, one finds that the option price indeed becomes independent of the drift, and is given by the average pay-off over the so-called 'risk neutral' distribution of price changes.

The conclusion of this section is that although non Δ-hedging schemes are interesting as these can sometimes substantially reduce the extreme risks and the hedging costs, one must bear in mind that these alternative strategies lead to a residual exposure to the drift of the underlying. This risk becomes the dominant one for long maturity options.

15.2.2 The price of delta-hedged options

Suppose that we choose to hedge the option *à la* Black-Scholes, i.e., imposing $\phi = \Delta = \partial \mathcal{C}/\partial x$. The fair price of the option must be the solution of the following self-consistent equation:

$$\mathcal{C}(x_0, x_s, N) = \int \mathcal{Y}(x' - x_s) P_m(x', N | x_0, 0) \, dx'$$

$$- m_1 \sum_{k=0}^{N-1} \int \frac{\partial \mathcal{C}(x, x_s, N-k)}{\partial x} P_m(x, k | x_0, 0) \, dx, \tag{15.19}$$

with \mathcal{Y} representing the (arbitrary) pay-off of the option.

This equation can be solved by making the following ansatz for \mathcal{C}:

$$\mathcal{C}(x, x_s, N-k) = \int \Omega(x' - x_s, N-k) P_m(x', N | x, k) \, dx' \tag{15.20}$$

where Ω is an unknown kernel which we try to determine. Using the fact that the option price only depends on the price difference between the strike price and the present price of the underlying, this gives:

$$\int \Omega(x' - x_s, N) P_m(x', N | x_0, 0) \, dx' = \int \mathcal{Y}(x' - x_s) P_m(x', N | x_0, 0) \, dx'$$

$$+ m_1 \sum_{k=0}^{N-1} \int P_m(x, k | x_0, 0) \times \frac{\partial}{\partial x_s} \int \Omega(x' - x_s, N-k) P_m(x', N | x, k) \, dx' \, dx. \tag{15.21}$$

Now, using the Chapman–Kolmogorov equation:

$$\int P_m(x', N | x, k) P_m(x, k | x_0, 0) \, dx = P_m(x', N | x_0, 0), \tag{15.22}$$

one obtains, after a few manipulations, the following equation for Ω (after changing $k \to N - k$):

$$\Omega(x' - x_s, N) + m_1 \sum_{k=1}^{N} \frac{\partial \Omega(x' - x_s, k)}{\partial x'} = \mathcal{Y}(x' - x_s). \tag{15.23}$$

The solution to this equation is $\Omega(x' - x_s, k) = \mathcal{Y}(x' - x_s - m_1 k)$. Indeed, if this is the case, one has:

$$\frac{\partial \Omega(x' - x_s, k)}{\partial x'} \equiv -\frac{1}{m_1} \frac{\partial \Omega(x' - x_s, k)}{\partial k} \tag{15.24}$$

and therefore:

$$\mathcal{Y}(x' - x_s - m_1 N) - \sum_{k=1}^{N} \frac{\partial \mathcal{Y}(x' - x_s - m_1 k)}{\partial k} \simeq \mathcal{Y}(x' - x_s) \tag{15.25}$$

where the last equality holds in the small τ, large N limit, when the sum over k can be approximated by an integral.[†]

Therefore, the price of the option is given by:

$$C = \int \mathcal{Y}(x' - x_s - m_1 N) P_m(x', N | x_0, 0) \, dx'. \tag{15.26}$$

Now, using the fact that $P_m(x', N | x_0, 0) \equiv P_0(x' - m_1 N, N | x_0, 0)$, and changing variable from $x' \to x' - m_1 N$, one finally finds:

$$C = \int \mathcal{Y}(x' - x_s) P_0(x', N | x_0, 0) \, dx', \tag{15.27}$$

thereby showing that the pricing kernel Q defined in Eq. (15.10) is equal to P_0, and thus *independent* of m, whenever the chosen hedge is the Δ.

Therefore, the Δ-hedge has the virtue of hedging against drift risk. The price of the option is given by the average pay-off over the 'detrended' probability distribution.

Note that, interestingly, the equality $Q = P_0$ holds independently of the particular pay-off function \mathcal{Y}, which gives to Q a general status. We shall actually show in the next section that the same result also holds for path dependent options.

Furthermore, since the pricing kernel Q is now positive everywhere, the negative option price paradox disappears. This means that *if* the seller of the option was 100% confident about the value of the average future return, he could afford to sell strongly out of the money options at a zero price and still make money on average. However, in these conditions, the drift risk would become the major source of risk, and his strategy should get closer to the Δ, so that the fair-price of the option becomes positive again.

[†] The resulting error is of the order of $m^2 \tau / D$, which is extremely small in practice for $\tau = 1$ day.

15.2.3 A general option pricing formula

Let us reformulate the above calculation in terms of an 'instantaneous' wealth balance, assuming Δ-hedging.[†] Between time k and $k+1$, the wealth of the writer of the option changes by:

$$\delta W_k = \phi_k \delta x_k - (\mathcal{C}_{k+1} - \mathcal{C}_k) = \frac{\partial \mathcal{C}_k}{\partial x_k} \delta x_k - (\mathcal{C}_{k+1} - \mathcal{C}_k). \tag{15.28}$$

(We have set the interest rate to zero.) Now, write $\delta x_k = m_k + \xi_k$, where m_k is the (possibly time and price dependent) conditional drift at time k and ξ_k a random variable of zero conditional mean. This means that conditional to all present information, one decomposes the next price change into a *known* part m_k and a random, unbiased part ξ_k.

If both m_k and the time between k and $k+1$ are sufficiently small, one can write:

$$\mathcal{C}_{k+1}(x_{k+1}) \approx \mathcal{C}_{k+1}(x_k + \xi_k) + m_k \frac{\partial \mathcal{C}_k}{\partial x_k}. \tag{15.29}$$

Injecting this last expression in Eq. (15.28) and setting the conditional average of δW_k to zero leads to:

$$\langle \delta W_k \rangle_k = 0 = -\langle \mathcal{C}_{k+1}(x_k + \xi_k) \rangle_k + \mathcal{C}_k(x_k). \tag{15.30}$$

This leads to the fundamental equation for option pricing in the general case:

$$\mathcal{C}_k(x_k) = \langle \mathcal{C}_{k+1}(x_k + \xi_k) \rangle_k, \tag{15.31}$$

where $\langle \cdots \rangle_k$ means averaging over ξ_k. This equation holds for any type of options (possibly path dependent), and for any model of price fluctuations, provided one uses the Δ-hedge. To first order in the drift, this equation tells us that the price of the option today is the average of tomorrow's price over the distribution of 'detrended' price returns, ξ_k, where the local drift is set to zero. In more technical terms, Eq. (15.31) states that the option price is a *martingale* (i.e. a stochastic process such that the average future value is equal to today's value), over the 'detrended' stochastic process where the local average conditional return is set to zero.

One therefore sees that although the probability distribution to be used to price option is not exactly the 'true' one, it is in practice very close to it, because the conditional drift m_k is usually very small. In particular, it shares with it the same tails and volatility fluctuations. This contrasts with what is sometimes thought by those educated within the strict mathematical finance orthodoxy: that the pricing kernel Q has nothing to do whatsoever with the true, historical probability distribution.

> *Other pricing kernels Q have been proposed in the literature. An interesting one, based on the theory of optimal gambling, suggests to choose the following distribution of*

[†] The following presentation was suggested to us by Andrew Matacz.

returns:

$$Q(\eta) = \frac{P(\eta)}{1 + \lambda^* \eta}, \qquad (15.32)$$

where $\lambda^ > 0$ obeys the following equation:*

$$\int_{-\infty}^{\infty} \frac{\eta}{1 + \lambda^* \eta} P(\eta) \, d\eta = 0. \qquad (15.33)$$

Then, by construction, $Q(\eta) > 0$, and $\int \eta Q(\eta) d\eta = 0$. The only property that one needs to check in order to interpret $Q(\eta)$ as a probability distribution is that $\int Q(\eta) d\eta = 1$, which follows from:

$$\int_{-\infty}^{\infty} (1 + \lambda^* \eta) Q(\eta) \, d\eta \equiv \int_{-\infty}^{\infty} P(\eta) \, d\eta = 1, \qquad (15.34)$$

and the fact that $\int \eta Q(\eta) d\eta = 0$. Therefore, $Q(\eta)$ has again a zero conditional mean. See Aurell et al. (2000).

15.3 Pricing and hedging in the presence of temporal correlations (*)

15.3.1 A general model of correlations

Let us now suppose that the price increments are correlated in time. The model for correlated price increments we will consider is the following. We first define an auxiliary set of *uncorrelated iid random variables* $\{\xi_k\}$, distributed according to an arbitrary probability distribution $P(\xi)$. The joint distribution is then given by

$$P(\{\xi_k\}) = \prod_k P_0(\xi_k). \qquad (15.35)$$

We will assume that $P_0(\xi) = P_0(-\xi)$, and define $D_1 = D\tau$ as the variance of the ξ_k. We now construct the set of correlated price increments $\{\delta x_k\}$ by writing

$$\delta x_k = \sum_\ell M_{k,\ell} \xi_\ell + m_1, \qquad (15.36)$$

with

$$M_{k,\ell} = \delta_{k,\ell} + \frac{1}{2D_1} c_{k-\ell}. \qquad (15.37)$$

The correlation coefficients c_k are assumed to satisfy

$$c_0 = 0; \qquad c_{-k} = c_k. \qquad (15.38)$$

In the sequel, we assume that both the c_k's and m_1 are small, and will work only to first order in both m_1 and c.

Because $\langle \xi \rangle = 0$, m_1 is the average *unconditional* drift of the price process. Moreover, to first order in c, the coefficients D_1 and c denote, respectively, the *unconditional* local variance and correlation of the price increments. Note that because of the correlations, the

long-time volatility D_N, given by:

$$D_N = ND_1 + \sum_{k=1}^{N-1} 2(N-k)c_k, \tag{15.39}$$

is different from the uncorrelated result, ND_1. One can furthermore show that the *conditional* drift and variance m_k and D_k are given by

$$m_{k-1} = m_1 + \sum_{\ell < k} c_{k-\ell} \left(\frac{1}{2D_1} \delta x_\ell - \frac{1}{2} \frac{\partial \log P_0}{\partial \delta x_\ell} \right), \tag{15.40}$$

$$D_k = D_1. \tag{15.41}$$

15.3.2 Derivative pricing with small correlations

Applying the general mean-variance pricing procedure to the above correlated process, one can derive the following formula for an arbitrary pay-off \mathcal{Y}, in the limit of small correlations:

$$\mathcal{C} = \langle \mathcal{Y} \rangle_0 + m_1 \sum_{k=1}^{N} \langle F(\delta x_k) \mathcal{Y} \rangle_0 + \frac{1}{2D_1} \sum_{k=1}^{N} \sum_{\ell < k} c_{k-\ell} \langle F(\delta x_k) \delta x_\ell \mathcal{Y} \rangle_0, \tag{15.42}$$

where

$$F(u) = -\frac{\partial \log P_0}{\partial u} - \frac{1}{D_1} u, \tag{15.43}$$

and the notation $\langle \ldots \rangle_0$ means that we are averaging over the *uncorrelated* process P_0, and not the 'correlated' (historical) price increment distribution.

Let us first consider the purely Gaussian case, where $\log P_0(u) = -u^2/2D_1 - 1/2 \log(2\pi D_1)$. This leads to $F(u) \equiv 0$. Therefore, all corrections brought about by the drift and correlations strictly vanish, and the price can be calculated by assuming that the process is correlation-free. The fact that the drift disappears was already shown in Section 15.2.2 above, and we see that this result can be extended to the conditional drift. (As will be clear in the next chapter, this can also be understood directly within the Ito continuous time framework.)

In other words, the price of the option is given by the Black–Scholes price, with a volatility given by the small scale volatility D_1 (that corresponds to the hedging time scale), and *not* with the volatility corresponding to the terminal distribution, D_N/N. When price increments are correlated (anticorrelated), D_1 is smaller (larger) than D_N/N, and the price of the contract is lower (higher) than the objective average pay-off. This is due to the fact that the hedging strategy is on average making (or losing) money because of the correlations. A rise of the price of the underlying leads to an increase of the hedge, which is followed (on average) by a further increase of the price. The opposite is true in the presence of anticorrelations.

For an uncorrelated non-Gaussian process, the effect of a non-zero drift m does not vanish in general, a situation already discussed in details above. The same is true with correlations: the correction no longer disappears and one cannot in general price the option

by considering a process with the same statistics of price increments but no correlations.[†] Note that the correction term involves all δx_ℓ with $\ell < 0$. This means that, in the presence of correlations, the price of the option does not only depend on the current price of the underlying, but also on all past price increments.

15.3.3 The case of delta-hedging

Let us conclude this section by discussing other possible hedging schemes. Instead of choosing the optimal hedge that minimizes the variance, one can follow the (sub–optimal) Black–Scholes Δ-hedge. As emphasized above, this leads in general to a very small risk increase, and has the advantage of removing the exposure to the drift m. One might therefore expect that Δ-hedging also allows one to get rid of the correlations. However, the following calculation shows that this is not the case.

Up to first order in c, the price of the contract is given by

$$\mathcal{C} = \langle \mathcal{Y} \rangle - \sum_{k=0}^{N-1} \langle m_k \phi_{BS,k} \rangle \tag{15.44}$$

$$= \langle \mathcal{Y} \rangle + \sum_{k=0}^{N-1} \left\langle m_k \frac{\partial \log P_0}{\partial \delta x_{k+1}} y \right\rangle \tag{15.45}$$

$$= \langle \mathcal{Y} \rangle_0 + \frac{1}{2} \sum_{k=1}^{N} \sum_{\ell<k} c_{k-\ell} \left\langle F(\delta x_k) \frac{\partial \log P_0}{\partial \delta x_\ell} y \right\rangle_0 . \tag{15.46}$$

Note that, in this case, the drift m indeed disappears to first order even if $F(u)$ is not zero; the correlations, on the other hand, do affect the option price even within a Δ-hedging scheme (except in the Gaussian case).

15.4 Conclusion

15.4.1 Is the price of an option unique?

As we shall see in the next chapter, the use of Ito's differential calculus rule immediately leads to the two main results of Black and Scholes, namely: the existence of a riskless hedging strategy, and the fact that the value of the average trend disappears from the final expressions. These two results are however not valid as soon as the hypothesis underlying Ito's stochastic calculus are violated (continuous-time, Gaussian statistics). The approach based on the global wealth balance, presented in the above sections, is by far less elegant but much more general. It allows one to understand the very peculiar nature of the limit considered by Black and Scholes.

As we have already discussed, the existence of a non-zero residual risk (and more precisely of a negatively skewed distribution of the optimized wealth balance) necessarily means that the bid and ask prices of an option will be different, because the market makers will try to

[†] This conclusion however depends on the particular form of the above model of correlations. See Cornalba et al. (2002) for details.

compensate for part of this risk. On the other hand, if the average return m is not zero, the fair price of the option explicitly depends on the optimal strategy ϕ^*, and thus of the chosen measure of risk. For example, as was the case for portfolios (see Chapter 12), the optimal strategy corresponding to a minimal variance of the final result is different from the one corresponding to a minimum value-at-risk.

> The price of an option therefore depends on the operator, on his definition of risk and on his ability to hedge this risk. In the Black–Scholes model, the price is uniquely determined since all definitions of risk are equivalent (and are all zero!).

This property is often presented as a major advantage of the continuous time Gaussian models. Nevertheless, it is clear that it is precisely the existence of an ambiguity in the price that justifies the very existence of option markets.[†] A market can only exist if *some* uncertainty remains. In this respect, it is interesting to note that new markets continually open (for example: weather derivatives), where more and more sources of uncertainty become tradable. Option markets correspond to a risk transfer: buying or selling a call are not identical operations (recall the skew in the final wealth distribution – Fig. 14.1 and the discussion of Section 14.3.3), except in the Black–Scholes world where options would actually be useless, since they would be equivalent to holding a certain number of the underlying asset (given by the Δ). This is actually at the heart of 'portfolio insurance' which substitutes the Δ hedge strategy to real put options.

15.4.2 Should one always optimally hedge?

The question of hedging is in fact related to the previous one, since the price of the option depends on the chosen hedge. It is also ambiguous because the optimal hedge depends both on the very definition of risk, and on the identification of the different sources of risk. We have seen above that when the drift is unknown, the sub-optimal Δ-hedge may be preferable to the optimal variance hedge. The case of anticorrelated underlying fluctuations is also quite interesting. Take for example a mean reverting model, where price changes are Gaussian, but follow an Ornstein-Uhlenbeck process (see Section 4.3.1). We have seen in Section 15.3.2 that for Gaussian correlated increments, the price of the option is given by the average pay-off using the uncorrelated price distribution (see also Section 16.2.2). As we noted, this price is higher than the 'true' average pay-off of the option, because the hedging strategy loses money on average, due to the anti-correlations. When the maturity of the option tends to infinity the problem becomes acute since the price of the option (given

[†] This ambiguity is related to the residual risk, which, as discussed above, comes both from the presence of price 'jumps', and from the very uncertainty on the parameters describing the distribution of price changes ('volatility risk').

by the standard Black–Scholes formula) tends to infinity, while the 'true' payoff of the option is bounded, since the price itself is bounded. Nobody would agree to pay an infinite amount of money for an option with a limited pay-off just because the writer of the option has to face infinite hedging costs! As the rehedging time τ increases, the residual risk seen by the writer of the option increases but the option price (which depends on the volatility measured on the time between rehedging) decreases. The sum of the fair-price plus a risk premium therefore reaches a minimum as a function of τ, and this leads to an acceptable trade-off between the risk taken by the writer and the price that an investor will agree to pay to cover the hedging costs of the writer. Again, the real price of an option cannot be determined without considering the risk issue.

In fact, the presence of anti-correlations in the price process play a very similar role to *transaction costs*, that we will discuss in more details in Section 17.1.4. In a similar way, as the rehedging time τ increases, the residual risk increases but the hedging costs decrease. The writer of the option cannot expect to transfer his transaction costs to the option buyer, and will have to choose a reasonable risk target.

15.5 Summary

- The drift only influences the price of an option because of the difference between the chosen hedging strategy and the Δ-hedge. The drift has therefore no influence on the option price for Gaussian increments since the Δ-hedge is in this case the optimal hedge, but is relevant in general for non-Gaussian increments.
- Conversely, the Δ-hedge removes any exposure to drift risk, which can become an important source of risk for long maturity options.
- Most importantly for applications, the price of a Δ-hedged option is equal to its (actualized) average pay-off, over the 'detrended' distribution of returns, where any conditional drift is subtracted.
- In the presence of temporal correlations that lead to a non zero conditional drift, the price of the option is given, in the Gaussian case, by the average pay-off as if the process was uncorrelated, but with a volatility computed on the rehedging time scale.
- The price of an option is not unique; it is the price at which two investors with different risk constraints and different perceptions of future volatility agree to trade.

15.6 Appendix E: optimal strategy in the presence of a bias

We give here, without giving the details of the computation, the general equation satisfied by the optimal hedging strategy in the presence of a non-zero average return m, when the price fluctuations are arbitrary but uncorrelated. Assuming that $\langle \delta x_k \delta x_\ell \rangle - m_1^2 = D\tau \delta_{k,\ell}$, and introducing the unbiased variables $\chi_k \equiv x_k - m_1 k$, one gets for the optimal strategy

$\phi_k^*(\chi)$ the following (involved) integral equation:

$$D\tau\phi_k^*(\chi) - \int_{\chi_s}^{\infty}(\chi' - \chi_s)\frac{\chi' - \chi}{N-k}P_0(\chi', N|\chi, k)\,d\chi' =$$

$$-m_1\left\{\int_{\chi_s}^{\infty}(\chi' - \chi_s)[P_0(\chi', N|\chi_0, 0) - P_0(\chi', N|\chi, k)]\,d\chi'\right.$$

$$+ \sum_{\ell=k+1}^{N-1}\int\left[\frac{\chi' - \chi}{\ell - k} + m_1\right]\phi_\ell^*(\chi')P_0(\chi', \ell|\chi, k)\,d\chi'$$

$$\left.+ \sum_{\ell=0}^{k-1}\int\left[\frac{\chi - \chi'}{k - \ell} + m_1\right]\phi_\ell^*(\chi')\frac{P_0(\chi', \ell|\chi_0, 0)P_0(\chi, k|\chi', \ell)}{P_0(\chi, k|\chi_0, 0)}\,d\chi' - m_1\overline{\phi^*}\right\}, \qquad (15.47)$$

with $\chi_s = x_s - m_1 N$ and

$$\overline{\phi^*} \equiv \sum_{k=0}^{N-1}\int\phi_k^*(\chi')P_0(\chi', k|\chi_0, 0)\,d\chi'. \qquad (15.48)$$

In the limit $m_1 = 0$, the right-hand side vanishes and one recovers the optimal strategy determined above. For small m, Eq. (15.47) is a convenient starting point for a perturbative expansion in m_1.

Using Eq. (15.47), one can establish a simple relation for $\overline{\phi^*}$, which fixes the correction to the 'Bachelier price' coming from the hedge: $C = \langle\max(x_N - x_s, 0)\rangle - m_1\overline{\phi^*}$.

$$\overline{\phi^*} = \sum_{k=0}^{N-1}\int_{\chi_s}^{\infty}(\chi' - \chi_s)\frac{\chi' - \chi}{N-k}P_0(\chi', N|\chi, k)P_0(\chi, k|\chi_0, 0)\,d\chi'$$

$$- m_1\sum_{\ell=k+1}^{N-1}\int\frac{\chi' - \chi}{\ell - k}\phi_\ell^*(\chi')P_0(\chi', \ell|\chi_0, 0)\,d\chi'. \qquad (15.49)$$

Replacing $\phi_\ell^*(\chi')$ by its corresponding equation $m_1 = 0$, one obtains the correct value of $\overline{\phi^*}$ to order m_1, and thus the value of C to order m_1^2 included.

• FURTHER READING

M. Baxter, A. Rennie, *Financial Calculus, An Introduction to Derivative Pricing*, Cambridge University Press, Cambridge, 1996.

J. C. Hull, *Futures, Options and Other Derivatives*, Prentice Hall, Upper Saddle River, NJ, 1997.

D. Lamberton, B. Lapeyre, *Introduction au calcul stochastique appliqué à la finance*, Ellipses, Paris, 1991.

M. Musiela, M. Rutkowski, *Martingale Methods in Financial Modelling*, Springer, New York, 1997.

P. Wilmott, *Derivatives, The Theory and Practice of Financial Engineering*, John Wiley, 1998.

P. Wilmott, *Paul Wilmott Introduces Quantitative Finance*, John Wiley, 2001.

- SPECIALIZED PAPERS

E. Aurell, S. Simdyankin, *Pricing risky options simply*, International Journal of Theoretical and Applied Finance, **1**, 1 (1998).

E. Aurell, R. Baviera, O. Hammarlid, M. Serva, A. Vulpiani, *Growth optimal investment and pricing of derivatives*, Int. Jour. Theo. Appl. Fin., **3**, 1 (2000).

L. Cornalba, J.-P. Bouchaud, M. Potters, *Option pricing and Hedging with temporal correlations*, International Journal of Theoretical and Applied Finance, **5**, 307 (2002).

M. Schweizer, *Risky options simplified*, International Journal of Theoretical and Applied Finance, **2**, 59 (1999).

16

Options: the Black and Scholes model

The heresies we should fear are those which can be confused with orthodoxy.
(Jorge Luis Borges, The Theologians.)

16.1 Ito calculus and the Black-Scholes equation

After having discussed at length the peculiarities of the continuous time Gaussian limit in the previous chapters, it is interesting to describe how option pricing theory is usually introduced using Ito's stochastic differential calculus. This framework is extremely powerful and allows one to obtain rather quickly some of the results that were painfully derived in the previous chapters, such as the optimal hedge or the independence of the price on the drift. Once the limitations of the Black-Scholes model and the subtleties of the continuous limit are fully understood, the 'blind' use of Ito calculus to obtain a first approximation to the price of derivatives becomes justified. However, we are convinced that in order to go beyond Black-Scholes and study more realistic models, one should be prepared to abandon Ito calculus and the world of partial differential equations on which most of mathematical finance relies.

16.1.1 The Gaussian Bachelier model

We first assume that the price itself (and not its logarithm) follows a continuous time random walk, with drift m and diffusion constant D. We also assume that the interest rate r is equal to zero. Consider now the variation between times t and $t + dt$ of the value of the portfolio of the writer of an option (worth C) who also holds $\phi(x, t)$ underlying assets as a hedge:

$$\frac{df}{dt} = -\frac{dC}{dt} + \phi \frac{dx}{dt}. \tag{16.1}$$

The first term comes with a minus sign because the writer is 'short' the option (which means that he has sold the option). Therefore he loses (gains) if the option increases (decreases) in value. The second term reflects the variation of value of his hedge.

Now, we assume that $x(t)$ is a continuous time random walk. We may thus apply Ito's formula, Eq. (3.22) to dC/dt. One finds:

$$\frac{df}{dt} = \phi \frac{dx}{dt} - \left[\frac{\partial C(x, t)}{\partial t} + \frac{\partial C(x, t)}{\partial x} \frac{dx}{dt} + \frac{D}{2} \frac{\partial^2 C(x, t)}{\partial x^2} \right]. \tag{16.2}$$

One immediately sees in the above formula that if one chooses ϕ such that $\phi = \phi^* \equiv \partial C(x, t)/\partial x$, the coefficient of the only random term in the above expression, namely dx/dt, vanishes. The evolution of the wealth of the writer is then known with certainty!

Now, writing options cannot be profitable with certainty, otherwise arbitragers would massively write options, pushing the price down until the profit disappears. Conversely, this wealth variation cannot be negative with certainty, for in that case, arbitragers would do the reverse operation (buy and hedge) driving up the price of options. The only remaining possibility is that the wealth of the writer must be constant.

Note that we assumed that the interest rate is zero, so we could ignore return on capital and cost of borrowing. Setting $df/dt = 0$ leads to a partial differential equation (PDE) for the price C:

$$\frac{\partial C(x, t)}{\partial t} = -\frac{D}{2} \frac{\partial^2 C(x, t)}{\partial x^2}. \tag{16.3}$$

This PDE is called the diffusion (or heat) equation. However, because of the minus sign, the interpretation of this diffusion equation is different from the one that appears in physical contexts, where it describes, for example, the way the spatial density of pollutants evolves in time. Here, because of the minus sign, time flows 'backwards': in order to make sense, the above PDE must be supplemented by a boundary condition in the future, and not by an 'initial condition'. In the present financial context, this terminal boundary condition makes perfect sense, since the price of the option at maturity ($t = T$) is perfectly known in all circumstances and is equal to the pay-off of the option: $C(x, T) = \mathcal{Y}(x)$, where, for example $\mathcal{Y}(x) = \max(x - x_s, 0)$ for European options, with x_s the strike price.

Before showing how this PDE can be solved for an arbitrary pay-off function, let us note the following: (i) the solution of the PDE determines the price of the option everywhere in the half plane $x, t < T$, where the point of interest $x = x_0, t = 0$ is only a special point. In other words, the option price $C(x, t)$ is a *field* that cannot be determined on a single point without knowing its surroundings; (ii) since the mean return m does not appear in the equation, it will not be involved in the solution either. The price and the hedging strategy are therefore immediately seen to be completely independent of the average return in this framework, a result obtained after rather cumbersome considerations in the previous chapters.

16.1.2 Solution and Martingale

In order to find the option price for an arbitrary pay-off and get some insight into the meaning of the obtained solution, let us first examine the following *forward* diffusion equation:

$$\frac{\partial P(x', t'|x, t)}{\partial t'} = \frac{D}{2} \frac{\partial^2 P(x', t'|x, t)}{\partial x'^2}, \tag{16.4}$$

with boundary conditions:

$$P(x', t' = t|x, t) = \delta(x' - x). \tag{16.5}$$

The solution to this diffusion equation is well known to be the (zero drift) Gaussian distribution:

$$P(x, t'|x, t) = P_G(x', t'|x, t) = \frac{1}{\sqrt{2\pi D(t' - t)}} \exp\left(-\frac{(x' - x)^2}{2D(t' - t)}\right). \tag{16.6}$$

Since $P_G(x', t'|x, t)$ only depends on $x' - x$ and on $t' - t$, it also obeys the equation:

$$\frac{\partial P(x', t'|x, t)}{\partial t} = -\frac{D}{2} \frac{\partial^2 P(x', t'|x, t)}{\partial x^2}, \tag{16.7}$$

Now, let us consider the following quantity:

$$\mathcal{C}_G(x, t) = \int \mathcal{Y}(x') P_G(x', T|x, t) \, dx' \tag{16.8}$$

which is the average of the pay-off over the probability distribution P_G. Take the derivative of $\mathcal{C}_G(x, t)$ with respect to t and use the above Eq. (16.7) to find:

$$\frac{\partial \mathcal{C}_G(x, t)}{\partial t} = \int \mathcal{Y}(x') \frac{\partial P_G(x', T|x, t)}{\partial t} \, dx' = -\frac{D}{2} \int \mathcal{Y}(x') \frac{\partial^2 P_G(x', T|x, t)}{\partial x^2} \, dx'$$

$$= -\frac{D}{2} \frac{\partial^2 \mathcal{C}_G(x, t)}{\partial x^2}. \tag{16.9}$$

Hence, $\mathcal{C}_G(x, t)$ obeys the pricing PDE, Eq. (16.3). Now, using the boundary condition $P_G(x', T|x, T) = \delta(x' - x)$, it is readily seen that $\mathcal{C}_G(x, t)$ also obeys the correct terminal boundary condition:

$$\mathcal{C}_G(x, T) = \int \mathcal{Y}(x') P_G(x', T|x, T) \, dx' = \int \mathcal{Y}(x') \delta(x' - x) \, dx' = \mathcal{Y}(x). \tag{16.10}$$

Hence, we find that $\mathcal{C}(x, t) = \mathcal{C}_G(x, t)$: the price of the option is equal to average of the pay-off over the probability distribution P_G that describes the undrifted evolution of the underlying price x.

Now, using the (Chapman-Kolmogorov) chain property of $P_G(x', T|x, t)$, namely:

$$P_G(x', T|x, t) = \int P_G(x', T|x'', t'') P_G(x'', t''|x, t) dx'' \qquad (t \leq t'' \leq T), \tag{16.11}$$

we also establish the following **martingale** property of the price:

$$\mathcal{C}_G(x, t) = \int \mathcal{C}_G(x'', t'') P_G(x'', t''|x, t) dx'' = \langle \mathcal{C}_G(x'', t'') \rangle_0, \tag{16.12}$$

which means that the price of the option today is equal to the average of the option price anytime in the future, provided one uses the undrifted distribution of future prices (hence the notation $\langle \ldots \rangle_0$. As discussed in Section 15.2.2, this is a more general property that only relies on Δ-hedging. In this case, one also has that $\mathcal{C}(x, t) = \langle \mathcal{C}(x'', t'') \rangle_0$ for any $t'' > t$. As explained in Section 17.2.4, this property allows to price more complicated options, such as American options.

16.1.3 Time value and the cost of hedging

Let us discuss an apparent contradiction in the above discussion: we just stated that the price of the option is such that it does not vary on average. Yet, Bachelier's PDE, Eq. (16.3), shows that the partial derivative of $C(x, t)$ with respect to time is negative, or in traders language, that theta is negative:

$$\frac{d\langle C \rangle}{dt} = 0 \quad \text{but} \quad \Theta = \frac{\partial C}{\partial t} < 0. \tag{16.13}$$

Equation (16.3) in fact only tells us that, conditional to the fact that the price x does not vary at all between t and $t + dt$, the price of the option C does decay. This variation can be seen as the cost of the hedging strategy due to the back and forth motion of the price between t and $t + dt$.

If we place ourselves in case where $x(t + dt) = x(t)$, we can first suppose that between t and $t + dt/2$, the price increases by a small amount $dx > 0$. The Black-Scholes hedging strategy is such that the quantity of stocks held by the writer must increase by $d\phi = (\partial \phi / \partial x)dx = (\partial^2 C / \partial x^2)dx$. But since between $t + dt/2$ and $t + dt$ the price goes back to x, this extra investment causes a loss, since it corresponds to buying high and selling low. The loss is equal to $d\phi dx$, which must also be the depreciation of the option between t and $t + dt$. Therefore:

$$\frac{\partial C(x, t)}{\partial t}dt = -d\phi dx = -\frac{\partial^2 C(x, t)}{\partial x^2}(dx)^2. \tag{16.14}$$

Since only $(dx)^2$ enters this expression, one sees that the hedging strategy is experiencing a loss in both cases $dx > 0$ and $dx < 0$. But since for a Brownian motion between t and $t + dt/2$, $(dx)^2 = Ddt/2$, the above expression is seen to reproduce Eq. (16.3).

16.1.4 The log-normal Black-Scholes model

Now, let us turn to classical Black-Scholes case where the logarithm of the price is a continuous time random walk, and the interest rate is non zero. Using again Ito's formula to compute df/dt, one now finds:

$$\frac{df}{dt} = \phi \frac{dx}{dt} - \left[\frac{\partial C(x, t)}{\partial t} + \frac{\partial C(x, t)}{\partial x} \frac{dx}{dt} + \frac{\sigma^2 x^2}{2} \frac{\partial^2 C(x, t)}{\partial x^2} \right]. \tag{16.15}$$

Again, the choice $\phi = \phi^* \equiv \partial C(x, t)/\partial x$, makes all uncertainty vanish. However, since the interest rate is non zero, the wealth differential should now equate the (zero risk) interest on the total value of the portfolio $\phi x - C(x, t)$. Therefore, one must have:

$$\frac{df}{dt} = r[\phi x - C(x, t)] = r\left[\frac{\partial C(x, t)}{\partial x}x - C(x, t) \right], \tag{16.16}$$

which leads finally to the celebrated Black-Scholes partial differential equation:

$$\frac{\partial C}{\partial t} + rx\frac{\partial C}{\partial x} + \frac{\sigma^2 x^2}{2}\frac{\partial^2 C}{\partial x^2} = rC, \tag{16.17}$$

where we have dropped the set of variables (x, t) on which C depends. The boundary condition is again, obviously, $C(x, t = T) \equiv \mathcal{Y}(x)$, i.e. the value of the option at expiry $t = T$.

The solution of this partial differential equation can be obtained using a variety of different methods, including path integral methods (see Chapter 3). The explicit solution reads, as already mentioned in Section 13.3.3:

$$C(x_0, t = 0) = x_0 \mathcal{P}_{G>} \left(\frac{y_-}{\sigma \sqrt{T}} \right) - x_s e^{-rT} \mathcal{P}_{G>} \left(\frac{y_+}{\sigma \sqrt{T}} \right), \tag{16.18}$$

where

$$y_\pm = \log(x_s/x_0) - rT \pm \sigma^2 T/2 \tag{16.19}$$

and $\mathcal{P}_{G>}(u)$ is the cumulative normal distribution defined by Eq. (2.36). Since the drift of the underlying does not appear in Eq. (16.17), it does not appear in the price of the option (nor in the hedging strategy) either. Again, the price of the option can be written as the discounted average pay-off over the undrifted distribution of price returns, which is such that the average terminal price is equal to the forward: $\langle x(T) \rangle_0 = x(t = 0)e^{rT}$.

16.1.5 General pricing and hedging in a Brownian world

The Black-Scholes protocol can be generalized to much more complex options or any type of derivatives, such as bonds (see Chapter 19). Suppose one has to price an option that depends on the terminal price of a set of (possibly correlated) assets $\{x_i\}, i = 1, \ldots, M$, all assumed to follow a geometric Brownian motion. The price of the option therefore depends both on time and on the price of all assets: $C(x_1, \ldots x_M; t)$. Using Ito's formula for several Gaussian noises (see Eq. (3.23)), one finds, for the variation of the portfolio of the option seller that hedges using a fraction ϕ_i of stocks of type i:

$$\frac{df}{dt} = \sum_{i=1}^{M} \phi_i \frac{dx_i}{dt} - \left[\frac{\partial C}{\partial t} + \sum_{i=1}^{M} \frac{\partial C}{\partial x_i} \frac{dx_i}{dt} + \sum_{i,j=1}^{M} \frac{\sigma_{ij}^2 x_i x_j}{2} \frac{\partial^2 C}{\partial x_i \partial x_j} \right], \tag{16.20}$$

where σ_{ij} is the covariance matrix of the returns. Setting the hedges to their Black-Scholes Δ value, i.e.

$$\phi_i \equiv \frac{\partial C}{\partial x_i}, \tag{16.21}$$

leads again to a zero risk portfolio, that must therefore yield a return equal to $r[\sum_i \phi_i x_i - C]$. The PDE fixing the price of the option is then a multidimensional diffusion equation:

$$\frac{\partial C}{\partial t} + r \sum_{i=1}^{M} x_i \frac{\partial C}{\partial x_i} + \sum_{i,j=1}^{M} \frac{\sigma_{ij}^2 x_i x_j}{2} \frac{\partial^2 C}{\partial x_i \partial x_j} = rC. \tag{16.22}$$

Interestingly, the general solution of this equation can again be written as the discounted average pay-off over the undrifted distribution of price returns, which is such that the average terminal price of all assets is equal to the forwards: $\langle x_i(T) \rangle_0 = x_i(t = 0)e^{rT}$.

The Black-Scholes machinery is very general and very beautiful: using hedges that are the linear sensitivity of the derivative to all assets, one finds that the resulting portfolio is riskless, and that the price of the derivative is simply the discounted average pay-off over the undrifted distribution of price returns. This machinery can be applied to many cases, including derivatives on the yield curve. In practice, however, the beauty of many theories often distract their followers from common sense, and the limitations of the Black-Scholes protocol should always be carefully taken into account.

16.1.6 The Greeks

The way C_{BS} varies with the current price x_0, the maturity T and the volatility defines the 'Greeks', because they are denoted by Greek letters. For the sake of completeness, we give here the explicit formulas of the Greeks in the Black-Scholes. First, the hedging ratios:

$$\Delta = \frac{\partial C}{\partial x_0} = \mathcal{P}_{G>}\left(\frac{y_-}{\sigma\sqrt{T}}\right);$$

(16.23)

(where y_- is defined in Eq. (16.19)) and

$$\Gamma = \frac{\partial \Delta}{\partial x_0} = \frac{1}{x_0\sigma\sqrt{2\pi T}}\exp\left(-\frac{y_-^2}{2\sigma^2 T}\right).$$

(16.24)

The Vega \mathcal{V}, that is actually not a Greek letter and should not exist in the Black-Scholes world where the volatility is known and constant, is given by:

$$\mathcal{V} = \frac{\partial C}{\partial \sigma} = x_0\sqrt{\frac{T}{2\pi}}\exp\left(-\frac{y_-^2}{2\sigma^2 T}\right);$$

(16.25)

and finally, in the limit $rT \ll 1$, the 'time value' Θ of the option (see Section 16.1.3):

$$\Theta = -\frac{\partial C}{\partial T} \approx -\frac{\sigma}{2T}\mathcal{V}.$$

(16.26)

16.2 Drift and hedge in the Gaussian model (*)

16.2.1 Constant drift

In the continuous-time Gaussian case, the solution of Eqs. (15.47) and (15.48) for the optimal strategy ϕ^* happens to be completely independent of the drift m as is clear from the above approach using Ito's calculus. Let us explicitly show how the compensation between the increase in average pay-off and the gain due to the hedging strategy occurs in this case. Using the explicit form of the hedging strategy, one has:

$$\phi^*(x, t) = -\int_{x_s}^{\infty} \frac{x' - x_s}{\sqrt{2\pi D(T - t)}}\frac{\partial}{\partial x'}\exp\left[-\frac{(x' - x)^2}{2D(T - t)}\right]dx'.$$

(16.27)

The average profit induced by the hedge is thus:

$$m_1\overline{\phi^*} = m\int_0^T \int \frac{1}{\sqrt{2\pi Dt}}\phi^*(x, t)\exp\left[-\frac{(x - x_0 - mt)^2}{2Dt}\right]dx\, dt.$$

(16.28)

Performing the Gaussian integral over x, one finds (setting $u = x' - x_s$ and $u_0 = x_0 - x_s$):

$$m_1\overline{\phi^*} = -m \int_0^T \int_0^\infty \frac{u}{\sqrt{2\pi DT}} \frac{\partial}{\partial u} \exp\left[-\frac{(u - u_0 - mt)^2}{2DT}\right] du\, dt$$

$$= \int_0^T \int_0^\infty \frac{u}{\sqrt{2\pi DT}} \frac{\partial}{\partial t} \exp\left[-\frac{(u - u_0 - mt)^2}{2DT}\right] du\, dt,$$

or else:

$$m_1\overline{\phi^*} = \int_0^\infty \frac{u}{\sqrt{2\pi DT}} \left\{\exp\left[-\frac{(u - u_0 - mT)^2}{2DT}\right] - \exp\left[-\frac{(u - u_0)^2}{2DT}\right]\right\} du. \quad (16.29)$$

The price of the option in the presence of a non-zero return $m \neq 0$ is thus given, in the Gaussian case, by:

$$\mathcal{C}_m(x_0, x_s, T) = \langle\max(x - x_s, 0)\rangle_m - m_1\overline{\phi^*}$$

$$= \int_0^\infty \frac{u}{\sqrt{2\pi DT}} \exp\left[-\frac{(u - u_0 - mT)^2}{2DT}\right] du - m_1\overline{\phi^*}$$

$$= \int_0^\infty \frac{u}{\sqrt{2\pi DT}} \exp\left[-\frac{(u - u_0)^2}{2DT}\right] du \quad (16.30)$$

$$\equiv \mathcal{C}_{m=0}(x_0, x_s, T), \quad (16.31)$$

(cf. Eq. (13.43)). Hence, the compensation between increased average pay-off and gain of the hedging strategy is *exact* in a Gaussian model, but only approximate in a more general model. (see Section 15.1.)

16.2.2 Price dependent drift and the Ornstein-Uhlenbeck paradox

Let us now study, in the context of Ito's stochastic differential calculus, a case where the disappearance of the drift from the option price becomes somewhat paradoxical. Suppose that the price follows an Ornstein-Uhlenbeck process:

$$\frac{dx}{dt} = -\Omega(x - \overline{x}) + \xi(t), \quad (16.32)$$

where $\xi(t)$ is a Gaussian white noise of zero mean, with $\langle\xi(t)\xi(t')\rangle = D\delta(t - t')$. For large maturities T, the distribution of terminal price becomes stationary, and independent of the current price:

$$P(x, T|x_0, 0) = \frac{1}{\sqrt{2\pi\sigma^2}} \exp -\left(\frac{(x - \overline{x})^2}{2\Delta^2}\right), \quad (16.33)$$

with $\Delta^2 = D/2\Omega$. Hence, the potential excursions of the price are bounded; the probability that $x - \overline{x}$ becomes much larger than Δ rapidly becomes negligible.

Now, suppose that we want to price an option on such an underlying using the Ito technique. Since the drift $-\Omega(x - \overline{x})$ never appears in Ito's formula, the partial differential

equation giving the option price is *exactly* the same as Eq. 16.3. For an at-the-money option, the price is therefore $C = \sqrt{DT/2\pi}$, *independent* of the mean reversion strength Ω.

However, the price fluctuations will never much exceed Δ, and therefore the profit of the buyer of this option will never be much larger than Δ. On the other hand, if T is large, the price of the option is much larger than this potential profit!

The difference comes from the fact that the mean-reverting anticorrelations in the price process plays the role of transaction costs for the writer of the option that hedges according to Black-Scholes. The mean reverting effect means that on average the hedger buys high and sells low. For long maturity options, his risk minimization strategy becomes infinitely costly and therefore unrealistic. (See also the discussion in Chapter 15.)

16.3 The binomial model

The **binomial model** for price evolution – the famous **Cox–Ross–Rubinstein model** (Cox et al. (1979)) – shares with the continuous-time Gaussian model the zero risk property. This model is very much used in practice, due to its easy numerical implementation. Furthermore, the zero risk property appears in the clearest fashion. Suppose indeed that between $t_k = k\tau$ and t_{k+1}, the price difference can only take two values: $\delta x_k = \delta x_{1,2}$. For this very reason, the option value itself can only evolve along two paths only, to be worth (at time t_{k+1}) $C_{1,2}^{k+1}$. Consider now the hedging strategy where one holds a fraction ϕ of the underlying asset, and a quantity B of bonds, with a risk-free interest rate ρ. If one chooses ϕ and B such that:

$$\phi_k(x_k + \delta x_1) + B_k(1 + \rho) = C_1^{k+1}, \tag{16.34}$$

$$\phi_k(x_k + \delta x_2) + B_k(1 + \rho) = C_2^{k+1}, \tag{16.35}$$

or else:

$$\phi_k = \frac{C_1^{k+1} - C_2^{k+1}}{\delta x_1 - \delta x_2}, \qquad B_k(1 + \rho) = \frac{\delta x_1 C_2^{k+1} - \delta x_2 C_1^{k+1}}{\delta x_1 - \delta x_2}, \tag{16.36}$$

one sees that in the two possible cases ($\delta x_k = \delta x_{1,2}$), the value of the hedging portfolio is *strictly equal* to the option value. The option value at time t_k is thus equal to $C^k(x_k) = \phi_k x_k + B_k$, *independently* of the probability p [resp. $1 - p$] that $\delta x_k = \delta x_1$ [resp. δx_2]. One then determines the option value by iterating this procedure from time $k + 1 = N$, where C^N is known, and equal to $\max(x_N - x_s, 0)$, or more generally $\mathcal{Y}(x_N)$. The price of the option is in this case completely independent of the 'historical' probability p. This unfortunate property has led many authors to write that the 'risk neutral' probability that one should use to price options is in general totally unrelated to the 'true' historical probabilities, which is an absurd statement that is only valid in oversimplified situations. It is, for example, easy to see that as soon as δx_k can take three or more values, it is impossible to find a perfect strategy.[†]

The Ito process can be obtained as the continuum limit of the binomial tree. But even in its discrete form, the binomial model shares some important properties with the Black–Scholes model. The independence of the premium on the transition probability p is the analogue of

[†] See e.g. Aurell & Simdyankin (1998).

the independence of the premium on the excess return m in the Black–Scholes model. The magic of zero risk in the binomial model can therefore be understood as follows. Consider the quantity $s^2 = (\delta x_k - a)^2$ for a fixed value of a; s^2 is in general a random number, but for a precise value of a, namely $a = (\delta x_1 - \delta x_2)/2$, s^2 is non-fluctuating ($s^2 \equiv (\delta x_1 - \delta x_2)^2/4$). Because there are only two possible outcomes for δx, a proper choice of a can remove all the fluctuations in s^2, this is the origin of "the hedge that stole the risk." The Ito process shares this property: in the continuum limit quadratic quantities do not fluctuate. For example, the quantity

$$S^2 = \lim_{\tau \to 0} \sum_{k=0}^{T/\tau - 1} [x((k+1)\tau) - x(k\tau) - m\tau]^2 , \tag{16.37}$$

is equal to DT with probability one when $x(t)$ follows an Ito process (see Fig. 3.1). In a sense, continuous-time Brownian motion represents a very weak form of randomness since quantities such as S^2 can be known with certainty. But it is precisely this property that allows for zero risk in the Black–Scholes world.

16.4 Summary

- The Ito formula allows one to see immediately how perfect hedging is possible in a world where uncertainty is taken to be a continuous time Gaussian process.
- The same Ito formula also immediately shows that any drift term, that can be time and price dependent, completely disappears from the option price. As emphasized in previous chapters, the same result is only approximately true when the statistics of price returns is non Gaussian, *if* the hedging strategy is the (possibly suboptimal) Δ-hedge.
- The option price can be written, for an arbitrary pay-off, as the average pay-off over the undrifted distribution of prices (with possible corrections coming from interest rate effects).
- The binomial model, where the price change can only take two values, is another model where complete elimination of risk is possible.
- A riddle for traders – Black-Scholes in Greek:
 When I hedge my option
 I can't lose with Delta
 What I lose with Theta
 I will gain with Gamma
 And Greeks have no Vega.

- FURTHER READING

L. Bachelier, *Théorie de la spéculation*, Gabay, Paris, 1995.

M. Baxter, A. Rennie, *Financial Calculus, An Introduction to Derivative Pricing*, Cambridge University Press, Cambridge, 1996.

J. P. Fouque, G. Papanicolaou, R. Sircar, *Derivatives in Financial Markets with Stochastic Volatility*, Cambridge University Press, 2000.

J. C. Hull, *Futures, Options and Other Derivatives*, Prentice Hall, Upper Saddle River, NJ, 1997.

M. Musiela, M. Rutkowski, *Martingale Methods in Financial Modelling*, Springer, New York, 1997.

D. Lamberton, B. Lapeyre, *Introduction au calcul stochastique appliqué à la finance*, Ellipses, Paris, 1991.

R. C. Merton, *Continuous-Time Finance*, Blackwell, Cambridge, 1990.

M. Musiela, M. Rutkowski, *Martingale Methods in Financial Modelling*, Springer, New York, 1997.

I. Nelken (ed.), *The Handbook of Exotic Options*, Irwin, Chicago, IL, 1996.

N. Taleb, *Dynamical Hedging: Managing Vanilla and Exotic Options*, Wiley, New York, 1998.

P. Wilmott, J. N. Dewynne, S. D. Howison, *Option Pricing: Mathematical Models and Computation*, Oxford Financial Press, Oxford, 1993.

P. Wilmott, *Derivatives, The theory and practice of financial engineering*, John Wiley, 1998.

P. Wilmott, *Paul Wilmott Introduces Quantitative Finance*, John Wiley, 2001.

- SPECIALIZED PAPERS: SOME CLASSICS

F. Black, M. Scholes, *The pricing of options and corporate liabilities*, Journal of Political Economy, **3**, 637 (1973).

J. C. Cox, S. A. Ross, M. Rubinstein, *Option pricing: a simplified approach*, Journal of Financial Economics, **7**, 229 (1979).

R. C. Merton, *Theory of rational option pricing*, Bell Journal of Economics and Management Science, **4**, 141 (1973).

- SPECIALIZED PAPERS

E. Aurell, S. Simdyankin, *Pricing risky options simply*, International Journal of Theoretical and Applied Finance, **1**, 1 (1998).

E. Aurell, R. Baviera, O. Hammarlid, M. Serva, A. Vulpiani, *Growth optimal investment and pricing of derivatives*, Int. Jour. Theo. Appl. Fin., **3**, 1 (2000)

M. Schweizer, *Risky options simplified*, International Journal of Theoretical and Applied Finance, **2**, 59 (1999).

- SOME RECENT READS

R. Cont, P. Tankov, *Financial Modelling with Jump Processes*, Chapman & Hall / CRC Press, 2003.

J. Gatheral, *The Volatility Surface: A Practitioner's Guide*, Wiley Finance, 2006.

P. Henry-Labordère, *Analysis, Geometry, and Modeling in Finance: Advanced Methods in Option Pricing*, Chapman & Hall / CRC Press, forthcoming.

17

Options: some more specific problems

This chapter can be skipped at first reading.
(J.-P. Bouchaud, M. Potters, *Theory of Financial Risk and Derivative Pricing*.)

17.1 Other elements of the balance sheet

We have until now considered the simplest possible option problem, and tried to extract the fundamental ideas associated to its pricing and hedging. In reality, several complications may appear, either in the very definition of the option contract (see next section), or in the elements that must be included in the wealth balance – for example dividends or transaction costs – that we have neglected up to now.

17.1.1 Interest rate and continuous dividends

The influence of interest rates (and continuous dividends) can be estimated using different models for the statistics of price increments.

These models give slightly different answers; however, *in a first approximation*, they provide the following answer for the option price:

$$C(x_0, x_s, T, r) = e^{-rT} C(x_0 e^{rT}, x_s, T, r = 0), \tag{17.1}$$

where $C(x_0, x_s, T, r = 0)$ is simply given by the average of the pay-off over the terminal undrifted price distribution.

In the presence of a non-zero continuous dividend d, the quantity $x_0 e^{rT}$ should be replaced by $x_0 e^{(r-d)T}$, i.e. in all cases the present price of the underlying should be replaced by the corresponding *forward* price – see Eq. (4.23).

Let us present three different models for stock returns that all lead to the above approximate formula; but with slightly different corrections to it in general. These three models offer alternative ways to understand the influence of a non-zero interest rate (and/or dividend) on the price of the option.

Rate and dividend induced drift of the price

In this model, one assumes that the price evolves according to $x_{k+1} - x_k = (\rho - d\tau)x_k + \delta x_k$, where $d\tau$ is the dividend over a unit time period τ, and δx_k are uncorrelated random variables of zero mean. Note that this 'dividend' is, in the case of currency options, the interest rate corresponding to the underlying currency (and ρ is the interest rate of the reference currency).

This model is convenient because if $\langle \delta x_k \rangle$ is zero, the average cost of the hedging strategy is zero, since it precisely includes the terms $x_{k+1} - x_k - (\rho - \delta)x_k$, where $\delta \equiv d\tau$. However, the terminal distribution of x_N must be constructed by noting that:

$$x_N = x_0(1 + \rho - \delta)^N + \sum_{k=0}^{N-1} \delta x_k (1 + \rho - \delta)^{N-k-1}. \tag{17.2}$$

Thus the terminal distribution is, for large N and small $\rho - \delta$, a function of the difference $x(T) - x_0 e^{(r-d)T}$ between the price and the futures. Furthermore, even if the δx_k are independent random variables of variance $D\tau$, the variance $c_2(T)$ of the terminal distribution is given by:

$$c_2(T) = \frac{D}{2(r-d)} \left[e^{2(r-d)T} - 1 \right], \tag{17.3}$$

which is equal to $c_2(T) = DT(1 + (r - d)T)$ when $(r - d)T \to 0$. There is thus an increase of the effective volatility to be used in the option pricing formula within this model. This increase depends on the maturity; its relative value is, for $T = 1$ year, $r - d = 5\%$, equal to $(r - d)T/2 = 2.5\%$. (Note indeed that $c_2(T)$ is the *square* of the volatility.) Up to this small change of volatility, the above rule, Eq. (17.1) is therefore appropriate.

Independence between price increments and interest rates/dividends

The above model is perhaps not very realistic since it assumes a direct, systematic relation between price changes and interest rates/dividends. It might be more appropriate to consider the case where $x_{k+1} - x_k = \delta x_k$ are uncorrelated random variables of zero mean. Reality is presumably in between these two limiting models. Now the terminal distribution is a function of the difference $x_N - x_0$, with no correction to the variance brought about by interest rates. However, in the present case, the cost of the hedging strategy cannot be neglected and reads:

$$\langle \Delta W_S \rangle = -(\rho - \delta) \sum_{k=0}^{N-1} \langle x_k \phi_k^{N*}(x_k) \rangle. \tag{17.4}$$

Writing $x_k = x_0 + \sum_{j=0}^{k-1} \delta x_j$, we find that $\langle \Delta W_S \rangle$ is the sum of a term proportional to x_0 and the remainder. The first contribution therefore reads:

$$\langle \Delta W_S \rangle_1 = -(\rho - \delta)x_0 \sum_{k=0}^{N-1} \int P(x, k|x_0, 0)\phi_k^{N*}(x)\,dx. \tag{17.5}$$

This term thus has exactly the same shape as in the case of a non-zero average return treated above, Eq. (15.2), with an effective mean return $m_{1\text{eff}} = -(\rho - \delta)x_0$. If we choose

for ϕ^* the suboptimal Δ-hedge (which we know does only induce minor errors), then, as shown above, this hedging cost can be reabsorbed in a change of the terminal distribution, where $x_N - x_0$ becomes $x_N - x_0 - m_{1\text{eff}}N$, or else, to first order: $x_N - x_0 \exp[(\rho - \delta)N]$. To order $(\rho - \delta)N$, this again corresponds to replacing the current price of the underlying by its forward price. For the second type of terms we write, for $j < k$:

$$\langle \delta x_j \phi_k^{N*}(x_k) \rangle = \int P(x_j, j|x_0, 0) \, dx_j \int P(x_k, k|x_j, j) \frac{x_k - x_j}{k - j} \phi_k^{N*}(x_k) \, dx_k. \tag{17.6}$$

This quantity is not easy to calculate in the general case. Assuming that the process is Gaussian allows us to use the identity:

$$P(x_k, k|x_j, j) \frac{x_k - x_j}{k - j} = -D\tau \frac{\partial P(x_k, k|x_j, j)}{\partial x_k}; \tag{17.7}$$

one can therefore integrate over x_j to find a result independent of j:

$$\langle \delta x_j \phi_k^{N*}(x_k) \rangle = D\tau \frac{\partial}{\partial x_0} \int P(x_k, k|x_0, 0) \phi_k^{N*}(x_k) \, dx_k, \tag{17.8}$$

or, using the expression of $\phi_k^{N*}(x_k)$ in the Gaussian case:

$$\langle \delta x_j \phi_k^{N*}(x_k) \rangle = D\tau \frac{\partial}{\partial x_0} \int_{x_s}^{\infty} P(x', N|x_0, 0) \, dx' = D\tau P(x_s, N|x_0, 0). \tag{17.9}$$

The contribution to the cost of the strategy is then obtained by summing over k and over $j < k$, finally leading to:

$$\langle \Delta W_S \rangle_2 \simeq -\frac{N^2}{2}(\rho - \delta)D\tau P(x_s, N|x_0, 0). \tag{17.10}$$

This extra cost is maximum for at-the-money options, and leads to a relative increase of the call price given, for Gaussian statistics, by:

$$\frac{\delta C}{C} = \frac{N}{2}(\rho - \delta) = \frac{T}{2}(r - d), \tag{17.11}$$

exactly of the same order as for the previous model, see Eq. (17.3). This correction is again quite small for short maturities. (For $r = 5\%$ annual, $d = 0$ and $T = 100$ days, one finds $\delta C/C \leq 1\%$.)

Multiplicative model

Let us finally assume that $x_{k+1} - x_k = (\rho - \delta + \eta_k)x_k$, where the η_k's are independent random variables of zero mean. Therefore, the average cost of the strategy is again zero. Furthermore, if the elementary time step τ is small, the terminal value x_N can be written as:

$$\log\left(\frac{x_N}{x_0}\right) = \sum_{k=0}^{N-1} \log(1 + \rho - \delta + \eta_k) \simeq N(\rho - \delta) + y; \qquad y = \sum_{k=0}^{N-1} \eta_k, \tag{17.12}$$

thus $x_N = x_0 e^{(r-d)T+y}$. Introducing the distribution $P(y)$, the price of the option can be written as:

$$C(x_0, x_s, T, r, d) = e^{-rT} \int P(y) \max\left(x_0 e^{(r-d)T+y} - x_s, 0\right) dy, \tag{17.13}$$

which is obviously of the general form, Eq. (17.1), which is exact in this case.

We thus conclude that the simple rule, Eq. (17.1) above, is, for many purposes, sufficient to account for interest rates and (continuous) dividends effects. But a more accurate formula, including corrections of order $(r - d)T$, depends somewhat on the chosen model.

17.1.2 Interest rate corrections to the hedging strategy

It is also important to estimate the interest rate corrections to the optimal hedging strategy. Here again, one could consider the above three models, that is, the systematic drift model, the independent model or the multiplicative model. In the former case, the result for a general terminal price distribution is rather involved, mainly due to the appearance of the conditional average detailed in Appendix D, Eq. (14.69). In the case where this distribution is Gaussian, the result is simply Black–Scholes' Δ-hedge, i.e. the optimal strategy is the derivative of the option price with respect to the price of the underlying contract (see Eq. (14.22)). As discussed in Chapter 15, this simple rule leads in general to a suboptimal strategy, but the relative increase of risk is rather small. The Δ-hedge procedure, on the other hand, has the advantage that the price is independent of the average return (see Appendix D).

In the 'independent' model, where the price change is unrelated to the interest rate and/or dividend, the order ρ correction to the optimal strategy is easier to estimate in general. From the general formula, Eq. (15.47), one finds:

$$\phi_k^*(x) = (1+\rho)^{1+k-N}\phi_k^{*0}(x) - \frac{r-d}{D}xC(x, x_s, N-k)$$

$$+ \frac{r-d}{2D}x \sum_{\ell<k} \int \frac{x-x'}{k-\ell} \frac{P(x,k|x',\ell)P(x',\ell|x_0,0)}{P(x,k|x_0,0)} \phi_\ell^{*0}(x')\,dx'$$

$$+ \frac{r-d}{2D} \sum_{\ell>k} \int \frac{x'-x}{\ell-k} x' P(x',\ell|x,k)\phi_\ell^{*0}(x')\,dx', \tag{17.14}$$

*where ϕ_k^{*0} is the optimal strategy for $r = d = 0$.*

Finally, for the multiplicative model, the optimal strategy is given by a much simpler general expression:

$$\phi_k^*(x) = \frac{(1+\rho)^{1+k-N}}{x\sigma^2(N-k)} \times \int_{x_s}^{\infty}(x'-x_s)\log\left(\frac{x'}{x(1+\rho-\delta)^{N-k}}\right) P(x',N|x,k)\,dx'. \tag{17.15}$$

17.1.3 Discrete dividends

More generally, for an arbitrary dividend d_k (per share) at time t_k, the extra term in the wealth balance reads:

$$\Delta W_D = \sum_{k=1}^{N} \phi_k^N(x_k)d_k. \tag{17.16}$$

Very often, this dividend occurs once a year, at a given date k_0: $d_k = d_0\delta_{k,k_0}$. In this case, the corresponding share price decreases immediately by the same amount (since the value of the company is decreased by an amount equal to d_0 times the number of outstanding shares):

$x \to x - d_0$. Therefore, the net balance $d_k + \delta x_k$ associated to the dividend is zero. For the same reason, the probability $P_{d_0}(x, N | x_0, 0)$ is given, for $N > k_0$, by $P_{d_0=0}(x + d_0, N | x_0, 0)$. The option price is then equal to: $C_{d_0}(x, x_s, N) \equiv C(x, x_s + d_0, N)$. If the dividend d_0 is not known in advance with certainty, this last equation should be averaged over a distribution $P(d_0)$ describing as well as possible the probable values of d_0. A possibility is to choose the distribution of least information (maximum entropy) such that the average dividend is fixed to \overline{d}, which in this case reads:

$$P(d_0) = \frac{1}{\overline{d}} \exp(-d_0/\overline{d}). \tag{17.17}$$

17.1.4 Transaction costs

The problem of transaction costs is important: the rebalancing of the optimal hedge as time passes induces extra costs that must be included in the wealth balance as well. These 'costs' are of different nature – some are proportional to the number of traded shares, whereas another part is fixed, independent of the total amount of the operation. Let us examine the first situation, assuming that the rebalancing of the hedge takes place at every time step τ. The corresponding cost is then equal to:

$$\delta W_{\text{trk}} = \nu |\phi_{k+1}^* - \phi_k^*|, \tag{17.18}$$

where ν is a measure of the transaction costs. In order to keep the discussion simple, we shall assume that:

- the most important part in the variation of ϕ_k^* is due to the change of price of the underlying, and not from the explicit time dependence of ϕ_k^* (this is justified in the limit where $\tau \ll T$);
- δx_k is small enough such that a Taylor expansion of ϕ^* is acceptable;
- the kurtosis effects on ϕ^* are neglected, which means that the simple Black–Scholes Δ-hedge is acceptable and does not lead to large hedging errors.

One can therefore write that:

$$\delta W_{\text{trk}} = \nu \left| \delta x_k \frac{\partial \phi_k^*(x)}{\partial x} \right| = \nu |\delta x_k| \frac{\partial \phi_k^*(x)}{\partial x}, \tag{17.19}$$

since $\partial \phi_k^*(x)/\partial x$ is positive. The average total cost associated to rehedging is then, after a straightforward calculation:[†]

$$\langle \Delta W_{\text{tr}} \rangle = \left\langle \sum_{k=0}^{N-1} \delta W_{\text{trk}} \right\rangle = \nu \langle |\delta x| \rangle N P(x_s, N | x_0, 0). \tag{17.20}$$

The order of magnitude of $\langle |\delta x| \rangle$ is given by $\sigma_1 x_0$: for an at-the-money option, $P(x_s, N | x_0, 0) \sim (\sigma_1 x_0 \sqrt{N})^{-1}$; hence, finally, $\langle \Delta W_{\text{tr}} \rangle \sim \nu \sqrt{N}$. It is natural to compare $\langle \Delta W_{\text{tr}} \rangle$ to the option price itself, which is of order $C \sim \sigma_1 x_0 \sqrt{N}$:

$$\frac{\langle \Delta W_{\text{tr}} \rangle}{C} \propto \frac{\nu}{\sigma_1 x_0}. \tag{17.21}$$

[†] One should add to the following formula the cost associated with the initial value of the hedge ϕ_0^*, equal to $\nu x_0 \phi_0^*$.

This part of the transaction costs is in general proportional to x_0: taking for example $\nu = 10^{-4}x_0$, $\tau = 1$ day and a daily volatility of $\sigma_1 = 1\%$, one finds that the transaction costs represent 1% of the option price. On the other hand, for higher transaction costs (say $\nu = 10^{-2}x_0$), a daily rehedging becomes absurd, since the ratio, Eq. (17.21), is of order 1. The rehedging frequency should then be lowered, such that the volatility on the scale of τ increases, such as to become larger than ν and to justify the cost of hedging. Remember that a small error ε in the hedging strategy only affects the risk to second order ε^2. Note also that in practice, one often hedges several call and put options simultaneously. Since for some the hedge increases, and for others it decreases, part of the trades cancel out and will not be executed. Therefore, the total cost of hedging of an option book can be much smaller than inferred from the above estimate.

In summary, the transaction costs are higher when the trading time is smaller. On the other hand, decreasing of τ allows one to diminish the residual risk (Fig. 14.4). This analysis suggests that the optimal trading frequency should be chosen such that the transaction costs are comparable to the residual risk.

17.2 Other types of options: 'puts' and 'exotic options'

17.2.1 'Put-Call' parity

A 'put' contract is a sell option, defined by a pay-off at maturity given by $\max(x_s - x, 0)$. A put protects its owner against the risk that the shares he owns drops below the strike price x_s. Puts on stock indices like the S&P 500 are very popular. The price of a European put will be noted $C^\dagger[x_0, x_s, T]$ (we reserve the notation \mathcal{P} for a probability). This price can be obtained using a very simple 'parity' (or no arbitrage) argument. Suppose that one simultaneously buys a call with strike price x_s, and a put with the same maturity and strike price. At expiry, this combination is therefore worth:

$$\max(x(T) - x_s, 0) - \max(x_s - x(T), 0) \equiv x(T) - x_s. \tag{17.22}$$

Exactly the same pay-off can be achieved by buying the underlying now, and selling a bond paying x_s at maturity. Therefore, the above call+put combination must be worth:

$$C[x_0, x_s, T] - C^\dagger[x_0, x_s, T] = x_0 - x_s e^{-rT}, \tag{17.23}$$

which allows one to express the price of a put knowing that of a call. If interest rate effects are small ($rT \ll 1$), this relation reads:

$$C^\dagger[x_0, x_s, T] \simeq C[x_0, x_s, T] + x_s - x_0. \tag{17.24}$$

Note in particular that at-the-money ($x_s = x_0$), the two contracts have the same value, which is obvious by symmetry (again, in the absence of interest rate effects).

17.2.2 'Digital' options

More general option contracts can stipulate that the pay-off is not the difference between the value of the underlying at maturity x_N and the strike price x_s, but rather an arbitrary

function $\mathcal{Y}(x_N)$ of x_N. For example, a 'digital' option is a pure bet, in the sense that it pays a fixed premium \mathcal{Y}_0 whenever x_N exceeds x_s. Therefore:

$$\mathcal{Y}(x_N) = \mathcal{Y}_0 \quad (x_N > x_s); \quad \mathcal{Y}(x_N) = 0 \quad (x_N < x_s). \tag{17.25}$$

The price of the option in this case can be obtained following the same lines as above. In particular, in the absence of bias (i.e. for $m = 0$) the fair price is given by:

$$\mathcal{C}_{\mathcal{Y}}(x_0, N) = \langle \mathcal{Y}(x_N) \rangle = \int \mathcal{Y}(x) P_0(x, N | x_0, 0) \, dx, \tag{17.26}$$

whereas the optimal strategy is still given by the general formula, Eq. (13.33). In particular, for Gaussian fluctuations, one always finds the Black–Scholes recipe:

$$\phi_{\mathcal{Y}}^*(x) \equiv \frac{\partial \mathcal{C}_{\mathcal{Y}}[x, N]}{\partial x}. \tag{17.27}$$

The case of a non-zero average return can be treated as above, in Section 15.1.1. To first order in m, the price of the option is given by:

$$\mathcal{C}_{\mathcal{Y},m} = \mathcal{C}_{\mathcal{Y},m=0} - \frac{mT}{D\tau} \sum_{n=3}^{\infty} \frac{c_{n,1}}{(n-1)!} \int \mathcal{Y}(x') \frac{\partial^{n-1}}{\partial x^{n-1}} P_0(x', N | x, 0) \, dx', \tag{17.28}$$

which reveals, again, that in the Gaussian case, the average return disappears from the final formula. In the presence of non-zero kurtosis, however, a (small) systematic correction to the fair price appears. Note that $\mathcal{C}_{\mathcal{Y},m}$ can be written as an average of $\mathcal{Y}(x)$ using the effective, 'risk neutral' probability $Q(x)$ introduced in Section 15.1.1. If the Δ-hedge is used, this risk neutral probability is simply P_0, and the value of the drift disappears from the option price – see Section 15.2.2.

17.2.3 'Asian' options

The problem is slightly more complicated in the case of the so-called 'Asian' options. The pay-off of these options is calculated not on the value of the underlying stock at maturity, but on a certain average of this value over a certain number of days preceding maturity. This procedure is used to prevent an artificial rise of the stock price precisely on the expiry date, a rise that could be triggered by an operator having an important long position on the corresponding option. The contract is thus constructed on a fictitious asset, the price of which being defined as:

$$\tilde{x} = \sum_{i=0}^{N} w_i x_i, \tag{17.29}$$

where the $\{w_i\}$'s are some weights, normalized such that $\sum_{i=0}^{N} w_i = 1$, which define the averaging procedure. The simplest case corresponds to:

$$w_N = w_{N-1} = \cdots = w_{N-K+1} = \frac{1}{K}; \quad w_k = 0 \quad (k < N - K + 1), \tag{17.30}$$

where the average is taken over the last K days of the option life. One could however consider more complicated situations, for example an exponential average ($w_k \propto s^{N-k}$).

The wealth balance then contains the modified pay-off: $\max(\tilde{x} - x_s, 0)$, or more generally $\mathcal{Y}(\tilde{x})$. The first problem therefore concerns the statistics of \tilde{x}. As we shall see, this problem is very similar to the case encountered in Chapter 14 where the volatility is time dependent. Indeed, one has:

$$\sum_{i=0}^{N} w_i x_i = \sum_{i=0}^{N} w_i \left(\sum_{\ell=0}^{i-1} \delta x_\ell + x_0 \right) = x_0 + \sum_{k=0}^{N-1} \gamma_k \delta x_k, \tag{17.31}$$

where

$$\gamma_k \equiv \sum_{i=k+1}^{N} w_i. \tag{17.32}$$

Said differently, everything goes as if the price did not vary by an amount δx_k, but by an amount $\delta y_k = \gamma_k \delta x_k$, distributed as:

$$\frac{1}{\gamma_k} P_1 \left(\frac{\delta y_k}{\gamma_k} \right). \tag{17.33}$$

In the case of Gaussian fluctuations of variance $D\tau$, one thus finds:[†]

$$P(\tilde{x}, N | x_0, 0) = \frac{1}{\sqrt{2\pi \tilde{D} N \tau}} \exp \left[-\frac{(\tilde{x} - x_0)^2}{2\tilde{D} N \tau} \right], \tag{17.34}$$

where

$$\tilde{D} = \frac{D}{N} \sum_{k=0}^{N-1} \gamma_k^2. \tag{17.35}$$

More generally, $P(\tilde{x}, N | x_0, 0)$ is the Fourier transform of

$$\prod_{k=0}^{N-1} \hat{P}_1(\gamma_k z). \tag{17.36}$$

This information is sufficient to fix the option price (in the limit where the average return is very small) through:

$$\mathcal{C}_{\text{asi}}[x_0, x_s, N] = \int_{x_s}^{\infty} (\tilde{x} - x_s) P(\tilde{x}, N | x_0, 0) \, d\tilde{x}. \tag{17.37}$$

In order to fix the optimal hedging strategy, one must however calculate the following quantity:

$$P(\tilde{x}, N | x, k) \langle \delta x_k \rangle |_{(x,k) \to (\tilde{x}, N)}, \tag{17.38}$$

conditioned to a certain terminal value for \tilde{x} (cf. Eq. (14.15)). The general calculation is given in Appendix D. For a small kurtosis, the optimal strategy reads:

$$\phi_k^{N*}(x) = \frac{\partial \mathcal{C}_{\text{asi}}[x, x_s, N - k]}{\partial x} + \frac{\kappa_1 D\tau}{6} \gamma_k^2 \frac{\partial^3 \mathcal{C}_{\text{asi}}[x, x_s, N - k]}{\partial x^3}. \tag{17.39}$$

[†] The case of a multiplicative process is more involved: see, e.g. H. Gemam, M. Yor, Bessel processes, Asian options and perpetuities, *Mathematical Finance*, **3**, 349 (1993).

Note that if the instant of time 'k' is outside the averaging period, one has $\gamma_k = 1$ (since $\sum_{i>k} w_i = 1$), and the formula, Eq. (14.24), is recovered. If on the contrary k gets closer to maturity, γ_k diminishes as does the correction term.

17.2.4 'American' options

We have up to now focused our attention on 'European'-type options, which can only be exercised on the day of expiry. In reality, most traded options on organized markets can be exercised at any time between the issue date and the expiry date: by definition, these are called 'American' options. It is obvious that the price of an American option must be greater or equal to the price of a European option with the same maturity and strike price, since the contract is *a priori* more favourable to the buyer. The pricing problem is therefore more difficult, since the writer of the option must first determine the optimal strategy that the buyer can follow in order to fix a fair price. Now, in the absence of dividends, the optimal strategy for the buyer of a call option is to keep it until the expiry date, thereby converting *de facto* the option into a European option. Intuitively, this is due to the fact that the average $\langle \max(x_N - x_s, 0) \rangle$ grows with N, hence the average pay-off is higher if one waits longer. The argument can be made more convincing as follows. Let us define a 'two-shot' option, of strike x_s, which can only be exercised at times N_1 and $N_2 > N_1$ only.[†] At time N_1, the buyer of the option may choose to exercise a fraction $f(x_1)$ of the option, which in principle depends on the current price of the underlying x_1. The remaining part of the option can then be exercised at time N_2. What is the average profit $\langle \mathcal{G} \rangle$ of the buyer at time N_2?

Considering the two possible cases, one obtains:

$$\langle \mathcal{G} \rangle = \int_{x_s}^{+\infty} (x_2 - x_s)\,dx_2 \int P(x_2, N_2|x_1, N_1)[1 - f(x_1)]P(x_1, N_1|x_0, 0)\,dx_1 +$$
$$\int_{x_s}^{+\infty} f(x_1)(x_1 - x_s)e^{r\tau(N_2-N_1)}P(x_1, N_1|x_0, 0)\,dx_1, \tag{17.40}$$

which can be rewritten as:

$$\langle \mathcal{G} \rangle = C[x_0, x_s, N_2]e^{r\tau N_2} +$$
$$\int_{x_s}^{\infty} f(x_1) \times P(x_1, N_1|x_0, 0)(x_1 - x_s - C[x_1, x_s, N_2 - N_1])e^{r\tau(N_2-N_1)}\,dx_1. \tag{17.41}$$

The last expression means that if the buyer exercises a fraction $f(x_1)$ of his option, he pockets immediately the difference $x_1 - x_s$, but loses *de facto* his option, which is worth $C[x_1, x_s, N_2 - N_1]$.

The optimal strategy, such that $\langle \mathcal{G} \rangle$ is maximum, therefore consists in choosing $f(x_1)$ equal to 0 or 1, according to the sign of $x_1 - x_s - C[x_1, x_s, N_2 - N_1]$. Now, this difference is always negative, whatever x_1 and $N_2 - N_1$. This is due to the put–call parity relation

[†] Options that can be exercised at certain specific dates (more than one) are called 'Bermudan' options.

(cf. Eq. (17.24)):

$$C^\dagger[x_1, x_s, N_2 - N_1] = C[x_1, x_s, N_2 - N_1] - (x_1 - x_s) - x_s \left(1 - e^{-r\tau(N_2 - N_1)}\right). \qquad (17.42)$$

Since $C^\dagger \geq 0$, $C[x_1, x_s, N_2 - N_1] - (x_1 - x_s)$ is also greater or equal to zero.

The optimal value of $f(x_1)$ is thus zero; said differently the buyer should wait until maturity to exercise his option to maximize his average profit. This argument can be generalized to the case where the option can be exercised at any instant N_1, N_2, \ldots, N_n with n arbitrary.

Note however that choosing a non-zero f increases the total probability of exercising the option, but reduces the average profit! More precisely, the total probability of reaching x_s before maturity is twice the probability of exercising the option at expiry (if the distribution of δx is even, see Section 10.4). OTC American options are therefore favourable to the writer of the option, since some buyers might be tempted to exercise before expiry.

It is interesting to generalize the problem and consider the case where the two strike prices x_{s1} and x_{s2} are different at times N_1 and N_2, in particular in the case where $x_{s1} < x_{s2}$. The average profit, Eq. (17.41), is then equal to (for $r = 0$):

$$\langle \mathcal{G} \rangle = C[x_0, x_{s2}, N_2] + \int_{x_s}^{\infty} f(x_1) P(x_1, N_1 | x_0, 0)$$

$$\times (x_1 - x_{s1} - C[x_1, x_{s2}, N_2 - N_1]) dx_1. \qquad (17.43)$$

The equation

$$x^* - x_{s1} - C[x^*, x_{s2}, N_2 - N_1] = 0 \qquad (17.44)$$

then has a non-trivial solution, leading to $f(x_1) = 1$ for $x_1 > x^$. The average profit of the buyer therefore increases, in this case, upon an early exercise of the option.*

American puts

Naively, the case of the American puts looks rather similar to that of the calls, and these should therefore also be equivalent to European puts. This is not the case for the following reason.[§] Using the same argument as above, one finds that the average profit associated to a 'two-shot' put option with exercise dates N_1, N_2 is given by:

$$\langle \mathcal{G}^\dagger \rangle = C^\dagger[x_0, x_s, N_2] e^{r\tau N_2} + \int_{-\infty}^{x_s} f(x_1) P(x_1, N_1 | x_0, 0)$$

$$\times (x_s - x_1 - C^\dagger[x_1, x_s, N_2 - N_1]) e^{r\tau(N_2 - N_1)} dx_1. \qquad (17.45)$$

Now, the difference $(x_s - x_1 - C^\dagger[x_1, x_s, N_2 - N_1])$ can be transformed, using the put-call parity, as:

$$x_s \left[1 - e^{-r\tau(N_2 - N_1)}\right] - C[x_1, x_s, N_2 - N_1]. \qquad (17.46)$$

This quantity may become positive if $C[x_1, x_s, N_2 - N_1]$ is very small, which corresponds to $x_s \gg x_1$ (puts deep in the money). The smaller the value of r, the larger should be the

[§] The case of American calls with non-zero dividends is similar to the case discussed here.

difference between x_1 and x_s, and the smaller the probability for this to happen. If $r = 0$, the problem of American puts is identical to that of the calls.

In the case where the quantity (17.46) becomes positive, an 'excess' average profit $\delta\mathcal{G}$ is generated, and represents the extra premium to be added to the price of the European put to account for the possibility of an early exercise. Let us finally note that the price of the American put C_{am}^\dagger is necessarily always larger or equal to $x_s - x$ (since this would be the immediate profit), and that the price of the 'two-shot' put is a lower bound to C_{am}^\dagger.

> The perturbative calculation of $\delta\mathcal{G}$ (and thus of the 'two-shot' option) in the limit of small interest rates is not very difficult. As a function of N_1, $\delta\mathcal{G}$ reaches a maximum between $N_2/2$ and N_2. For an at-the-money put such that $N_2 = 100$, $r = 5\%$ annual, $\sigma = 1\%$ per day and $x_0 = x_s = 100$, the maximum is reached for $N_1 \approx 80$ and the corresponding $\delta\mathcal{G} \approx 0.15$. This must be compared with the price of the European put, which is $C^\dagger \approx 4$. The possibility of an early exercise leads in this case to a 5% increase of the price of the option.
>
> More generally, when the increments are independent and of average zero, one can obtain a numerical value for the price of an American put C_{am}^\dagger by iterating backwards the following exact equation:
>
> $$C_{am}^\dagger[x, x_s, N] = \max(x_s - x, e^{-r\tau}\langle C_{am}^\dagger[x + \delta x, x_s, N + 1]\rangle) \tag{17.47}$$
>
> This equation means that the put is worth the average value of tomorrow's price if it is not exercised today ($C_{am}^\dagger > x_s - x$), or $x_s - x$ if it is immediately exercised. The averaging over the price increment must be performed with the undrifted distribution if the option is Δ-hedged, see Section 15.2.2.
>
> Using this procedure, one can calculate the price of a European, American and 'two-shot' option of maturity 100 days (Fig. 17.1). For the 'two-shot' option, the optimal value of N_1 as a function of the strike is shown in the inset.

17.2.5 'Barrier' options (*)

Let us now turn to another family of options, called 'barrier' options, which are such that if the price of the underlying x_k reaches a certain 'barrier' value x_b during the lifetime of the option, the option is lost. (Conversely, there are options that are only activated if the value x_b is reached.) This clause leads to cheaper options, which can be more attractive to the investor. Also, if $x_b > x_s$, the writer of the option limits his possible losses to $x_b - x_s$. What is the probability $P_b(x, N|x_0, 0)$ for the final value of the underlying to be at x, conditioned to the fact that the price has not reached the barrier value x_b?

In some cases, it is possible to give an exact answer to this question, using the so-called method of images. Let us suppose that for each time step, the price x can only change by an discrete amount, ± 1 tick. The method of images is explained graphically in Figure 17.2: one can notice that all the trajectories going through x_b between $k = 0$ and $k = N$ has a 'mirror' trajectory, with a statistical weight *precisely equal* (for $m = 0$) to the one of the trajectory one wishes to exclude. It is clear that the conditional probability we are looking for is obtained by subtracting the weight of these image trajectories:

$$P_b(x, N|x_0, 0) = P(x, N|x_0, 0) - P(x, N|2x_b - x_0, 0). \tag{17.48}$$

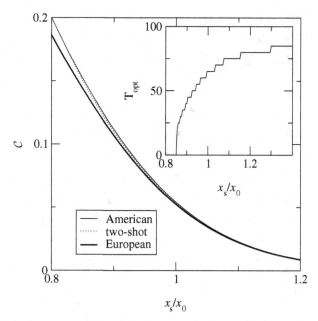

Fig. 17.1. Price of a European, American and 'two-shot' put option as a function of the strike, for a 100-days maturity and a daily volatility of 1% and $r = 1\%$. The top curve is the American price, while the bottom curve is the European price. In the inset is shown the optimal exercise time N_1 as a function of the strike for the 'two-shot' option.

In the general case where the variations of x are not limited to 0, ± 1, the previous argument fails, as one can easily be convinced by considering the case where δx takes the values ± 1 and ± 2. However, if the possible variations of the price during the time τ are small, the error coming from the uncertainty about the exact crossing time is small, and leads to an error on the price C_b of the barrier option on the order of $\langle |\delta x| \rangle$ times the total probability of ever touching the barrier. Discarding this correction, the price of barrier options reads:

$$C_b[x_0, x_s, N] = \int_{x_s}^{\infty} (x - x_s)[P(x, N|x_0, 0) - P(x, N|2x_b - x_0, 0)] \, dx$$

$$\equiv C[x_0, x_s, N] - C[2x_b - x_0, x_s, N],$$

$$(x_b < x_0), \, or \tag{17.49}$$

$$C_b[x_0, x_s, N] = \int_{x_s}^{x_b} (x - x_s) [P(x, N|x_0, 0) - P(x, N|2x_b - x_0, 0)] \, dx, \tag{17.50}$$

$(x_b > x_s)$; the option is worthless whenever $x_0 < x_b < x_s$.

One can also find 'double barrier' options, such that the price is constrained to remain within a certain channel $x_b^- < x < x_b^+$, or else the option vanishes. One can generalize the method of images to this case. The images are now successive reflections of the starting point x_0 in the two parallel 'mirrors' x_b^-, x_b^+.

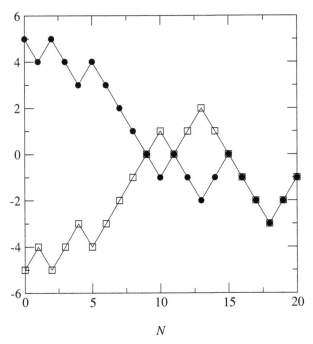

N

Fig. 17.2. Illustration of the method of images. A trajectory, starting from the point $x_0 = -5$ and reaching the point $x_{20} = -1$ can either touch or avoid the 'barrier' located at $x_b = 0$. For each trajectory touching the barrier, as the one shown in the figure (squares), there exists one single trajectory (circles) starting from the point $x_0 = +5$ and reaching the same final point – only the last section of the trajectory (after the last crossing point) is common to both trajectories. In the absence of bias, these two trajectories have exactly the same statistical weight. The probability of reaching the final point without crossing $x_b = 0$ can thus be obtained by subtracting the weight of the image trajectories. Note that the argument is not exact if jump sizes are not constant.

17.2.6 Other types of options

One can find many other types of options, which we shall not discuss further. Some options, for example, are calculated on the maximum value of the price of the underlying reached during a certain period. It is clear that in this case, a Gaussian or log-normal model is particularly inadequate, since the price of the option is directly governed by extreme events. Only an adequate treatment of the tails of the distribution can allow one to price this type of option correctly.

17.3 The 'Greeks' and risk control

The 'Greeks', which is the traditional name given by professionals to the derivative of the price of an option with respect to the price of the underlying, the volatility, etc., are often used for local risk control purposes. Indeed, if one assumes that the underlying asset does not vary too much between two instants of time t and $t + \tau$, one may expand the variation

of the option price in Taylor series:

$$\delta C = \Delta \delta x + \frac{1}{2}\Gamma(\delta x)^2 + V\delta\sigma + \Theta\tau, \tag{17.51}$$

where δx is the change of price of the underlying. If the option is hedged by simultaneously selling a proportion ϕ of the underlying asset, one finds that the change of the portfolio value is, to this order:

$$\delta W = (\Delta - \phi)\delta x + \frac{1}{2}\Gamma(\delta x)^2 + V\delta\sigma + \Theta\tau. \tag{17.52}$$

Note that the Black–Scholes (or rather, Bachelier) equation is recovered by setting $\phi^* = \Delta$, $\delta\sigma \equiv 0$, and by recalling that for a continuous-time Gaussian process, $(\delta x)^2 \equiv D\tau$ (see Chapter 15). In this case, the portfolio does not change with time ($\delta W = 0$), provided that $\Theta = -D\Gamma/2$, which is precisely Eq. (16.3) in the limit $\tau \to 0$.

In reality, due to the non-Gaussian nature of δx, the large risk corresponds to cases where $\Gamma(\delta x)^2 \gg |\Theta\tau|$. Assuming that one chooses to follow the Δ-hedge procedure (which is in general suboptimal, see Section 14.2.2), one finds that the fluctuations of the price of the underlying leads to an *increase* in the value of the portfolio of the buyer of the option (since $\Gamma > 0$). Losses can only occur if the implied volatility of the underlying decreases. If δx and $\delta\sigma$ are uncorrelated (which is in general not true), one finds that the 'instantaneous' variance of the portfolio is given by:

$$\langle(\delta W)^2\rangle = \frac{3 + \kappa_1}{4}\Gamma^2\langle\delta x^2\rangle^2 + V^2\langle\delta\sigma^2\rangle, \tag{17.53}$$

where κ_1 is the kurtosis of δx. For an at-the-money option of maturity T, one has:

$$\Gamma \sim \frac{1}{\sigma x_0\sqrt{T}} \qquad V \sim x_0\sqrt{T}. \tag{17.54}$$

Typical values are, on the scale of $\tau = $ one day, $\kappa_1 = 3$ and $\delta\sigma \sim \sigma$. The Γ contribution to risk is therefore on the order of $\sigma x_0\tau/\sqrt{T}$. This is equal to the typical fluctuations of the underlying contract multiplied by $\sqrt{\tau/T}$, or equivalently the price of the option reduced by a factor $N = T/\tau$. The Vega contribution is much larger for long maturities, since it is of order of the price of the option itself.

17.4 Risk diversification (*)

We have put the emphasis on the fact that for real world options, the Black–Scholes divine surprise – i.e. the fact that the risk is zero – does not occur, and a non-zero residual risk remains. One can ask whether this residual risk can be reduced further by including other assets in the hedging portfolio. Buying stocks other than the underlying to hedge an option can be called an 'exogenous' hedge. A related question concerns the hedging of a 'basket' option, the pay-off of which being calculated on a linear superposition of different assets. A rather common example is that of 'spread' options, which depend on the *difference* of the price between two assets (for example the difference between the Nikkei and the S&P 500, or between the British and German interest rates, etc.). An interesting conclusion is

that in the Gaussian case, any exogenous hedge increases the risk. An exogenous hedge is only useful in the presence of non-Gaussian effects. Another possibility is to hedge some options using different options; in other words, one can ask how to optimize a whole 'book' of options such that the global risk is minimum.

'Portfolio' options and 'exogenous' hedging

Let us suppose that one can buy M assets X^i, $i = 1, \ldots, M$, the price of which being x_k^i at time k. As in Chapter 12, we shall suppose that these assets can be decomposed over a basis of independent factors E^a:

$$x_k^i = \sum_{a=1}^{M} v_{ia} e_k^a. \tag{17.55}$$

The E^a are independent, of unit variance, and of distribution function P_a. The correlation matrix of the fluctuations, $\langle \delta x^i \delta x^j \rangle$ is equal to $\sum_a v_{ia} v_{ja} = [\mathbf{vv}^\dagger]_{ij}$.

One considers a general option constructed on a linear combination of all assets, such that the pay-off depends on the value of

$$\tilde{x} = \sum_i f_i x^i, \tag{17.56}$$

and is equal to $\mathcal{Y}(\tilde{x}) = \max(\tilde{x} - x_s, 0)$. The usual case of an option on the asset X^1 thus corresponds to $f_i = \delta_{i,1}$. A spread option on the difference $X^1 - X^2$ corresponds to $f_i = \delta_{i,1} - \delta_{i,2}$, etc. The hedging portfolio at time k is made of all the different assets X^i, with weight ϕ_k^i. The question is to determine the optimal composition of the portfolio, ϕ_k^{i}.*

Following the general method explained in Section 13.3.3, one finds that the part of the risk which depends on the strategy contains both a linear and a quadratic term in the ϕ's. Using the fact that the E^a are independent random variables, one can compute the functional derivative of the risk with respect to all the $\phi_k^i(\{x\})$. Setting this functional derivative to zero leads to:[†]

$$\sum_j [\mathbf{vv}^\dagger]_{ij} \phi_k^j = \int \hat{\mathcal{Y}}(z) \left[\prod_b \hat{P}_b \left(z \sum_j f_j v_{jb} \right) \right]$$

$$\times \sum_a \frac{v_{ia}}{\sum_j f_j v_{ja}} \frac{\partial}{\partial \mathrm{i} z} \log \hat{P}_a \left(z \sum_j f_j v_{ja} \right) dz. \tag{17.57}$$

Using the cumulant expansion of P_a (assumed to be even), one finds that:

$$\frac{\partial}{\partial \mathrm{i} z} \log \hat{P}_a \left(z \sum_j f_j v_{ja} \right) = \mathrm{i} z \left(\sum_j f_j v_{ja} \right)^2 - \mathrm{i} \frac{z^3}{6} \kappa_a \left(\sum_j f_j v_{ja} \right)^4 + \cdots. \tag{17.58}$$

The first term combines with

$$\sum_a \frac{v_{ia}}{\sum_j f_j v_{ja}}, \tag{17.59}$$

[†] In the following, i denotes the unit imaginary number, except when it appears as a subscript, in which case it is an asset label.

to yield:

$$iz \sum_{aj} v_{ia} v_{ja} f_j \equiv iz[\mathbf{vv}^\dagger \cdot f]_a,\tag{17.60}$$

which finally leads to the following simple result:

$$\phi_k^{i*} = f_i \mathcal{P}\left[\{x_k^i\}, x_s, N - k\right]\tag{17.61}$$

where $\mathcal{P}[\{x_k^i\}, x_s, N - k]$ is the probability for the option to be exercised, calculated at time k. In other words, in the Gaussian case ($\kappa_a \equiv 0$) the optimal portfolio is such that the proportion of asset i precisely reflects the weight of i in the basket on which the option is constructed. In particular, in the case of an option on a single asset, the hedging strategy is not improved if one includes other assets, even if these assets are correlated with the former.

However, this conclusion is only correct in the case of Gaussian fluctuations and does not hold if the kurtosis is non-zero.[§] In this case, an extra term appears, given by:

$$\delta\phi_k^{i*} = \frac{1}{6}\left[\sum_{ja} \kappa_a[\mathbf{vv}^\dagger]_{ij}^{-1} v_{ja} \sum_l f_l v_{la}^3\right] \frac{\partial P(\tilde{x}, x_s, N - k)}{\partial x_s}.\tag{17.62}$$

This correction is not, in general, proportional to f_i, and therefore suggests that, in some cases, an exogenous hedge can be useful. However, one should note that this correction is small for at-the-money options ($\tilde{x} = x_s$), since $\partial P(\tilde{x}, x_s, N - k)/\partial x_s = 0$.

Option portfolio

Since the risk associated with a single option is in general non-zero, the global risk of a portfolio of options ('book') is also non-zero. Suppose that the book contains p_i calls of 'type' i (i therefore contains the information of the strike x_{si} and maturity T_i). The first problem to solve is that of the hedging strategy. In the absence of volatility risk, it is not difficult to show that the optimal hedge (in the sense of variance minimization) for the book is the linear superposition of the optimal strategies for each individual option:[†]

$$\phi^*(x, t) = \sum_i p_i \phi_i^*(x, t).\tag{17.63}$$

The residual risk is then given by:

$$\mathcal{R}^{*2} = \sum_{i,j} p_i p_j C_{ij},\tag{17.64}$$

where the 'correlation matrix' C is equal to:

$$C_{ij} = \langle \max(x(T_i) - x_{si}, 0) \max(x(T_j) - x_{sj}, 0) \rangle$$

$$-\mathcal{C}_i \mathcal{C}_j - D\tau \sum_{k=0}^{N-1} \langle \phi_i^*(x, k\tau)\phi_j^*(x, k\tau)\rangle,\tag{17.65}$$

[§] The case of Lévy fluctuations is also such that an exogenous hedge is useless.

[†] Note however that this would *not* be true for other risk measures, such as the fourth moment of the wealth balance, or the value-at-risk.

where C_i is the price of the option i. If the constraint on the p_i's is of the form $\sum_i p_i = 1$, the optimum portfolio is given by:

$$p_i^* = \frac{\sum_j C_{ij}^{-1}}{\sum_{i,j} C_{ij}^{-1}} \tag{17.66}$$

(remember that by assumption the mean return associated to an option is zero). Very often, however, the constraint is rather of the type $\sum_i |p_i| = 1$, for which the problem becomes much more intricate, see Section 12.2.2.

Let us finally note that we have not considered, in the above calculation, the risk associated with volatility fluctuations, which is rather important in practice. It is common practice to try to hedge, if possible, this volatility risk using other types of options (for example, an exotic option can be hedged using a 'plain vanilla' option). A generalization of the Black–Scholes argument (assuming that option prices themselves follow a Gaussian process, which is far from being the case) suggests that the optimal strategy is to hold a fraction

$$\tilde{\Delta} = \frac{\partial C_2}{\partial \sigma} \bigg/ \frac{\partial C_1}{\partial \sigma} \tag{17.67}$$

of options of type 2 to hedge the volatility risk associated with an option of type 1. Using the formalism established in Chapter 14, one could work out the correct hedging strategy, taking into account the non-Gaussian nature of the price variations of options.

17.5 Summary

- In the presence of interest rates and dividends, the option price is given, in a first approximation, by the discounted value of the option price with zero interest rate and dividends, computed as if the current price was the futures price, see Eq. (17.1).
- In the presence of transaction costs, the rehedging time should be such that the volatility on this rehedging time is much larger than the fractional cost, otherwise the total cost of hedging would become comparable to the price of the option itself.
- The general method of wealth balance can be applied to different types of 'exotic' options. If these options are Δ-hedged, their price can be computed by replacing the true distribution of returns by the 'detrended' distribution ('risk neutral' valuation).

- FURTHER READING

J. C. Hull, *Futures, Options and Other Derivatives*, Prentice Hall, Upper Saddle River, NJ, 1997.

I. Nelken (ed.), *The Handbook of Exotic Options*, Irwin, Chicago, IL, 1996.

N. Taleb, *Dynamical Hedging: Managing Vanilla and Exotic Options*, Wiley, New York, 1998.

P. Wilmott, J. N. Dewynne, S. D. Howison, *Option Pricing: Mathematical Models and Computation*, Oxford Financial Press, Oxford, 1993.

P. Wilmott, *Derivatives, The Theory and Practice of Financial Engineering*, John Wiley, 1998.

P. Wilmott, *Paul Wilmott Introduces Quantitative Finance*, John Wiley, 2001.

18

Options: minimum variance Monte–Carlo

If uncertain, randomize!

(Mark Kac)

18.1 Plain Monte-Carlo

18.1.1 Motivation and basic principle

The calculation of the price of options can only be done analytically in some very special cases: the statistics of the underlying contract must be 'simple enough' (for example, the Black-Scholes log-normal process), and the option contract itself must not be too 'exotic'. For example, 'American' options, that can be exercised at any time, cannot be priced exactly, even in the simple framework of the Black-Scholes model.

On the other hand, as emphasized throughout this book, the fluctuations of the underlying assets are in general non-Gaussian, with complex volatility-volatility and return-volatility correlations. Furthermore, real world options often involve clauses that are difficult to deal with analytically. It can therefore be compulsory to rely on numerical methods to price and hedge these options. A versatile and commonly used method is the Monte-Carlo method, that we first explain in the context of 'plain vanilla' European options. Suppose that the interest rate r is zero, and that one chooses to Δ-hedge the option. In this case, the price of the option today is equal to the average pay-off of the option, over the detrended distribution of price returns (this price is called the 'risk-neutral' price, see the discussion in Section 15.2.2).

The most naive method to compute the option price would be to generate an ensemble of paths according to a certain (zero-drift) stochastic model, and to compute the average pay-off over this ensemble. Suppose for example that one chooses to describe the asset dynamics as a succession of independent random increments δx_k (*additive* model). The increments are all chosen with a common 'risk-neutral' distribution $Q(\delta x)$ – say a Student distribution, or any other distribution. A *path* is the set of the successive positions of the price, x_0, x_1, \ldots, x_N, such that $x_{k+1} - x_k = \delta x_k$, and that start from the current price, x_0. Each path requires N random variables δx_k to be constructed. The construction of these δx_k from a random number generator giving a uniform distribution in the interval [0, 1], such that the correct distribution $Q(\delta x)$ is obtained, is a standard problem discussed in details in Vetterling et al. (1993). Some simple cases are discussed in Appendix F.

An interesting alternative is to draw the price increments (or the returns) of the underlying stock using the exact historical distribution, simply by making a list of the historically observed δx's, subtracting their average value (such as to remove the drift) and choosing at random from this list the sequence of δx_k. In other words, one scrambles the list of returns in a random order, a method sometimes called 'Bootstrap Monte-Carlo'. (Note however that this procedure destroys the volatility correlations.)

Another possibility is to choose independent random returns η_k (*multiplicative* model), such that $x_{k+1} - x_k = \delta x_k = x_k \eta_k$. The η_k can again be drawn according to a given probability distribution $Q(\eta)$, or obtained by scrambling the true (detrended) returns of a given asset.

Now suppose that one wants to price a plain vanilla European option using M such paths, labeled by $\alpha = 1, \ldots, M$. For zero interest rate, one would then compute the option price as:

$$C_{MC} = \frac{1}{M} \sum_{\alpha=1}^{M} \max \left(x_N^\alpha - x_s, 0 \right),$$ (18.1)

where x_s is the strike. For large M, the above sum converges towards the average pay-off and gives the required option price. (The error on the price using this method is discussed below, see Section 18.2.3.) When the interest rate is non-zero, one can apply the following general formula:

$$C(x_0, x_s, T, r) = e^{-rT} C(x_0 e^{rT}, x_s, T, r = 0),$$ (18.2)

(see Eq. (17.1)) to convert the price obtained for $r = 0$ into the correct price. Remember however that this formula is approximate in the general case, and only becomes exact for a multiplicative model – see Section 17.1.1.

Of course, the above method is a little silly in the particular case at hand, where increments are assumed to be iid and the option pay-off only depends on the terminal price. It would be much more efficient to use the hypothesis of iid increments to compute numerically the exact form of $Q(x_N, N|x_0, 0)$, as the Nth convolution of $Q(\delta x)$. The average pay-off can then be computed from a numerical evaluation of a one-dimensional integral. However, if the option is price dependent, one needs to generate the whole path to compute its price. Take for example an Up and Out call option, such that the option becomes void as soon as the underlying stock price exceeds a certain value x_b. Using the M paths defined above, the price of this option will be given by:

$$C_{b,MC} = \frac{1}{M} \sum_{\alpha=1}^{M} \left(\prod_{k=1}^{N-1} \Theta\left(x_b - x_k^\alpha\right) \right) \max \left(x_N^\alpha - x_s, 0 \right),$$ (18.3)

where $\Theta(x > 0) = 1$ and $\Theta(x \le 0) = 0$. This formula involves the whole paths $x_0, x_1^\alpha, \ldots, x_N^\alpha$: the product of Θ function therefore enforces that x_k^α is always below x_b. In cases where the method of images can be used (see Section 17.2.5), an analytic formula for

barrier option can be given in terms of the plain vanilla prices. However, in the important cases where jumps are present, a Monte-Carlo method is needed.

Another case where option pricing is difficult to achieve analytically is when increments are not independent, e.g. when volatility clustering is present. Generating paths with a given structure of volatility correlations is not very difficult. For example, the Bacry-Muzy-Delour multifractal model can be readily simulated using the construction explained in Sections 2.5 and 7.3.1. One first constructs the volatility process by drawing independent Gaussian 'shocks' ξ_k with $k = -N', -N' - 1, \ldots N - 1$, and summing them using a memory kernel $K(\ell)$ much as in Eq. (2.76). One then draws a second set of independent Gaussian random variables ξ'_k to construct the return as:

$$\eta_k = \xi'_k \exp \left(\sum_{k'=-N'}^{k-1} K(k - k')\xi_k \right). \tag{18.4}$$

A difficulty here is that the memory kernel $K(\ell)$ only decays very slowly with ℓ, which requires N' to be rather large. A method based on Fourier transforms is then often preferable (see Chapter 2), although the strict causality of the process is lost.

18.1.2 Pricing the forward exactly

As shown in Section 15.2.2, the implicit use of the Δ-hedge means (at least when the time step τ is small enough) that the probability distribution that must be used to generate the paths and price the option must be of zero drift, or, when the interest rate is non zero, with a drift equal to the interest rate. Another, more precise way to state this is to consider the problem of pricing a forward contract, which is a particular option with a linear pay-off equal to x_N. Therefore, if one prices the forward as the average over all paths of the terminal price x_N, one should find exactly $x_0 e^{N\rho}$, where $\rho = r\tau$. In fact, this should be true for any intermediate price x_k: its average over all paths should equal $x_0 e^{k\rho}$.

The point here is that the numerical average of x_k over a set of random paths can a priori only reproduce this result approximately. In the case of an additive model, it is easy to impose *exactly* the price of the forward, by first computing the numerical average over the paths of the kth increment:

$$\langle \delta x_k \rangle_{MC} = \frac{1}{M} \sum_{\alpha=1}^{M} \delta x_k^\alpha, \tag{18.5}$$

and redefining new price increments as:

$$x_{k+1} - x_k = \rho x_k + \delta x_k - \langle \delta x_k \rangle_{MC}. \tag{18.6}$$

Of course, if $Q(\delta x)$ is chosen to be of zero mean, this shift is useless for large M. However, for finite M, the above procedure removes a systematic bias that would, for example, ruin the Put-Call parity relation. Indeed, the numerical difference between the Call and the Put

is equal to:

$$\mathcal{C}_{MC} - \mathcal{C}_{MC}^{\dagger} = \frac{e^{-rT}}{M} \sum_{\alpha=1}^{M} \max\left(x_N^{\alpha} - x_s, 0\right) - \max\left(x_s - x_N^{\alpha}, 0\right) \equiv \frac{e^{-rT}}{M} \sum_{\alpha=1}^{M} \left(x_N^{\alpha} - x_s\right),$$

$$(18.7)$$

which should *exactly* equal $x_0 - e^{-rT}x_s$. This is important because in order to gain precision, one usually prices the contract with the smallest price (see below): in-the-money Calls from out-of-the-money Puts (and vice-versa), using the Put-Call parity relation.

If $Q(\delta x)$ is even, a simpler way to impose this condition is to add to all generated paths their 'mirror image' obtained by changing all the signs of the increments: $\delta x_k^{\alpha} \to -\delta x_k^{\alpha}$. It is clear that in this case, the numerical average of x_k over the $2M$ paths is precisely equal to $x_0 e^{k\rho}$.

For a multiplicative model, where $x_{k+1} - x_k = x_k \eta_k$, one first computes all forward prices:

$$\langle x_k \rangle_{MC} = \frac{1}{M} \sum_{\alpha=1}^{M} x_k^{\alpha},$$

$$(18.8)$$

using the generated paths, and then redefine new returns η_k' such that:

$$1 + \eta_k' = \frac{\langle x_k \rangle_{MC}}{\langle x_{k+1} \rangle_{MC}} (1 + \eta_k) e^{\rho},$$

$$(18.9)$$

such that the new process has now $\langle x_k' \rangle_{MC} \equiv x_0 e^{k\rho}$ exactly for all k.

Therefore, we see that in order to price the simplest derivative, namely the forward, exactly, some treatment of the 'raw' random variables is needed to obtain a consistent approximate risk neutral measure with a finite number of paths.

18.1.3 Calculating the Greeks

Finite difference method

Let us now turn to the computation of the Greeks using the Monte-Carlo method. The Δ is equal to $\partial \mathcal{C}/\partial x_0$. Therefore, the most naive idea is to compute the option price using finite differences, using two nearby values of the initial underlying price, say x_0 and x_0', where $dx = x_0' - x_0$ is sufficiently small. One then obtains Δ as:

$$\Delta_{MC} \approx \frac{\mathcal{C}_{MC}(x_0') - \mathcal{C}_{MC}(x_0)}{dx}.$$

$$(18.10)$$

However, there is a very important point to notice: the paths used to compute both $\mathcal{C}_{MC}(x_0')$ and $\mathcal{C}_{MC}(x_0)$ must involve exactly the same random variables δx_k^{α}, otherwise, the difference between the two is dominated by the statistical error and does not reflect the true dependence

of the price on x_0.[†] The adequate choice of dx depends on the moneyness of the option: the larger the Gamma of the option, the smaller dx.

Choosing the optimal mesh size

The choice of the 'mesh' $dx = x_0' - x_0$ is guided by the following consideration: when computing $C_{MC}(x_0)$ and $C_{MC}(x_0')$, one finds the following contributions:

$$C_{MC}(x_0) = C(x_0) + \varepsilon_{MC} + \varepsilon_{Pr}, \tag{18.11}$$

and

$$C_{MC}(x_0') = C(x_0) + \Delta(x_0)dx + \frac{1}{2}\Gamma(x_0)(dx)^2 + \varepsilon_{MC}' + \varepsilon_{Pr}', \tag{18.12}$$

where the ε_{MC} are random error terms coming from the Monte-Carlo noise, and the ε_{Pr} are random error terms coming from the finite machine precision. In general, one has $\varepsilon_{MC} \gg \varepsilon_{Pr}$. However, if one chooses the very same paths to price the two options, $\varepsilon_{MC}' = \varepsilon_{MC}$ and this contribution drops out from the computation of the price difference. Therefore, the error on the estimated Δ reads:

$$\delta\Delta \sim \frac{1}{2}\Gamma(x_0)dx + \sqrt{2}\frac{\varepsilon_{Pr}}{dx}, \tag{18.13}$$

which reaches a minimum for $dx^* \sim \sqrt{\varepsilon_{Pr}/\Gamma}$. Let us estimate the above number. The relative numerical error is given by the typical machine precision (10^{-12}) times the square-root of the number of operations N_{op}. On the other hand, Γ is of the order of the option price C divided by the square of the absolute fluctuations of the terminal price (for at the money options): $\Gamma \sim C/(x_0^2\sigma^2 T)$. Using $C \sim \sqrt{\sigma^2 T}x_0$, one finally gets:

$$\frac{dx^*}{x_0} \sim 10^{-6}N_{op}^{1/4}\sigma\sqrt{T}, \tag{18.14}$$

and an optimal precision on Δ given by $\delta\Delta^* \sim 10^{-6}N_{op}^{1/4}$. For $N_{op} = 10^4$ and $\sigma\sqrt{T} = 10\%$, one finds $dx^*/x_0 \sim 10^{-6}$ and $\delta\Delta^* \sim 10^{-5}$. Note that for the same value of Γ, both dx^* and $\delta\Delta^*$ are proportional to \sqrt{C}, which means that one should compute out of the money calls and out of the money puts, and use the Put-Call parity relation to compute the complementary contract.

Using some analytical results

In some special cases, there is actually no need to take a numerical derivative to compute the Δ of the option. Suppose first that one deals with an additive model. In this case, the price of the option C can be written as \sqrt{DT} times a function of $(x_0 - x_s)/\sqrt{DT}$, where \sqrt{DT} is the variance of the terminal price. The derivative of C with respect to x_0 is then equal, up

[†] Similarly, the Gamma of the option can be computed using three nearby values of the initial price, again using the very same randomness to compute all three prices.

to a change of sign, to the derivative of C with respect to x_s. This last quantity is nothing but the probability of exercising the option, which is directly given by the Monte-Carlo simulation as:

$$\Delta(x_0)_{MC} \approx \frac{1}{M} \sum_{\alpha=1}^{M} \Theta\left(x_N^\alpha - x_s\right), \tag{18.15}$$

which does not require any finite difference scheme. Similarly, if the model is multiplicative, the option price can be written as $C = x_0 f(x_s/x_0)$. Therefore:

$$\Delta = \frac{\partial C}{\partial x_0} \equiv f - \frac{x_s}{x_0} f' = \frac{1}{x_0} \left[C - x_s \frac{\partial C}{\partial x_s} \right]. \tag{18.16}$$

(This relation is a special case of Euler's formula for homogeneous functions of several variables.) Again, $\partial C/\partial x_s$ is simply the probability that the option is exercised, and can be computed as above. Using Eq. (18.16), one can also directly compute Δ without introducing any small price difference dx. Unfortunately, this is not true for the Γ. In the additive model, one would naively get:

$$\Gamma(x_0)_{MC} \approx \frac{1}{M} \sum_{\alpha=1}^{M} \delta\left(x_N^\alpha - x_s\right). \tag{18.17}$$

However, for a finite number of points M, one necessarily has to introduce a finite width to the above δ-function, and take for example a Gaussian of width dx.

The Vega

If one assumes a multiplicative model where price returns are iid, the Vega \mathcal{V} can be usefully defined as the derivative of the price with respect to the root mean square σ of the returns, for an arbitrary distribution of returns. This Vega can then be computed by estimating the price of the option using a set of random variables δx_k^α, and the very same set but scaled by a factor $1 + \epsilon$, with $\epsilon \ll 1$. Then,

$$\mathcal{V}_{MC} \approx \frac{\sigma}{\epsilon} \left(C_{MC}' - C_{MC}\right), \tag{18.18}$$

where C_{MC}' is the option price computed with the scaled δx_k^α.

18.1.4 Drawbacks of the method

There are however two serious problems with the simplest implementation of the Monte-Carlo method explained above:

- First, the numerical value of C_{MC} only converges towards its true value when $M \to \infty$. More precisely, the error on C_{MC} decreases as \mathcal{R}_0/\sqrt{M}, where \mathcal{R}_0 is the bare risk computed in Section 14.2.2 precisely as the variance of the pay-off function. Typically, this variance is on the order of the option price for an at-the-money option. A precision of 1% on the price therefore requires 10^4 different paths. The situation is much worse for out of the money options, for which \mathcal{R}_0 can easily be ten times larger than the option price.

• More fundamentally, this method assumes that one Δ-hedges the option, and therefore that the distribution to be used to generate the paths is not the objective, historical distribution P, but the 'risk-neutral' probability distribution Q, obtained by removing any drifts from the historical process P (cf. Section 15.2.2). This is why this method is called 'Risk-neutral Monte-Carlo pricing'. However, this method does not account directly for the cost of the hedging strategy, and does not allow to determine the truly optimal hedge, which is general is *not* the Δ-hedge. Furthermore, the formula Eq. (18.2) that allows one to account for the effect of a non zero interest rate is only approximate in the general case; the corrections come from the precise value of the cost of hedging.

The above two problems are actually deeply related. If one could simulate the evolution of the whole wealth balance, including the hedging strategy, the cost of hedging strategy would be accounted for. Furthermore, finding the optimal strategy would reduce the variance of the wealth balance, and therefore the numerical error on \mathcal{C}_{MC}. Said differently, the financial risk arising from the imperfect replication of the option by the hedging strategy is directly related to the variance of the Monte-Carlo simulation. The residual risk \mathcal{R} is typically five to ten times smaller than \mathcal{R}_0, which means that for the same level of precision, the number of trajectories needed in the Monte-Carlo would be up to a hundred times smaller. The Monte-Carlo method that we now describe is directly inspired from the ideas presented in this book. It allows one to determine simultaneously a numerical estimate of the price of the derivative, but also of the optimal hedge (which may be different from the Black-Scholes Δ-hedge for non Gaussian statistics) and of the residual risk. This method is furthermore of pedagogical interest, since it illustrates in a very concrete way the basic mechanisms involved in option pricing and hedging that were discussed in the previous chapters.

18.2 An 'hedged' Monte-Carlo method

18.2.1 Basic principle of the method

The idea is really to think as an option trader, who has sold a stock option and tries to hedge it. His profit and losses are 'marked-to-market', which means that he has to report his position every day. His position at time k is short one option, and long a certain number ϕ_k of underlying stocks. His concern is that the net value of his position does not vary too much (or preferably increases, if possible) between today and tomorrow. Without necessarily formulating it consciously, he envisages the probable variations of the stock price and considers the impact of these different possibilities on his position. More precisely, his position between today and tomorrow varies as:[†]

$$\delta W_k = -\mathcal{C}_{k+1}(x_{k+1}) + \mathcal{C}_k(x_k) + \phi_k(x_k)[x_{k+1} - x_k], \tag{18.19}$$

where we assume for simplicity that the option price \mathcal{C}_k only depends on the underlying price x_k (and of course on k), but not, for example, on the volatility. How should he price

[†] We assume that the interest rate is zero. When the interest rate r is non zero, the change reads: $\delta W_k = -e^{-\rho}\mathcal{C}_{k+1}(x_{k+1}) + \mathcal{C}_k(x_k) + \phi_k(x_k)[x_k - e^{-\rho}x_{k+1}]$, where $\rho = r\tau$ is the interest rate over the elementary time step τ.

and hedge his option such that the distribution of δW_k is as sharply peaked as possible? (This presentation obviously parallels that of Section 13.1.2.) Within a quadratic measure of risk, the price and the hedging strategy at time k are such that the variance of the wealth change between k and $k + 1$ is minimized.[†] The local 'risk' \mathcal{R}_k is defined as:

$$\mathcal{R}_k^2 = \langle (\mathcal{C}_{k+1}(x_{k+1}) - \mathcal{C}_k(x_k) + \phi_k(x_k)[x_k - x_{k+1}])^2 \rangle, \tag{18.20}$$

where $\langle ... \rangle$ means an average over the objective (true) probability distribution of price changes. The option trader is concerned with what can really happen, and thinks in terms of *real* probabilities. The Monte-Carlo method does this in a systematic way; and gives an estimate of the local risk as:

$$\mathcal{R}_k^2 \approx \frac{1}{M} \sum_{\alpha=1}^{M} \left(\mathcal{C}_{k+1}(x_{k+1}^\alpha) - \mathcal{C}_k(x_k^\alpha) + \phi_k(x_k^\alpha)[x_k^\alpha - x_{k+1}^\alpha] \right)^2, \tag{18.21}$$

where x_k^α is the price at time k along the αth trajectory.

The functional minimization of \mathcal{R}_k with respect to both $\mathcal{C}_k(x_k)$ and $\phi_k(x_k)$ gives equations that allow one determine the price and hedge, provided \mathcal{C}_{k+1} is known. Since the terminal price of the option is known, one can work recursively backwards in time. The problem is to formulate this in a way that is numerically tractable.[‡]

18.2.2 A linear parameterization of the price and hedge

The idea is to decompose the functions $\mathcal{C}_k(x)$ and $\phi_k(x)$ over a set of L appropriately chosen basis functions $C_a(x)$ and $F_a(x)$ (which might be themselves k-dependent):[§]

$$\mathcal{C}_k(x) = \sum_{a=1}^{L} \gamma_a^k C_a(x) \qquad \phi_k(x) = \sum_{a=1}^{L} \varphi_a^k F_a(x). \tag{18.22}$$

In other words, one solves the above minimization problem within the *variational space* spanned by the functions $C_a(x)$ and $F_a(x)$. The unknown quantities are the $2L$ coefficients γ_a^k, φ_a^k. This leads to a major simplification since the quantity to minimize is a quadratic function of these variables, which is very easy to solve explicitly. (This would not be true for other measures of risk.) These coefficients must be such that:

$$\mathcal{R}_k^2 \approx \frac{1}{M} \sum_{\alpha=1}^{M} \left(\mathcal{C}_{k+1}(x_{k+1}^\alpha) - \sum_{a=1}^{L} \gamma_a^k C_a(x_k^\alpha) + \sum_{a=1}^{L} \varphi_a^k F_a(x_k^\alpha)[x_k^\alpha - x_{k+1}^\alpha] \right)^2, \tag{18.23}$$

is minimized, assuming that the function \mathcal{C}_{k+1} is already known. Altogether, there are N minimization problems (one for each $k = 0, \dots, N - 1$) which are solved working backwards in time with $\mathcal{C}_N(x) = \mathcal{Y}(x)$ the known final pay-off function. Expression (18.23)

[†] Of course, as already discussed several times, one could choose other measures for the risk, combining other moments of the distribution of δW_k, or based on value-at-risk. This would however lead to a much less tractable problem from a numerical point of view.

[‡] See Potters et al. (2001).

[§] This method was proposed in Longstaff and Schwartz (2001), in a somewhat different context.

is messy because of the different indices, but it is clear that this is a quadratic form of the γ_a^k, φ_a^k; the minimization therefore leads to a linear system of equations for these quantities, that one estimates using the generated trajectories. Interestingly, the value of \mathcal{R}_k at the minimum is the minimum residual risk (within the variational space spanned by the chosen basis functions C_a, F_a).

Although in general the optimal strategy is not equal to the Black-Scholes Δ-hedge, the difference between the two is often small, and only leads to a second order increase of the risk (see the discussion in Section 14.4). Therefore, one can choose to work within a smaller variational space where the hedging strategy is restricted to be exactly the Δ-hedge. This means that:

$$\varphi_a^k \equiv \gamma_a^k \qquad F_a(x) \equiv \frac{dC_a(x)}{dx}. \tag{18.24}$$

This will lead to exact results only for Gaussian processes, but reduces the computation cost by a factor two.

Of course, a correct choice of the basis functions $F_a(x)$, $C_a(x)$ is crucial to obtain fast convergence as a function of the number L of such functions. A simple choice is to take these basis function to be piecewise linear for F_a (and therefore piecewise quadratic for C_a for Δ-hedge):

$$F_a(x) = 0 \quad \text{for } x \leq x_a^-; \qquad F_a(x) = \frac{x - x_a^-}{x_a^+ - x_a^-} \quad \text{for } x_a^- \leq x \leq x_a^+;$$

$$F_a(x) = 1 \quad \text{for } x \geq x_a^+; \tag{18.25}$$

with breakpoints x_a^\pm such that the intervals are covering the real axis (i.e. $x_a^+ = x_{a+1}^-$). A natural choice is such that the number of Monte-Carlo paths falling in each interval is constant, equal to $M/(L+2)$.

Note that it is useful to impose $\sum_{a=1}^{L} \varphi_a^k = \sum_{a=1}^{L} \gamma_a^k = 1$ exactly (note that the problem remains quadratic), in such a way that the price of the option tends to that of the underlying, and the hedging strategy to 1 for deep in-the-money option, as it should.

18.2.3 The Black-Scholes limit

As a first illustration, one can choose the paths to be the realizations of a (discretized) geometric random walk, and price an at-the-money three month European option, on an asset with 30% annualized volatility and a drift equal to the risk-free rate which we set to 5% per annum. The number of time intervals N is chosen to be 20. The initial stock and strike price are $x_0 = x_s = 100$; the corresponding Black-Scholes price is $C_0^{BS} = 6.58$. The number of basis functions is $L = 8$. One runs 500 simulations containing $M = 500$ paths each, from which one can extract both the average price and standard deviation on the price.

An example of the result of the optimization of the parameters γ_k^a is plotted in Fig. (18.1). Each data point corresponds to one trajectory of the Monte-Carlo at one instant of time k, and represents the quantity:

$$e^{-\rho}C_{k+1}(x_{k+1}) + \phi_k(x_k)[x_k - e^{-\rho}x_{k+1}], \tag{18.26}$$

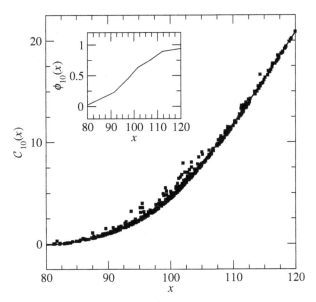

Fig. 18.1. Option price as a function of underlying price for a within the hedged Monte-Carlo simulation. Square symbols correspond to the option price on the next step $k = 11$, corrected by the hedge, Eq. (18.26) for individual Monte-Carlo trajectories. The full line corresponds to the fitted option price $\mathcal{C}_{10}(x)$ at the tenth hedging step ($k = 10$) of a simulation of length $N = 20$. Inset: Hedge as a function of underlying price at the tenth step of the same simulation.

as a function of x_k. The full line represents the result of the least-squared fit, which is by definition $\mathcal{C}_k(x_k)$. Note that some points deviate from the best fit, which means that the hedging error (or the risk) is non zero. We show in the inset the corresponding hedge ϕ_k, that was constrained in this case to be the Δ-hedge.

One obtains the following numerical results. For the standard (un-hedged) Monte-Carlo scheme, Eq. (18.1), one obtains as an average over the 500 simulations $\mathcal{C}_0^{MC} = 6.68$ with a standard deviation of 0.44. For the hedged scheme, one finds $\mathcal{C}_0^{HMC} = 6.55$ but with a standard deviation of 0.06, seven times smaller than with the un-hedged Monte-Carlo. This variance reduction is illustrated in Fig. 18.2, where the histogram of the 500 different MC results both for the un-hedged case (full bars) and for the 'optimally hedged' case (OHMC, dotted bars) are shown.

Now set the drift to 30% annual. The Black-Scholes price, as is well known, is unchanged. A naive un-hedged Monte-Carlo scheme with the objective probabilities gives a completely wrong price of 10.72, 60% higher than the correct price, with a standard deviation of 0.56. On the other hand, the the hedged Monte-Carlo indeed produces the correct price (6.52) with a standard deviation of 0.06.

Therefore, one checks explicitly that in the case of a geometric random walk, the hedge indeed compensates the drift exactly and reproduces the usual Black-Scholes results, as it should.

Fig. 18.2. Histogram of the option price as obtained of 500 MC simulations with different seeds. The dotted histogram corresponds to the unhedged scheme (SMC for standard MC) and the full histogram to the hedged Monte-Carlo. The dotted line indicates the exact Black-Scholes price. Note that on average both methods give the correct price, but that the hedged Monte-Carlo has an error that is more than seven times smaller.

18.3 Non-Gaussian models and purely historical option pricing

Obviously, the above method can be used with a realistic model of price changes, with fat tails, volatility correlations and leverage correlations, provided one can generate the corresponding ensemble of trajectories efficiently. A possible application is to calibrate the parameters of such a model using market prices of some plain vanilla options, and then use the same paths to price exotic options or unquoted options (see below).

But more interestingly, the method allows one to use purely historical data to price derivatives, short-circuiting the modelling of the underlying asset fluctuations. Within the hedged Monte-Carlo method, one can directly use 'chunks' of the historical time series of the asset to generate the paths. This preserves exactly not only the correct statistics of price increments, but also all the subtle volatility and leverage correlations. Moreover, when several assets are involved in the option, this method also preserves the cross-correlations between these assets. The fact that a rather small number of paths is needed to reach good accuracy means that the length of the historical time series needs not be very large, at least for options of not too long maturities.

As an illustration, one can price a one month (21 business days) option on Microsoft Corp., hedged daily, with zero interest rates, still assuming that the option price only depends on the underlying price and not on the volatility (but see below), and keep the simple Δ-hedge. We

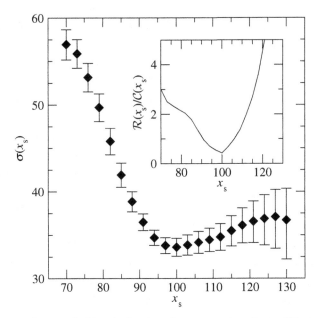

Fig. 18.3. Smile curve for a purely historical OHMC of a one-month option on Microsoft (volatility as a function of strike price). The error bars are estimated from the residual risk. The inset shows this residual risk as a function of strike normalized by the 'time-value' of the option (i.e. by the call or put price, whichever is out-of-the-money). The minimum value of the normalized risk is 0.42.

use here 2000 paths of length 21 days, extracted from the time series of Microsoft during the period May 1992 to May 2000. The initial price is always normalized to 100. From the numerically determined option prices, one extracts an implied Black-Scholes volatility by inverting the Black-Scholes formula and plot it as a function of the strike, in order to construct an implied volatility smile. The result is shown in Fig. 18.3. Since we now only perform a single MC simulation, the error bars shown are obtained from the residual risk of the optimally hedged options. The residual risk itself, divided by the call or the put option price (respectively for out of the money and in the money call options), is given in the inset. We find that the residual risk is around 42% of the option premium at the money, and rapidly reaches 100% when one goes out of the money. These risk numbers are comparable to those quoted in Chapter 14, and are much larger than the residual risk that one would get from discrete time hedging effects in a Black-Scholes world.

The obtained smile has a shape quite typical of those observed on option markets. However, it should be emphasized that we have neglected here the possible dependence of the option price on the local value of the volatility. One could let the function C_k depend not only on x_k but also on the value of some filtered past volatility σ_k. This would allow option prices to depend explicitly on the volatility, an obviously welcome possibility.

It is also interesting to price some exotic options within the same framework. We compare for example in Table 18.1 the price of a barrier option (Down and Out call) on Microsoft Corp. using the optimally hedged method and using the standard Black-Scholes formula

Table 18.1. Results for a Down and Out Call option on Microsoft, with a barrier at $95, for different maturities and strike. The spot price is normalized to $100.

Maturity/Strike	Implied Black-Scholes	OHMC
two weeks/100	$ 2.45	$ 2.39
two weeks/105	$ 0.86	$ 0.84
one month/100	$ 3.10	$ 3.08
one month/105	$ 1.61	$ 1.61

for barrier options with the implied volatility corresponding to the same strike, as shown in Fig. 19.3. As could be expected, the correct price is lower than the Black-Scholes price, reflecting the presence of jumps in the underlying that increases the probability to hit the barrier. Interestingly, however, the difference becomes small as the maturity increases.

18.4 Discussion and extensions. Calibration

The above hedged Monte-Carlo scheme closely follows the ideas and methodology exposed in the previous chapters. The inclusion of the optimal hedging strategy allows one to reduce the financial risk associated to option trading, and for the very same reason the variance of the Monte-Carlo pricing scheme as compared to the standard Monte-Carlo schemes. Reduced variance Monte-Carlo techniques for option pricing were discussed previously in the literature, and bear some similarity with the present method. The general idea is to add to the observable one wants to average some so called 'control variates' which have by construction a zero average value but such that the resulting sum has smaller fluctuations. The profit and loss of some appropriate hedging strategy is an obvious candidate for such control variates, as was demonstrated in Clewlow & Caverhill (1994). However, the hedge proposed in their work is only an approximate hedge (for example, the Δ-hedge corresponding to an option not too dissimilar to the one under consideration, and for which an analytical formula is known), and not the optimal hedge for the particular option and underlying under consideration.

The explicit accounting of the hedging cost naturally converts the objective probability into the 'risk-neutral' one. This allows a consistent use of purely historical time series to price derivatives and obtain their residual risk. There are many potential extensions of the above scheme. For example, with some modifications and extra numerical cost, the method presented here could be used to deal with transaction costs, or with non quadratic risk measures (VaR hedging).[†]

Another very important extension is to consider cases where the option price depends *a priori* on several variables, for example an option that depends on the behaviour of two underlyings, or, as mentioned above, the case where the dependence of the option price on

[†] B. Pochart, Ph. D. Thesis, in preparation.

the current value of the volatility must be included. Another very important case is that of options on the interest rate curve. In these cases, the hedging strategy will include several different assets, for example, short term options to Vega-hedge the volatility of long term options. The practical difficulty here is to find a suitable linear parameterization of functions of two or more variables.

Let us finally mention a very interesting idea to calibrate the results of a Monte-Carlo simulation on a set of option prices that are observed on markets.[†] The idea is to define a certain set of M 'possible' paths, $x_1^\alpha, x_2^\alpha, \ldots, x_N^\alpha$, with $\alpha = 1, \ldots, M$, with *a priori* unknown weights p_α. Suppose that one observes the market price C_i, $i = 1, \ldots, K$ of a certain number K of different options, with different maturities and strikes. The weights p_α are then determined such that the prices of these options are *exactly* reproduced, subject to the constraint that the information entropy associated to the p_α is maximum. Since one has, very generally,

$$C_i = \sum_{\alpha=1}^{M} p_\alpha f_i\left(x_1^\alpha, x_2^\alpha, \ldots, x_N^\alpha\right), \tag{18.27}$$

where f_i is a certain pay-off function of the path, the entropy maximization program can be written by introducing Lagrange parameters ζ_i:

$$\frac{\partial}{\partial p_\beta}\left[-\sum_{\alpha=1}^{M} p_\alpha \log(p_\alpha) + \sum_{i=1}^{K} \zeta_i \sum_{\alpha=1}^{M} p_\alpha f_i(\{x^\alpha\}) + \zeta_0 \sum_{\alpha=1}^{M} p_\alpha\right], \tag{18.28}$$

where the ζ_i are determined such that the prices of the K options have the correct value, and ζ_0 is such that $\sum_{\alpha=1}^{M} p_\alpha = 1$. The solution of the above problem is well known to be the Gibbs measure:

$$p_\alpha = \frac{1}{Z} \exp \sum_{i=1}^{K} \zeta_i f_i(\{x^\alpha\}). \tag{18.29}$$

The trick, as usual in statistical physics, is to compute the 'partition function' Z as a function of the ζ_i, as:

$$Z = \sum_{\alpha=1}^{M} \exp \sum_{i=1}^{K} \zeta_i f_i(\{x^\alpha\}), \tag{18.30}$$

and then to compute the ζ_i such that the following quantity:

$$G = \log Z - \sum_{i=1}^{K} \zeta_i C_i \tag{18.31}$$

is minimized, which leads to:

$$C_i \equiv \frac{\partial \log Z}{\partial \zeta_i}, \tag{18.32}$$

[†] See Avellaneda et al. (2001).

as it should. Once the ζ_i (and therefore the weights p_α) are determined, one can use the same paths with their corresponding weights to compute other option prices, not priced by the market, or to check the consistency between the different option prices.

18.5 Summary

- The 'Risk-Neutral' Monte-Carlo method prices options by computing the average pay-off over paths generated using the 'risk-neutral' distribution of returns. Care must be paid in order to price the futures contract exactly. The Greeks can be computed from a finite-difference scheme, using exactly the same random paths to price the nearby options.
- The hedged Monte-Carlo method aims to accomplish in a systematic way what an option trader does intuitively: price and hedge his option, such that the possible variations of his position due the potential fluctuations of the underlying varies as little as possible. In terms of Monte-Carlo pricing, including the optimal hedge reduces considerably the numerical error on the price.
- The proper accounting of the hedging strategy allows one to use consistently historical paths to price options. In the case of log-normal price fluctuations, one sees explicitly that the hedge converts the historical distribution (possibly with drift) into the 'risk-neutral' (un-drifted) distribution, and one recovers exactly the Black-Scholes prices.
- Using 'chunks' of historical price series allows one to retain all subtle correlations in the volatility and leads to very realistic option prices. The method can easily be extended to price exotic options.

18.6 Appendix F: generating some random variables

Suppose the one draws a random variable u, uniform in the interval $[0, 1]$ and a sign variable $\varepsilon = \pm 1$ with equal probability. One can transform these random variables, such as to obtain:

- An exponential random variable: $z = \varepsilon \log u$, such that $P(z) = e^{-|z|}/2$;
- A power-law variable: $z = \varepsilon(u^{-1/\mu} - 1)$, such that $P(z) = \mu(1 + |z|)^{-\mu-1}/2$;
- A Student variable with $\mu = 4$: $z = \varepsilon\sqrt{2}t/\sqrt{1 - t^2}$ where $t = 2\cos((\arccos(1 - 2u) + 4\pi)/3)$, such that $P(z) = 3/(2 + z^2)^{5/2}$.

In the above formulas, the Student variable has unit variance, while the exponential and power law variables should be multiplied by $1/\sqrt{2}$ and $\sqrt{(\mu - 2)(\mu - 1)/2}$ respectively to have unit variance, the last case being only valid if the variance exists (i.e. when $\mu > 2$).

More generally, if one has an explicit expression for the cumulative distribution $P_>(z)$ one wants to simulate, the random variable z can be obtained from a uniformly distributed variable u by solving the equation $P_>(z) = u$. This equation must often be solved numerically. This procedure can be used to generate $\mu = 3$ Student variables, using Eq. (1.40).

There are sometimes special tricks. For example, with two variables u, v independently and uniformly chosen in the interval $[0, 1]$, one can generate two independent Gaussian variables z_1, z_2, such that:

$$z_1 = \cos(2\pi v)\sqrt{-\log u}; \qquad z_2 = \sin(2\pi v)\sqrt{-\log u}. \tag{18.33}$$

With these two variables, one can also generate Lévy stable variables z as follows: set $u' = \pi(u - 1/2)$ and $v' = -\log v$. Then construct z as:

$$z = A_{\beta,\mu} \frac{\sin(\mu(u' + B_{\beta,\mu}))}{\cos^{1/\mu} u'} \left(\frac{\cos(u' - \mu(u' + B_{\beta,\mu}))}{v'} \right)^{1-\mu/\mu}, \tag{18.34}$$

with:

$$A_{\beta,\mu} = \cos^{-1/\mu} \left(\arctan \left(\beta \tan \frac{\pi\mu}{2} \right) \right), \tag{18.35}$$

and

$$B_{\beta,\mu} = \frac{1}{1 - |1 - \mu|} \arctan \left(\beta \tan \frac{\pi\mu}{2} \right). \tag{18.36}$$

Then, z is a Lévy stable distribution with an tail index equal to μ, an asymmetry coefficient equal to β, and with $a_\mu = 1$ (see Eq. (1.24)). For $\mu = 2$ the above equations boil down to the previous method to generate Gaussians. This method is due to Chambers et al. (1976).

• FURTHER READING

L. Clewlow, Ch. Strickland, *Implementing Derivatives Models*, Wiley Series in Financial Engineering, 1998.

B. Dupire, *Monte Carlo – Methodologies and Applications for Pricing and Risk Management*, Risk Publications, London, 1998.

J. C. Hull, *Futures, Options and Other Derivatives*, Prentice Hall, Upper Saddle River, NJ, 1997.

W. T. Vetterling, S. A. Teukolsky, W. H. Press, B. P. Flannery, *Numerical recipes in C: the art of scientific computing*, Cambridge University Press, 1993.

P. Wilmott, J. N. Dewynne, S. D. Howison, *Option Pricing: Mathematical Models and Computation*, Oxford Financial Press, Oxford, 1993.

P. Wilmott, *Derivatives, The Theory and Practice of Financial Engineering*, John Wiley, 1998.

P. Wilmott, *Paul Wilmott Introduces Quantitative Finance*, John Wiley, 2001.

• SPECIALIZED PAPERS

M. Avellaneda, R. Buff, C. Friedmann, N. Grandechamp, L. Kruk, J. Newman, *Weighted Monte-Carlo: A new technique for calibrating asset-pricing models*, Int. Jour. Theo. Appl. Fin., **4**, 91 (2001).

J. Chambers, C. Mallows, B. Struck, *A method for simulating stable random variables*, J. Am. Stat. Ass., **71**, 340 (1976).

L. Clewlow, A. Caverhill, *On the simulation of contingent claims*, Journal of Derivatives, **2**, 66 (1994), and RISK Magazine, **7**, 63 (1994).

J.-P. Laurent, H. Pham, *Dynamic programming and mean-variance hedging*, Finance and Stochastics, **3**, 83 (1999).

F. A. Longstaff, E. S. Schwartz, *Valuing American options by simulation: a simple least-squares approach*, Review of Financial Studies, **14**, 113 (2001).

M. Potters, J.-P. Bouchaud, D. Šestović, *Hedged Monte-Carlo: low variance derivative pricing with objective probabilities*, Physica A, **289**, 517, 2001, *Hedge your Monte Carlo*, Risk Magazine, **14**, 133 (March 2001).

19

The yield curve

Time flies like an arrow, fruit flies like a banana.

(Groucho Marx)

19.1 Introduction

The case of the interest rate curve is particularly complex and interesting, since it is not the random motion of a point, but rather the consistent history of a whole curve (corresponding to different loan maturities) which is at stake. Indeed, interest rates corresponding to all possible maturities (from one week to thirty years) are 'floating', that is, fixed by the market.[†] When money lenders are more numerous, money is cheaper to borrow, and the corresponding rate goes down. Conversely, the rates are high whenever money lenders are uncertain about the future and ask for a substantial yield on their loans.

The need for a consistent description of the whole interest rate curve is driven by the importance, for large financial institutions, of asset liability management and by the rapid development of interest rate derivatives (options, swaps, options on swaps, etc.[‡]). Present models of the interest rate curve fall into two categories: the first one is the Vasicek model and its variants, which focuses on the dynamics of the short-term interest rate, from which the whole curve is reconstructed.[§] The second one, initiated by Heath, Jarrow and Morton takes the full forward rate curve as dynamic variables, driven by (one or several) continuous-time Brownian motion, multiplied by a maturity-dependent scale factor. Most models are however primarily motivated by their mathematical tractability rather than by their ability to describe the data. For example, the fluctuations are often assumed to be Gaussian, thereby neglecting 'fat tail' effects.

Our aim in this chapter is to discuss the first category of interest rate models in the spirit of the previous chapters. Several ideas, introduced in the context of futures and options, will turn out to be useful for bonds as well. We then present an empirical study of the forward rate curve (FRC), where we isolate several important qualitative features

[†] All maturities are fixed by money markets, except the shortest one (day to day), which is fixed by central banks that try to monitor the way money is injected in the economy.

[‡] A swap is a contract where one exchanges fixed interest rate payments with floating interest rate payments (Section 5.2.6).

[§] For a compilation of the most important theoretical papers on the interest rate curve, see: Hughston (1996).

that a good model should be asked to reproduce, and discuss how badly the simplest models fail. Finally, some intuitive arguments are proposed to relate the series of observations reported below.

19.2 The bond market

A 'zero-coupon' bond is a contract that pays \$1 at a certain time in the future. The notation which we will use for the price of this contract at time t is $B(t, \theta)$, where θ is the maturity of the bond; the contract therefore expires at date $T = t + \theta$. Bonds on markets are usually more complex contracts since they also pay a certain amount every year, or every six months, traditionally called the 'coupon'. The price of the bond must therefore also allow for this cash flow at regular dates in the future, on top of the 'nominal' terminal payment.

Bonds of different maturities coexist in the market. What should the fair value of these different contracts be? The simplest assumption is that all these different bonds are sensitive to a unique source of uncertainty, namely the value of the short term (say daily) interest rate $r(t)$, which is every day subject to change. A naive 'fair-game' argument suggests that the value $B(t, \theta)$ of the bond today should be such that the compounded short rate interest on this value allows the seller of the bond to receive, one average, \$1 at maturity. One is thus led to a Bachelier-like price for bonds:

$$\left\langle \prod_{t'=t}^{T-\tau} (1 + r(t')\tau) \right\rangle B_B(t, \theta) \equiv 1, \tag{19.1}$$

which reads, in the continuous time limit $\tau \to 0$:

$$B_B(t, \theta) = \left[\left\langle \exp \left(\int_t^T r(u) \, du \right) \right\rangle \right]^{-1} \qquad T = t + \theta. \tag{19.2}$$

In the case where the short term rate is a constant equal to r_0, one finds, as expected:

$$B_B(t, \theta) = e^{-r_0 \theta}, \tag{19.3}$$

which is the exact result. If, on the other hand, the short term rate is random, one can wonder whether Eq. (19.2) is correct, or if as in the case of futures and options, a correction term coming from the possibility of hedging comes into play.

19.3 Hedging bonds with other bonds

19.3.1 The general problem

Consider a market where bonds of M different maturities dates T_n, $n = 1, \ldots, M$ are traded. The seller of a bond of a certain maturity T_N can try to hedge his position by trading simultaneously bonds of other maturities. Let us call ϕ_n the number of bonds of maturity T_n, with by definition $\phi_N = -1$ (one bond of maturity T_N sold). Between time t and time $t + \tau$, the value of $B_n(t) \equiv B(t, T_n - t)$ changes by an amount $\delta B_n \equiv B_n(t + \tau) - B_n(t)$.

The change of the value δS of the whole portfolio, compared to the risk-free short rate, is thus given by:

$$\delta S = \sum_{n=1}^{M} \phi_n \left(\delta B_n - r(t) \tau B_n(t) \right). \tag{19.4}$$

If one assumes that bond prices only depend on the short rate $r(t)$, the random part $\delta_r B_n$ of change of price of a bond can only come from the change of the short rate itself between time t and $t + \tau$:

$$\delta_r B_n = B_n(r + \delta r) - B_n(r). \tag{19.5}$$

The variance of δS can therefore be expressed in terms of the covariance matrix of the bond price changes much in the same way as for usual portfolios, see Chapter 12:

$$D_S \equiv \langle (\delta S)^2 \rangle = \sum_{n,m=1}^{M} \phi_n \phi_m C_{nm} \qquad C_{nm} \equiv \langle \delta_r B_n \cdot \delta_r B_m \rangle_c. \tag{19.6}$$

The minimization of D_S with the constraint $\phi_N = -1$ leads to:

$$\sum_{m \neq N} C_{nm} \phi_m^* = C_{nN} \longrightarrow \phi_n^* = \sum_{m \neq N} (\tilde{C})_{nm}^{-1} C_{mN}, \tag{19.7}$$

where \tilde{C} is the matrix C with the row and column associated to N suppressed. We now show that the matrix C_{nm} can be given a more explicit form if the spot rate has Gaussian fluctuations.

19.3.2 The continuous time Gaussian limit

We first introduce the Fourier transform of $B_n(r)$, defined as:

$$B_n(r) = \int e^{izr} \hat{B}_n(z) \frac{dz}{2\pi}. \tag{19.8}$$

Let us first compute the average of the product $B_n(r + \delta r) B_m(r + \delta r)$ over the random variable δr:

$$\langle B_n(r + \delta r) B_m(r + \delta r) \rangle = \int e^{i(z+z')(r+\delta r)} \hat{B}_n(z) \hat{B}_m(z') P(\delta r) \frac{dz}{2\pi} \frac{dz'}{2\pi} d\delta r. \tag{19.9}$$

Using the definition of the characteristic function $\hat{P}(z)$ of $P(\delta r)$, one finds:

$$\langle B_n(r + \delta r) B_m(r + \delta r) \rangle = \int e^{i(z+z')r} \hat{B}_n(z) \hat{B}_m(z') \hat{P}(z + z') \frac{dz}{2\pi} \frac{dz'}{2\pi}. \tag{19.10}$$

The calculation of C_{nm} involves three other similar terms; the final result reads:

$$C_{nm} = \int e^{i(z+z')r} \hat{B}_n(z) \hat{B}_m(z') [\hat{P}(z + z') - \hat{P}(z) - \hat{P}(z') + 1] \frac{dz}{2\pi} \frac{dz'}{2\pi}. \tag{19.11}$$

Let us first look at the case where $P(\delta r)$ is Gaussian, with a variance equal to $\sigma_1^2 \ll 1$. In this case, $\hat{P}(z) = \exp(-\sigma_1^2 z^2/2)$, and to first order in σ_1^2 one finally obtains:

$$C_{nm} = \sigma_1^2 \int e^{i(z+z')r} z z' \hat{B}_n(z) \hat{B}_m(z') \frac{dz}{2\pi} \frac{dz'}{2\pi} \equiv \sigma_1^2 \frac{\partial B_n}{\partial r} \frac{\partial B_m}{\partial r}. \tag{19.12}$$

We find the very important result that in the Gaussian case, the matrix C_{nm} is a *projector* on the vector $\partial B_n/\partial r$. Therefore, in this case, Eq. (19.7) has many solutions, since one can add to a certain solution ϕ_n^* any vector orthogonal to $\partial B_n/\partial r$ without changing the overall risk. A special family of solutions correspond to all ϕ_m^*, $m \neq N$, equal to zero except one. In this case, one chooses to hedge the bond B_N with a single other bond B_n, with:

$$\phi_n^* = \frac{\frac{\partial B_N}{\partial r}}{\frac{\partial B_n}{\partial r}}. \tag{19.13}$$

This hedge is the analogue of the Δ-hedge in the case of options, and has an obvious interpretation: the change of value of B_N is, to first order in δr, given by $(\partial B_N/\partial r)\delta r$; this change is, in the continuous time limit, exactly compensated by the quantity $\phi_n^*(\partial B_n/\partial r)\delta r$. In this limit, the hedging strategy becomes perfect, as for options in the Black-Scholes world.

The degeneracy of the hedging problem in the continuous time, Gaussian case, which tells us that one can hedge optimally using any other bond, is lifted as soon as non-Gaussian or discrete time effects exist. Indeed, to the kurtosis order, one finds (assuming that $P(\delta r)$ is symmetric) that C_{nm} can no longer be written as a projector, and reads:

$$C_{nm} = \sigma_1^2 \frac{\partial B_n}{\partial r} \frac{\partial B_m}{\partial r} + \frac{(\kappa+3)\sigma_1^4}{6} \left[\frac{\partial^3 B_n}{\partial r^3} \frac{\partial B_m}{\partial r} + \frac{3}{2} \frac{\partial^2 B_n}{\partial r^2} \frac{\partial^2 B_m}{\partial r^2} + \frac{\partial B_n}{\partial r} \frac{\partial^3 B_m}{\partial r^3} \right]. \tag{19.14}$$

As was the case with options, the optimal hedge in the presence of non-Gaussian fluctuations ceases to be the Δ-hedge. In the case of bonds, non-Gaussian effects in fact determine uniquely the optimal hedge, i.e. the maturities that should be used for hedging. One could of course extend the above calculation to the case where bond prices depend on other (possibly non-Gaussian) factors, such as the long-term interest rate.

19.4 The equation for bond pricing

Now we know the optimal hedging strategy, we can find an equation that sets the bond prices using a fair-game argument, once again on the full balance equation that includes the hedging strategy. The fair-game argument becomes a strict absence of arbitrage opportunities in the continuous time, Gaussian world, as was the case for options. When the risk is non-zero, a risk premium proportional to the residual risk $\sqrt{\langle \delta S^2 \rangle}$ could be added to the balance equation. In the present case, the set of equations fixing the fair-game price of all bonds simultaneously reads:

$$-\langle \tilde{\delta} B_N \rangle + \sum_{n \neq N} \phi_n^* \langle \tilde{\delta} B_n \rangle = 0, \tag{19.15}$$

where we have set $\tilde{\delta} B_n = \delta B_n - r(t)\tau B_n$. A trivial solution of the above equations is:

$$\langle \tilde{\delta} B_n(t) \rangle = 0 \quad \forall n \longrightarrow \langle \delta B_n(t) \rangle = r(t)\tau B_n(t). \tag{19.16}$$

Using the explicit form of ϕ_n^*, given by Eq. (19.7), one can actually show, using simple linear algebra (detailed in Appendix G), that the general solution to the bond pricing equations reads:

$$\langle \tilde{\delta} B_n \rangle + \lambda \sigma \frac{\partial B_1}{\partial r} \frac{\partial B_n}{\partial r} = 0, \tag{19.17}$$

where λ is an arbitrary parameter, possibly time dependent, but *independent of maturity*. The extra factor σ was added for later convenience.

The shortest maturity bond B_1 is trivial to price since there is no uncertainty about future short rates: $B_1 = \exp(-r\tau)$, and $\partial B_1/\partial r \approx -\tau$. Therefore the fundamental bond pricing equation takes the following form:

$$\langle \delta B_n(t) \rangle = r(t)\tau B_n(t) + \lambda \sigma \tau \frac{\partial B_n(t)}{\partial r}, \qquad B_n(t = T_n) \equiv 1. \tag{19.18}$$

Let us first show that in the continuous time limit, and for Gaussian fluctuations of $r(t)$, the above equation reproduces the form quoted in the literature (see e.g. Hull (1997), Hughston (1996)). We first write, to lowest order in τ:

$$\delta B_n(t) \simeq \frac{\partial B_n}{\partial t}\tau + \frac{\partial B_n}{\partial r}\langle \delta r \rangle + \frac{1}{2}\frac{\partial^2 B_n}{\partial r^2}\langle \delta r^2 \rangle. \tag{19.19}$$

Now, setting $\langle \delta r \rangle = m\tau$ and $\langle \delta r^2 \rangle = \sigma_1^2 = \sigma^2 \tau$, Eq. (19.18) becomes a partial differential equation in the limit $\tau \to 0$:[†]

$$\frac{\partial B_n}{\partial t} + (m - \lambda\sigma)\frac{\partial B_n}{\partial r} + \frac{\sigma^2}{2}\frac{\partial^2 B_n}{\partial r^2} = r(t)B_n(t), \tag{19.20}$$

where λ is often called the **market price of risk**. This interpretation comes from the following observation. Using Ito's lemma for the function $B_n(r, t)$ and $dr = mdt + \sigma dW$, one finds:

$$dB_n = \left[\frac{\partial B_n}{\partial t} + m\frac{\partial B_n}{\partial r} + \frac{\sigma^2}{2}\frac{\partial^2 B_n}{\partial r^2}\right]dt + \frac{\partial B_n}{\partial r}\sigma dW \equiv m_B B_n dt + \sigma_B B_n dW, \tag{19.21}$$

where the last part of the equation defines the average bond return m_B and the bond volatility σ_B. Using these quantities, Eq. (19.20) can be identically rewritten as:

$$m_B - r = \lambda\sigma_B, \tag{19.22}$$

which means that the average bond return should exceed the short term rate by an amount proportional to the volatility of the same bond, with a proportionality constant independent

[†] Note that both m and σ could depend both on time t and on the short rate r.

of the maturity of the bond. The risk of the bond, measured through σ_B, should lead to an increase of the return, hence the term 'market price of risk' for λ. Interestingly, λ is not fixed by the above theory: any value is in principle consistent with the fair-game/absence of arbitrage prescription.

19.4.1 A general solution

Interestingly, in the case where the short rate process is Markovian, Eq. (19.18) can be solved formally using a discrete time path integral method. Suppose first that the market price of risk λ is zero, and consider the following expression:

$$I_n(r, t) = \left\langle \prod_{t'=t}^{T_n - \tau} [1 - r(t')\tau] \right\rangle \Big|_t \qquad (19.23)$$

where the subscript $|_t$ means that the average is taken over all future histories of the short time rate, conditioned to all information available at time t. By construction, $I_n(r, T_n) \equiv 1$. Using the assumption that $r(t)$ follows a Markov process, the only relevant information at time t is in fact the value of $r(t)$. Therefore, Eq. (19.23) can be written as:

$$I_n(r, t) = [1 - r(t)\tau] \langle I_n(r + \delta r, t + \tau) \rangle_{\delta r}. \qquad (19.24)$$

Since by definition, $\langle \delta I_n \rangle \equiv \langle I_n(r + \delta r, t + \tau) \rangle_{\delta r} - I_n(r, t)$, one has:

$$\langle \delta I_n \rangle = r(t)\tau \langle I_n(r + \delta r, t + \tau) \rangle_{\delta r}. \qquad (19.25)$$

To first order in τ, one can furthermore replace $r(t)\tau \langle I_n(r + \delta r, t + \tau) \rangle_{\delta r}$ by $r(t)\tau I_n$. Hence Eqs. (19.25) and (19.18) are identical, which shows that the bond price B_n is in fact given by Eq. (19.23). So, in the limit $\tau \to 0$, one can write the bond price in the following form:

$$B_n(r, t) = \left\langle \exp\left(-\int_t^{T_n} r(u)\,du \right) \right\rangle \Big|_t. \qquad (19.26)$$

This formula is close to, but different from the Bachelier price, Eq. (19.2). The Bachelier formula says that the bond price is one over the average of e^{yT}, where $y = \int_t^T r(u)\,du / T$ is the yield. The above formula assumes that the instantaneous returns of all bonds is equal to the short rate, and gives the bond price as the average of e^{-yT}. Since the following bound is true:

$$\langle e^{-yT} \rangle \langle e^{yT} \rangle \geq 1, \qquad (19.27)$$

one concludes that the Bachelier price is a lower bound to the bond price. Said differently, selling a bond at price (19.26) and reinvesting continuously on a short term money market account is *on average* profitable (but risky).

When $\lambda \neq 0$, Eq. (19.26) continues to hold (in the small τ limit), but the average is taken assuming a fictitious 'trend' on the short rate, equal to $m - \lambda\sigma$, instead of the true trend m. When $\lambda > 0$, therefore, the price of bonds increases compared to the case $\lambda = 0$, since the effective future value of the short rate is, according to this procedure, reduced.

19.4.2 The Vasicek model

Let us now study a specific model for the short rate fluctuations. Since the short rate does not wander away infinitely far, it is reasonable to assume some sort of mean reverting term. The simplest model is:

$$r(t) = r_0 + \sum_{t'=-\infty}^{t-\tau} e^{-\Omega(t-t')}\xi_{t'}, \tag{19.28}$$

where r_0 is the 'reference' value for the short rate, Ω measures the strength of the mean reverting term, and ξ is a (possibly non-Gaussian) random noise. When the noise is Gaussian and the continuous time limit is taken, the above model defines the Ornstein-Uhlenbeck process which is called, in this context, the Vasicek model.

A simple property of Eq. (19.28) is that:

$$r(t_2) - r_0 = \sum_{t'=t_1}^{t_2-\tau} e^{-\Omega(t_2-t')}\xi_{t'} + e^{-\Omega(t_2-t_1)}\left(r(t_1) - r_0\right), \qquad (t_2 \geq t_1). \tag{19.29}$$

Hence, a quantity such as $\sum_{t'=t}^{T-\tau} r(t')\tau$, that appears in the exponential of Eq. (19.26), can be rewritten as:

$$\sum_{t'=t}^{T-\tau} r(t')\tau = r_0(T-t) + (r(t) - r_0)\,\tau \sum_{t'=t}^{T-\tau} e^{-\Omega(t'-t)} + \tau \sum_{t'=t}^{T-\tau}\sum_{t''=t}^{t'-\tau} e^{-\Omega(t'-t'')}\xi_{t''}. \tag{19.30}$$

Now, permuting the order of the sum over t' and t'', and evaluating the sums in the continuous time limit leads to:

$$\int_t^T r(u)\,du = r_0\theta + (r(t) - r_0)\frac{1 - e^{-\Omega\theta}}{\Omega} + \int_t^T \xi(T-t'')\frac{1 - e^{-\Omega t''}}{\Omega}\,dt'', \tag{19.31}$$

where $\theta = T - t$ is the maturity. The first two terms of this expression are not random, and can be pulled out from the average. The last term can be evaluated in general in terms of the characteristic function of the random variable ξ_t (see Section 19.4.4), but we will restrict here to the Vasicek model where ξ_t is Gaussian, of variance $\sigma^2\tau$. In this case, one finds:

$$\left\langle \exp\left(-\int_t^T \xi(T-t'')\frac{1 - e^{-\Omega t''}}{\Omega}\,dt''\right)\right\rangle = \exp\left(\frac{\sigma^2}{2}\int_t^T \left(\frac{1 - e^{-\Omega t''}}{\Omega}\right)^2 dt''\right). \tag{19.32}$$

This formula allows one to obtain an explicit expression for the bond price as a function of the present short term rate, the maturity of the bond, and the parameters of the Vasicek model (Ω, σ). Instead of writing down this formula, however, we will introduce the notion *forward rates*, in terms of which the result is much simpler.

19.4.3 Forward rates

The forward interest rate curve (FRC) at time t is fully specified by the collection of all **forward rates** $f(t, \theta)$, for different maturities θ. The forward rates are by definition such that they compound to give the bond price $B(t, \theta)$ exactly:

$$B(t, \theta) = \exp\left(-\int_0^\theta f(t, u)\, du\right),$$ (19.33)

$r(t) = f(t, \theta = 0)$ is called the short rate (spot rate). The forward rate $f(t, \theta)$ is the rate on a future loan, between times $t + \theta$ and $t + \theta + d\theta$, decided at time t. Using the above definition of forward rates, one sees that from the knowledge of all bond prices, one deduces the whole FRC as:

$$f(t, \theta) = -\partial \log B(t, \theta)/\partial \theta.$$ (19.34)

Noting that the derivative with respect to θ or with respect to T are identical for fixed t, one deduces very simply from the results of the previous section the FRC in the Vasicek model:

$$f(t, \theta) = r_0(1 - e^{-\Omega\theta}) + r(t)e^{-\Omega\theta} - \frac{\sigma^2}{2\Omega^2}(1 - e^{-\Omega\theta})^2.$$ (19.35)

This result will be further discussed and compared with empirical data in Section 19.6.1. Note that, in practice, the last term, proportional to σ^2, is quite small. When the market price of risk is non-zero, r_0 is replaced by $r_0 - \lambda\sigma/\Omega$.

19.4.4 More general models

Non-Gaussian Vasicek model

Using the definition of the cumulants c_n of $P(\delta r)$, the generalization of the formula for the FRC in the case of non Gaussian fluctuations is very simple and reads:

$$f(t, \theta) = r_0(1 - e^{-\Omega\theta}) + r(t)e^{-\Omega\theta} - \sum_{n=2}^\infty \frac{(-1)^n c_n}{n!\Omega^n}(1 - e^{-\Omega\theta})^n.$$ (19.36)

Heath-Jarrow-Morton and the dynamics of the whole FRC

If one takes the time derivative of Eq. (19.35) for a fixed maturity date $T = t + \theta$, one obtains a dynamical equation for the time evolution of the FRC which will lend itself to interesting generalizations of the Vasicek model. Using the fact that $\partial/\partial t|_T = -\partial/\partial\theta|_T$, this dynamical equation reads:

$$df(t, \theta)|_T = dr(t)e^{-\Omega\theta} + \Omega[r(t) - r_0]e^{-\Omega\theta}dt + \frac{\sigma^2}{\Omega}e^{-\Omega\theta}[1 - e^{-\Omega\theta}]dt.$$ (19.37)

In the continuous time limit, the Vasicek model corresponds to an Ornstein-Uhlenbeck model for the short rate, such that:

$$dr(t) = -\Omega[r(t) - r_0]\,dt + \sigma\,dW(t). \tag{19.38}$$

This allows one to transform Eq. (19.37) into:

$$df(t, \theta)|_T = \frac{\sigma^2}{\Omega} e^{-\Omega\theta}[1 - e^{-\Omega\theta}]\,dt + \sigma e^{-\Omega\theta}\,dW(t). \tag{19.39}$$

In the presence of a non-zero market price of risk, an extra $-\lambda e^{-\Omega\theta}dt$ term appears on the right-hand side.

Now, defining $\sigma(\theta) = \sigma e^{-\Omega\theta}$ as the volatility of the forward rate, and noting that

$$\int_0^\theta \sigma(\theta')\,d\theta' = \frac{\sigma}{\Omega}[1 - e^{-\Omega\theta}], \tag{19.40}$$

one can rewrite the above equation as:

$$df(t, \theta)|_T = m(\theta)\,dt + \sigma(\theta)\,dW(t); \quad m(\theta) = \sigma(\theta)\left[-\lambda + \int_0^\theta \sigma(\theta')\,d\theta'\right], \tag{19.41}$$

which shows that in the framework of the Vasicek model, the drift $m(\theta)$ of the forward rate is found to be directly related to the volatility $\sigma(\theta)$ of the same rate.

The interesting point now is to generalize the dynamical equation (19.41) to a more general form, possibly time dependent, of the volatility $\sigma(\theta)$, and write the so-called Heath-Jarrow-Morton equation:

$$df(t, \theta)|_T = m(t, \theta)\,dt + \sigma(t, \theta)\,dW(t) \quad T = t + \theta. \tag{19.42}$$

Since we consider a Gaussian model in continuous time, perfect hedging is possible, and the fair-game argument becomes equivalent to the constraint of absence of arbitrage opportunities. In the present context, this constraint leads to a relation between drift and volatility that generalizes the one found above, namely:

$$m(t, \theta) = \sigma(t, \theta)\left[-\lambda(t) + \int_0^\theta \sigma(t, \theta')\,d\theta'\right], \tag{19.43}$$

where the market price of risk itself can be time dependent, but, as emphasized above, independent of θ.

Although Eq. (19.42) is rather general and allows for a large class of maturity dependent volatilities for the forward rates, we are still in the framework of *single factor models*, where the factor is the short term rate. All HJM models are indeed equivalent to a certain model of short term rate dynamics. However, as the following sections will show, the description of real FRCs require at least two extra factors, namely the long term rate, and (say) the initial slope of the FRC. Generalized HJM models, where each forward rate is driven by several independent Gaussian noises, have to be considered. We prefer at this

stage to discuss the most important features of empirical data, that any plausible model should reproduce.

19.5 Empirical study of the forward rate curve

19.5.1 Data and notations

Our study is based on a data-set of daily prices of Eurodollar futures contracts on Libor USD interest rates, and similar contracts on rates of other countries.[†] The interest rate underlying the Eurodollar futures contract is a 90-day rate, earned on dollars deposited in a bank outside the U.S. by another bank. The interest in studying forward rates rather than yield curves is that one has a direct access to a 'derivative' (in the mathematical sense: $f(t, \theta) = -\partial \log B(t, \theta)/\partial \theta$), which obviously contains more precise information than the yield curve itself (defined from the logarithm of $B(t, \theta)$).

In practice, the futures markets price 3-months forward rates for *fixed* expiration dates, separated by 3-month intervals. Identifying 3-months futures rates with instantaneous forward rates, we have available a sequence of time series on forward rates $f(t, T_i - t)$, where T_i are fixed dates (March, June, September and December of each year). We can convert these into fixed maturity (multiple of 3-months) forward rates by a simple linear interpolation between the two nearest points such that $T_i - t \leq \theta \leq T_{i+1} - t$. Between 1990 and 1996, one has at least 15 different Eurodollar maturities for each market date. Between 1994 and 1999, the number of available maturities rises to 30 (as time grows, longer and longer maturity forward rates are being traded on future markets). Since we only consider daily data, our reference time scale will be $\tau = 1$ day. The variation of $f(t, \theta)$ between t and $t + \tau$ will be denoted as $\delta f(t, \theta)$:

$$\delta f(t, \theta) = f(t + \tau, \theta) - f(t, \theta). \tag{19.44}$$

19.5.2 Quantities of interest and data analysis

The description of the FRC has two, possibly interrelated, aspects:

(i) What is, at a given instant of time, the *shape* of the FRC as a function of the maturity θ?

(ii) What are the statistical properties of the increments $\delta f(t, \theta)$ between time t and time $t + \tau$, and how are they correlated with the shape of the FRC at time t?

The two basic quantities describing the FRC at time t are the value of the short-term interest rate $f(t, \theta_{\min})$ (where θ_{\min} is the shortest available maturity), and that of the short-term/long-term *spread* $s(t) = f(t, \theta_{\max}) - f(t, \theta_{\min})$, where θ_{\max} is the longest available

[†] In principle forward contracts and futures contracts are not strictly identical – they have different margin requirements – and one may expect slight differences, which we shall neglect in the following.

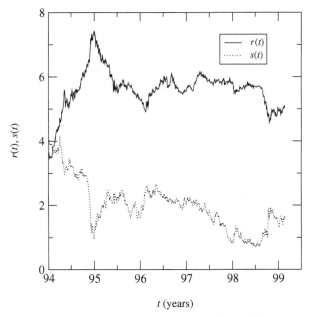

Fig. 19.1. The historical time series of the spot rate $r(t)$ from 1994 to 1999 (top curve) – actually corresponding to a 3-month future rate (dark line) and of the 'spread' $s(t)$ (bottom curve), defined with the longest maturity available over the whole period 1994–99 on future markets, i.e. $\theta_{max} = 9.5$ years.

maturity. The two quantities $r(t) \approx f(t, \theta_{min})$, $s(t)$ are plotted versus time in Figure 19.1;[†] note that:

- The volatility σ of the spot rate $r(t)$ is equal to $0.8\%/\sqrt{\text{year}}$.[‡] This obtained by averaging over the whole period.
- The spread $s(t)$ has varied between 0.6 and 4.0%. Contrarily to some European interest rates on the same period, $s(t)$ has always remained positive. (This however does not mean that the FRC is increasing monotonically, see below.)

Figure 19.2 shows the average shape of the FRC, determined by averaging the difference $f(t, \theta) - r(t)$ over time. Interestingly, this function is rather well fitted by a simple square-root law. This means that on average, the difference between the forward rate with maturity θ and the spot rate is equal to $a\sqrt{\theta}$, with a proportionality constant $a = 0.85\%/\sqrt{\text{year}}$ which turns out to be nearly identical to the spot rate volatility. A similar $\sqrt{\theta}$ law for the average shape of the FRC was also found for different rate markets (UK, Australia, Japan): see Matacz & Bouchaud (2000b). Again, the proportionality constant a is found to be

[†] We shall from now on take the 3-month rate as an approximation to the spot rate $r(t)$.

[‡] The dimension of r should really be % per year, but we conform here to the habit of quoting r simply in %. Note that this can sometimes be confusing when checking the correct dimensions of a formula.

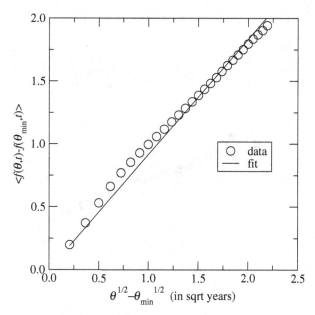

Fig. 19.2. The average FRC in the period 1994–99, as a function of the maturity θ. We have shown for comparison a *one parameter fit* with a square-root law, $a(\sqrt{\theta} - \sqrt{\theta_{min}})$.

comparable to the spot rate volatility. We shall propose a simple interpretation of this fact below.

Let us now turn to an analysis of the *fluctuations* around the average shape. These fluctuations are actually similar to that of a vibrating elastic string. The average deviation $\Delta(\theta)$ can be defined as:

$$\Delta^2(\theta) \equiv \left\langle \left[f(t, \theta) - r(t) - s(t)\sqrt{\frac{\theta}{\theta_{max}}} \right]^2 \right\rangle, \qquad (19.45)$$

and is plotted in Figure 19.3. The maximum of Δ is reached for a maturity of $\theta^* = 1$ year.

We now turn to the statistics of the *daily* increments $\delta f(t, \theta)$ of the forward rates, by calculating their volatility $\sigma(\theta) = \sqrt{\langle \delta f(t, \theta)^2 \rangle}$ and their excess kurtosis

$$\kappa(\theta) = \frac{\langle \delta f(t, \theta)^4 \rangle}{\sigma^4(\theta)} - 3. \qquad (19.46)$$

A very important quantity will turn out to be the following 'spread' correlation function:

$$C(\theta) = \frac{\langle \delta f(t, \theta_{min}) (\delta f(t, \theta) - \delta f(t, \theta_{min})) \rangle}{\sigma^2(\theta_{min})}, \qquad (19.47)$$

which measures the influence of the short-term interest fluctuations on the other modes of motion of the FRC, subtracting away any trivial overall translation of the FRC.

Figure 19.4 shows $\sigma(\theta)$ and $\kappa(\theta)$. Somewhat surprisingly, $\sigma(\theta)$, much like $\Delta(\theta)$ has a maximum around $\theta^* = 1$ year. The order of magnitude of $\sigma(\theta)$ is $0.05\%/\sqrt{day}$, or $0.8\%/\sqrt{year}$.

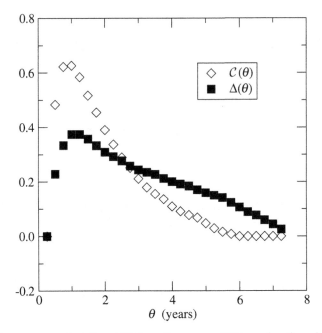

Fig. 19.3. Root mean square deviation $\Delta(\theta)$ from the average FRC as a function of θ. Note the maximum for $\theta^* = 1$ year, for which $\Delta \approx 0.38\%$. We have also plotted the correlation function $C(\theta)$ (defined by Eq. (19.47)) between the daily variation of the spot rate and that of the forward rate at maturity θ, in the period 1994–96. Again, $C(\theta)$ is maximum for $\theta = \theta^*$, and decays rapidly beyond.

The daily kurtosis $\kappa(\theta)$ is rather high (on the order of 5), and only weakly decreasing with θ.

Finally, $C(\theta)$ is shown in Figure 19.3; its shape is again very similar to those of $\Delta(\theta)$ and $\sigma(\theta)$, with a pronounced maximum around $\theta^* = 1$ year. This means that the fluctuations of the short-term rate are *amplified* for maturities around 1 year. We shall come back to this important point below.

19.6 Theoretical considerations (*)

19.6.1 Comparison with the Vasicek model

The explicit form of the FRC in the Vasicek model was computed in Section 19.4.3. The basic predictions of this model for $\lambda = 0$ are as follows:

- The FRC is a monotonous function of maturity, either increasing from $r(t)$ for $\theta = 0$ to r_0 for $\theta \to \infty$ when $r_0 > r(t)$, or decreasing if $r_0 < r(t)$.
- Since $\langle r_0 - r(t) \rangle = 0$, the average of $f(t, \theta) - r(t)$ is given by

$$\langle f(t, \theta) - r(t) \rangle = -\sigma^2/2\Omega^2(1 - e^{-\Omega\theta})^2, \tag{19.48}$$

and should thus be *negative*, at variance with empirical data. Note that in the limit $\Omega\theta \ll 1$, the order of magnitude of this (negative) term is very small: taking $\sigma = 1\%/\sqrt{\text{year}}$ and

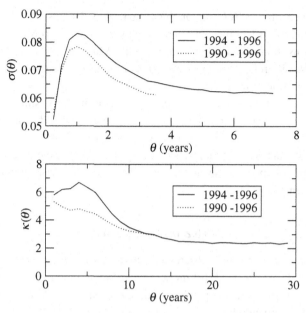

Fig. 19.4. The daily volatility and kurtosis as a function of maturity. Note the maximum of the volatility for $\theta = \theta^*$, while the kurtosis is rather high, and only very slowly decreasing with θ. The two curves correspond to the periods 1990–96 and 1994–96, the latter period extending to longer maturities.

$\theta = 1$ year, it is found to be equal to 0.005%, much smaller than the typical differences actually observed on forward rates.

- The volatility $\sigma(\theta)$ is monotonically decreasing as $\exp -\Omega\theta$, whereas the kurtosis $\kappa(\theta)$ is identically zero (because the noise ξ driving the short rate fluctuations is Gaussian).
- The correlation function $\mathcal{C}(\theta)$ is negative and is a monotonic decreasing function of its argument, in total disagreement with observations (Fig. 19.3).
- The variation of the spread $s(t)$ and of the spot rate should be perfectly correlated, which is not the case (Fig. 19.3): more than one factor is in any case needed to account for the deformation of the FRC.

An interesting extension of Vasicek's model designed to fit exactly the 'initial' FRC $f(t = 0, \theta)$ was proposed by Hull and White. It amounts to replacing the constants Ω and r_0 by time-dependent functions. For example, $r_0(t)$ represents the anticipated evolution of the 'reference' short-term rate itself with time. These functions can be adjusted to fit $f(t = 0, \theta)$ exactly. Interestingly, one can then derive the following relation:

$$\left\langle \frac{\partial r(t)}{\partial t} \right\rangle = \left\langle \frac{\partial f}{\partial \theta}(t, 0) \right\rangle, \tag{19.49}$$

up to a term of order σ^2 which turns out to be negligible, exactly for the same reason as explained above. On average, the second term (estimated by taking a finite difference estimate of the partial derivative using the first two points of the FRC) is definitely found to

be *positive*, and equal to 0.8%/year. On the same period (1990–96), however, the spot rate has *decreased* from 8 to 5%, instead of growing by $7 \times 0.8\% = 5.6\%$.

In simple terms, both the Vasicek and the Hull–White model mean the following: the FRC should basically reflect the market's expectation of the average evolution of the spot rate (up to a correction of the order of σ^2, but which turns out to be very small, see above). However, since the FRC is on average increasing with the maturity (situations when the FRC is 'inverted' are comparatively much rarer), this would mean that the market systematically expects the spot rate to rise, which it does not. It is hard to believe that the market would persist in error for such a long time. Hence, the upward slope of the FRC is not only related to what the market expects on average, but that a systematic risk premium is needed to account for this increase.

19.6.2 Market price of risk

A possibility is to assign the above discrepancy to the market price of risk, which we discuss in the context of the one factor HJM model, where we assume that the volatility $\sigma(\theta)$ and the market price of risk λ are independent of time t. From Eq. (19.42) above, one finds that the FRC at time t is given by:

$$f(t, \theta) = f(t_0, t - t_0 + \theta) + \int_{t_0}^{t} m(t + \theta - u)\,du + \int_{t_0}^{t} \sigma(t + \theta - u)\,dW(u), \qquad (19.50)$$

where t_0 is the initial time, and:

$$m(\theta) = \sigma(\theta)\left[\int_0^\theta \sigma(\theta')\,d\theta' - \lambda\right], \qquad (19.51)$$

The *average* FRC, over a certain time window of size \mathcal{T}, contains three terms. First is the contribution of the initial FRC. In our case the initial FRC was indeed somewhat steeper than the average FRC. Yet its contribution to the average FRC is roughly a factor of three smaller than the observed average. The second contribution comes from the $O(\sigma^2)$ term in Eq. (19.51). The size of this term is again very small, in any case at least a factor ten smaller than the observed average FRC. This term can therefore be neglected. More interesting is the contribution of the market price of risk λ, and is given by:

$$\langle f(t, \theta) - r(t)\rangle|_\lambda = \frac{\lambda}{\mathcal{T}} \int_0^{\mathcal{T}} (\mathcal{T} - u)(\sigma(u) - \sigma(u + \theta))du. \qquad (19.52)$$

For long time periods ($\mathcal{T} \gg \theta$), this contribution takes the form:

$$\langle f(t, \theta) - r(t)\rangle_\lambda \simeq \lambda\left[\int_0^\theta \sigma(u)\,du - \theta\sigma(\theta_{max})\right] + O(\sigma(\theta_{max})\theta^2/\mathcal{T}). \qquad (19.53)$$

We deduce that this contribution is *negative* (when $\lambda > 0$) for some initial region of the FRC if, as found empirically, $\sigma(\theta) > \sigma(\theta = 0)$ for all θ. In Figure 19.5 we show a plot of Eq. (19.53) where we use the empirical volatility $\sigma(\theta)$ and choose $\lambda = 4.4\sqrt{year}$ which gives a best fit to the average FRC. It is clear that this fit is very bad, in particular compared to

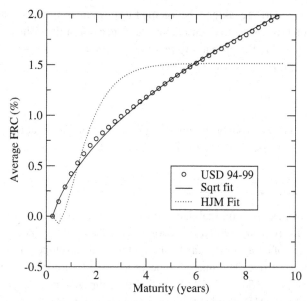

Fig. 19.5. Comparison between the empirical average FRC in the period 1994–99, as a function of the maturity θ and the one factor HJM model with a non zero market price of risk, Eq. (19.53), with the empirically determined $\sigma(\theta)$ (dotted line).

the simple square-root fit described above. The negative part of the average FRC predicted by the HJM model is even stronger for other rate markets.[†]

19.6.3 Risk-premium and the $\sqrt{\theta}$ law

The average FRC and value-at-risk pricing

The observation that on average the FRC follows a simple $\sqrt{\theta}$ law (i.e. $\langle f(t, \theta) - r(t) \rangle \propto \sqrt{\theta}$), with a prefactor precisely given by the short rate volatility suggests an intuitive, direct interpretation, related to the strong asymmetry, in bond markets, between lenders (investors) and borrowers.

At any time t, the market anticipates either a future rise, or a decrease of the spot rate. However, the average anticipated trend is, in the long run, zero, since the spot rate has bounded fluctuations. Hence, the *average* market's expectation is that the future spot rate $r(t)$ will be close to its present value $r(t = 0)$. In this sense, the average FRC should thus be flat. However, even in the absence of any trend in the spot rate, its *probable* change between now and $t = \theta$ is (assuming the simplest random walk behaviour) of the order of $\sigma\sqrt{\theta}$, where σ is the volatility of the spot rate. Money lenders agree at time t on a loan at rate $f(t, \theta)$, which will run between time $t + \theta$ and $t + \theta + d\theta$. These money lenders will themselves borrow money from central banks at the short-term rate prevailing at that date, i.e. $r(t + \theta)$. They will therefore lose money whenever $r(t + \theta) > f(t, \theta)$.

[†] see Matacz & Bouchaud (2000a).

Hence, money lenders are taking a bet on the future value of the spot rate and want to be sure not to lose their bet more frequently than, say, once out of five. Thus their price for the forward rate is such that the probability that the spot rate at time $t + \theta$, $r(t + \theta)$ actually exceeds $f(t, \theta)$ is equal to a certain number p:

$$\int_{f(t,\theta)}^{\infty} P(r', t + \theta | r, t) \, dr' = p, \tag{19.54}$$

where $P(r', t' | r, t)$ is the probability that the spot rate is equal to r' at time t' knowing that it is r now (at time t). Assuming that r' follows a simple random walk centred around $r(t)$ then leads to:[†]

$$f(t, \theta) = r(t) + a\sigma(0)\sqrt{\theta}, \qquad a = \sqrt{2} \, \text{erfc}^{-1}(2p), \tag{19.55}$$

which indeed matches the empirical data, with $p \approx 0.16$. Other rates (British, Australian, Japanese) show a similar $\sqrt{\theta}$ with a prefactor comparable to the volatility of the spot rate.

Hence, the shape of today's FRC can be thought of as an envelope for the probable future evolutions of the spot rate. The market appears to price future rates through a *Value-at-Risk* procedure (Eqs. (19.54) and (19.55)) rather than through an averaging procedure. As illustrated above by the HJM model, this averaging procedure leads to results that are quite inconsistent with the data.

The anticipated trend and the volatility hump

Let us now discuss, along the same lines, the shape of the FRC at a given instant of time, which of course deviates from the average square root law. For a given instant of time t, the market actually expects the spot rate to perform a biased random walk. We shall argue that a consistent interpretation is that the market estimates the trend $m(t)$ by extrapolating the past behaviour of the spot rate itself. Hence, the probability distribution $P(r', t + \theta | r, t)$ used by the market is not centred around $r(t)$ but rather around:

$$r(t) + \int_t^{t+\theta} m(t, t + u) \, du, \tag{19.56}$$

where $m(t, t')$ can be called the anticipated bias at time t', seen from time t.

It is reasonable to think that the market estimates m by extrapolating the recent past to the nearby future. Mathematically, this reads:

$$m(t, t + u) = m_1(t) \mathcal{Z}(u) \text{ where } m_1(t) \equiv \int_0^{\infty} K(v) \delta r(t - v) \, dv, \tag{19.57}$$

and where $K(v)$ is an averaging kernel of the past variations of the spot rate. One may call $\mathcal{Z}(u)$ the trend persistence function; it is normalized such that $\mathcal{Z}(0) = 1$, and describes how the present trend is expected to persist in the future. Equation (19.54) then gives:

$$f(t, \theta) = r(t) + a\sigma\sqrt{\theta} + m_1(t) \int_0^{\theta} \mathcal{Z}(u) \, du. \tag{19.58}$$

[†] This assumption is certainly inadequate for small times, where large kurtosis effects are present. However, on the scale of months, these non-Gaussian effects can be considered as small.

This mechanism is a possible explanation of why the three functions introduced above, namely $\Delta(\theta)$, $\sigma(\theta)$ and the correlation function $C(\theta)$ have similar shapes. Indeed, taking for simplicity an exponential averaging kernel $K(v)$ of the form $\epsilon \exp[-\epsilon v]$, one finds:[†]

$$\frac{dm_1(t)}{dt} = -\epsilon m_1 + \epsilon \frac{dr(t)}{dt} + \epsilon \xi(t), \tag{19.59}$$

where $\xi(t)$ is an independent noise of strength σ_ξ^2, added to introduce some extra noise in the determination of the anticipated bias. In the absence of temporal correlations, one can compute from the above equation the average value of m_1^2. It is given by:

$$\langle m_1^2 \rangle = \frac{\epsilon}{2} \left(\sigma^2(0) + \sigma_\xi^2 \right). \tag{19.60}$$

In the simple model defined by Eq. (19.58) above, one finds that the correlation function $C(\theta)$ is given by:[‡]

$$C(\theta) = \epsilon \int_0^\theta \mathcal{Z}(u) \, du. \tag{19.61}$$

Using the above result for $\langle m_1^2 \rangle$, one also finds:

$$\Delta(\theta) = \sqrt{\frac{\sigma^2(0) + \sigma_\xi^2}{2\epsilon}} C(\theta), \tag{19.62}$$

thus showing that $\Delta(\theta)$ and $C(\theta)$ are in this model simply proportional.

Turning now to the volatility $\sigma(\theta)$, one finds that it is given by:

$$\sigma^2(\theta) = [1 + C(\theta)]^2 \, \sigma^2(0) + C(\theta)^2 \sigma_\xi^2. \tag{19.63}$$

We thus see that the maximum of $\sigma(\theta)$ is indeed related to that of $C(\theta)$. Intuitively, the reason for the volatility maximum is as follows: a variation in the spot rate changes that market anticipation for the trend $m_1(t)$. But this change of trend obviously has a larger effect when multiplied by a longer maturity. For maturities beyond 1 year, however, the decay of the persistence function comes into play and the volatility decreases again. The relation Eq. (19.63) is tested against real data in Figure 19.6. An important prediction of the model is that the deformation of the FRC should be strongly correlated with the past trend of the spot rate, averaged over a time scale $1/\epsilon$ (see Eq. (19.59)). This correlation has been convincingly established recently, with $1/\epsilon \approx 100$ days.[§]

19.7 Summary

- Much like options, bonds can be hedged with other bonds. In the continuous time Gaussian limit, the optimal hedging portfolio is degenerate, in the sense that many different combinations of bonds are optimal. This degeneracy is lifted in the presence of non-Gaussian effects.

[†] This equation is quite similar to the second equation of the Hull-White II model, Hull & White (1994).
[‡] In reality, one should also take into account the fact that $a\sigma(0)$ can vary with time. This brings an extra contribution both to $C(\theta)$ and to $\sigma(\theta)$.
[§] see Matacz & Bouchaud (2000b).

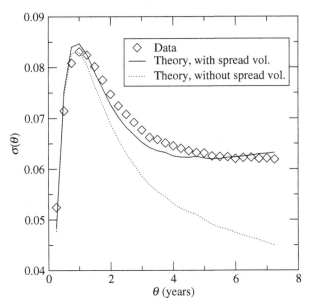

Fig. 19.6. Comparison between the theoretical prediction and the observed daily volatility of the forward rate at maturity θ, in the period 1994–96. The dotted line corresponds to Eq. (19.63) with $\sigma_\xi = \sigma(0)$, and the full line is obtained by adding the effect of the variation of the coefficient $a\sigma(0)$ in Eq. (19.58), which adds a contribution proportional to θ.

- A fair-game argument applied to an hedged portfolio of bonds allows one to obtain the price of bonds as a function of the short term interest rate. It is given by Eq. (19.26).
- The Vasicek model, or the more general Heath-Jarrow-Morton framework allows one to obtain a relation between the average shape of the forward rate curve and the maturity dependent volatility. This relation is not satisfied by empirical data, which suggests that some sort of value-at-risk premium sets the shape of the forward rate curve.
- The dynamics of the forward rate curve shows a number of interesting peculiarities, such as a humped shape volatility as a function of maturity, or a strong correlation between the initial slope of the curve and the past trend of the spot rate.

19.8 Appendix G: optimal portfolio of bonds

Consider the equation:

$$-\langle \tilde{\delta} B_N \rangle + \sum_{n \neq N} \phi_n^* \langle \tilde{\delta} B_n \rangle = 0, \tag{19.64}$$

where ϕ_n^* is the solution of the risk optimization problem, and is given by:

$$\phi_n^* = \sum_{m \neq N} (\tilde{C})_{nm}^{-1} C_{mN}, \tag{19.65}$$

with \tilde{C} is the matrix C with the row and column associated to N suppressed. Inject in Eq. (19.64) the ansatz $\langle \tilde{\delta} B_n \rangle = \lambda C_{nn_0}$. Then:

$$\sum_{n \neq N} \phi_n^* \langle \tilde{\delta} B_n \rangle = \lambda \sum_{m \neq N} C_{mN} \sum_{n \neq N} (\tilde{C})_{nm}^{-1} C_{nn_0} = \lambda \sum_{m \neq N} C_{mN} \delta_{mn_0}. \qquad (19.66)$$

Provided $n_0 \neq N$, one has:

$$\lambda \sum_{m \neq N} C_{mN} \delta_{mn_0} = \lambda C_{Nn_0} = \langle \tilde{\delta} B_N \rangle, \qquad (19.67)$$

and thus Eq. (19.64) is satisfied. Since n_0 must be different from all bonds traded in the markets, the only choice is $n_0 = 1$, corresponding to the short rate itself.

- FURTHER READING

M. Baxter, A. Rennie, *Financial Calculus, An Introduction to Derivative Pricing*, Cambridge University Press, Cambridge, 1996.

D. Brigo, F. Mercurio, *Interest Rate Models Theory and Practice: Theory and Practice*, Springer Finance, 2001.

L. Hughston (ed.), *Vasicek and Beyond*, Risk Publications, London, 1996.

L. Hughston (ed.), *New Interest Rate Models*, Risk Publications, London, 2001.

J. C. Hull, *Futures, Options and Other Derivatives*, Prentice Hall, Upper Saddle River, NJ, 1997.

M. Musiela, M. Rutkowski, *Martingale Methods in Financial Modelling*, Springer, New York, 1997.

R. Rebonato, *Interest-Rate Option Models: Understanding, Analyzing and Using Models for Exotic Interest-Rate Options*, Wiley Series in Financial Engineering, 1998.

- SPECIALIZED PAPERS

B. E. Baaquie, *Quantum Finance*, Cambridge University Press, 2004.

B. E. Baaquie and J.-P. Bouchaud, *Stiff field theory of interest rates and psychological time*, Wilmott Magazine, March 2004.

M. Baxter, *General Interest-Rate Models and the Universality of HJM*, in Mathematics of Derivative Securities, M. A. H. Dempster and S. R. Pliska. Cambridge University Press, 1997, p. 315.

J.-P. Bouchaud, N. Sagna, R. Cont, N. ElKaroui, M. Potters, *Phenomenology of the interest rate curve*, Applied Mathematical Finance, **6**, 209 (1999) and *Strings attached*, RISK Magazine, **11** (7), 56 (1998).

A. Brace, D. Gatarek, M. Musiela, *The market model of interest rate dynamics*, Math Finance, **7**, 127 (1997).

R. Cont, *Modelling the term structure of interest rates: an infinite dimensional approach*, to appear in International Journal of Theoretical and Applied Finance.

T. Di Matteo, T. Aste, *How does the Eurodollar interest rate behave?* Int. Jour. Theo. Appl. Fin., **5**, 107 (2002).

D. Heath, R. Jarrow and A. Morton, *Bond pricing and the term structure of interest rates: A new methodology for contingent claim valuation*, Econometrica, **60**, (1992), 77–105.

J. Hull and A. White, *Numerical procedures for implementing term structure models II: two-factor models*, Journal of Derivatives, **Winter**, 37 (1994).

D. P. Kennedy, *Characterizing Gaussian models of the term structure*, Mathematical Finance, **7**, 107 (1997).

A. Matacz, J.-P. Bouchaud, *Explaining the forward interest rate term structure*, International Journal of Theoretical and Applied Finance, **3**, 381 (2000a).

A. Matacz, J.-P. Bouchaud, *An empirical study of the interest rate curve*, International Journal of Theoretical and Applied Finance, **3**, 703 (2000b).

L. C. G. Rogers, *Which Model for Term-Structure of Interest Rates Should One Use?* in M. Davis, D. Duffie, W. Fleming and S. Shreve, eds. Mathematical Finance. IMA Vol. Math. Appl., **65** (New York: Springer-Verlag, 1995).

P. Santa-Clara, D. Sornette, *The dynamics of the forward rate curve with stochastic string shocks*, The Review of Financial Studies, **14**, 149 (2001).

20

Simple mechanisms for anomalous price statistics

In the future, all models will be right, but each one only for 15 minutes.
(E. Derman, after Andy Warhol, *Quantitative Finance (2001)*.)

20.1 Introduction

Physicists like using power-laws to fit data. The reason for this is that complex, collective phenomenon often give rise to power-laws which are furthermore *universal*, that is to a large degree independent of the microscopic details of the phenomenon. These power-laws emerge from collective action and transcend individual specificities. As such, they are unforgeable signatures of a collective mechanism. Examples in the physics literature are numerous. A well-known example is that of phase transitions, where a system evolves from a disordered state to an ordered state: many observables behave as universal power laws in the vicinity of the transition point (see e.g. Goldenfeld (1992)). This is related to an important property of power-laws, namely 'scale invariance': the characteristic length scale of a physical system at its critical point is infinite, leading to self-similar, scale-free fluctuations. Similarly, power-law distributions are scale invariant in the sense that the relative probability to observe an event of a given size s_0 and an event ten times larger is independent of the reference scale s_0. Another interesting example is fluid turbulence, where the statistics of the velocity field has scale invariant properties, to a large extent independent of the geometry, the nature of the fluid, of the power injected, etc. (see Frisch (1997)).

As reviewed in Chapters 6, 7, power-laws are also often observed in economic and financial data. For example, the tail of the distribution of high frequency stock returns is found to be well fitted by a power-law with an exponent $\mu \approx 3$. The temporal decay of volatility correlations is also found to be described by a slow power law. Compared to physics, however, much less efforts have yet been devoted to understanding these power-laws in terms of 'microscopic' (i.e. agent based) models and to relating the value of the exponents to some generic mechanisms. The aim of this chapter is to discuss several simple models which naturally lead to power-laws and could serve as a starting point for further developments. It should be stressed that none of the models presented here are intended to be fully realistic and complete, but are of pedagogical interest: they nicely illustrate how and when power-laws can arise. The role of models is not necessarily to provide an

exact description of reality. Their primary goal is to set a framework within which relevant questions, that are of wider scope than the model itself, can be formulated and answered.

20.2 Simple models for herding and mimicry

We first describe some very simple models for thick tails in the distribution of price increments in financial markets. An intuitive explanation is herding: if a large number of agents acting on markets coordinate their action, the price change is likely to be huge due to a large unbalance between buy and sell orders. However, this coordination can result from two rather different mechanisms.

- One is the feedback of past price changes onto themselves, which we will discuss in the following section. Since all agents are influenced by the very *same* price changes, this can induce non trivial collective behaviour: for example, an accidental price drop can trigger large sell orders, which lead to further downward moves.
- The second is direct influence between the traders, through exchange of information that leads to 'clusters' of agents sharing the same decision to buy, sell, or be inactive at any given instant of time.

20.2.1 Herding and percolation

A simple model of how herding affects the price fluctuations is based on the following ingredients. First, assume that the price increment δx depends linearly on the instantaneous offset between supply and demand. More precisely, if each operator in the market i wants to buy or sell a certain fixed quantity of the financial asset, one has:[†]

$$\delta x = \frac{1}{\lambda} \sum_i \varphi_i \equiv \frac{\Phi}{\lambda}, \qquad (20.1)$$

where φ_i can take the values -1, 0 or $+1$, depending on whether the operator i is selling, inactive, or buying, and λ is a measure of the market depth. Recent empirical analysis seems to confirm that the relation between price changes δx and order imbalance $\sum_i \varphi_i$, averaged over a time scale of several minutes, is indeed linear for small arguments, but bends down and perhaps even saturates for larger arguments.[‡]

Suppose now that the operators interact with one another in an heterogeneous manner: with a small probability p/\mathcal{N} (where \mathcal{N} is the total number of operators on the market), two operators i and j are 'connected', and with probability $1 - p/\mathcal{N}$, they ignore each other. The factor $1/\mathcal{N}$ means that on average, the number of operator connected to any particular

[†] This can alternatively be written for the relative price increment $\eta = \delta x / x$, which is more adapted to describe long time scales. On short time scales, however, an additive model is preferable. See the discussion in Chapter 8.

[‡] On this point, and on the possible relation with the value $\mu \approx 3$, see the recent paper by Gabaix et al. (2002).

one is equal to p. Suppose finally that if two operators are connected, they will come to agree on the strategy they should follow, i.e. $\varphi_i = \varphi_j$.

It is easy to understand that the population of operators clusters into groups sharing the same opinion. These clusters are defined such that there exists a connection between any two operators belonging to this cluster, although the connection can be indirect and follow a certain 'path' between operators. These clusters do not all have the same size, i.e. do not contain the same number of operators. If the size of cluster \mathcal{C} is called $S(\mathcal{C})$, one can write:

$$\delta x = \frac{1}{\lambda} \sum_{\mathcal{C}} S(\mathcal{C})\varphi(\mathcal{C}), \qquad (20.2)$$

where $\varphi(\mathcal{C})$ is the common opinion of all operators belonging to \mathcal{C}. The statistics of the price increments δx therefore reduces to the statistics of the size of clusters, a classical problem in percolation theory (Stauffer & Aharony (1994)). One finds that as long as $p < 1$ (less than one 'neighbour' on average with whom one can exchange information), then all $S(\mathcal{C})$'s are small compared to the total number of traders \mathcal{N}. More precisely, the distribution of cluster sizes takes the following form in the limit where $1 - p = \epsilon \ll 1$:

$$P(S) \propto_{S \gg 1} \frac{1}{S^{5/2}} \exp -\epsilon^2 S \qquad S \ll \mathcal{N}. \qquad (20.3)$$

When $p = 1$ (percolation threshold), the distribution becomes a pure power-law with an exponent $\mu = 3/2$, and the Central Limit Theorem tells us that in this case, the distribution of the price increments δx is precisely a pure symmetrical Lévy distribution of index $\mu = 3/2$ (assuming that $\varphi = \pm 1$ play identical roles, that is if there is no global bias pushing the price up or down). If $p < 1$, on the other hand, one finds that the Lévy distribution is truncated exponentially, leading to a larger effective tail exponent μ. If $p > 1$, a finite fraction of the \mathcal{N} traders have the same opinion: this leads to a crash. Very recently, a somewhat related model, where each agent probes the opinion of a pool of m randomly selected agents, was introduced by Eckmann et al. (2002). The agent then chooses either to conform to the majority opinion or to be contrarian if the majority is too strong. This interesting model leads to various types of market behaviour, from periodic to chaotic.

20.2.2 Avalanches of opinion changes

The above simple percolation model is interesting but has one major drawback: one has to assume that the parameter p is smaller than one, but relatively close to one such that Eq. (20.3) is valid, and non trivial statistics follows. One should thus explain why the value of p spontaneously stabilizes in the neighbourhood of the critical value $p = 1$. Furthermore, this model is purely static, and does not specify how the above clusters evolve with time. As such, it cannot address the problem of volatility clustering. Several extensions of this simple model have been proposed, in particular to increase the value of μ from $\mu = 3/2$ to $\mu \sim 3$ and to account for volatility clustering.

One particularly interesting model is inspired by the recent work of Dahmen & Sethna (1996), that describes the behaviour of random magnets in a time dependent magnetic field.

Transposed to market dynamics, this model describes the collective behaviour of a set of traders exchanging information, but having all different *a priori* opinions, say optimistic (buy) or pessimistic (sell). One trader can however change his mind and take the opinion of his neighbours if the herding tendency is strong, or if the strength of his *a priori* opinion is weak. More precisely, the opinion $\varphi_i(t)$ of agent i at time t is determined as:

$$\varphi_i(t) = \text{sign}\left(h_i(t) + \sum_{j=1}^{N} J_{ij}\varphi_j(t)\right), \tag{20.4}$$

where J_{ij} is a connectivity matrix describing how strongly agent j influences agent i (in the above percolation model, J_{ij} was either 0 or J), and $h_i(t)$ describes the *a priori* opinion of agent i: $h_i > 0$ means a propensity to buy, $h_i < 0$ a propensity to sell. We assume that h_i is a random variable with a time dependent mean $\overline{h}(t)$ and root mean square Δ. The quantity $\overline{h}(t)$ is the average optimism, for example confidence in long term economy growth ($\overline{h}(t) > 0$), or fear of recession ($\overline{h}(t) < 0$). Suppose that one starts at $t = 0$ from a 'euphoric' state, where $\overline{h} \gg \Delta, J$, such that $\varphi_i = 1$ for all i's. Here J denotes the order of magnitude of $\sum_j J_{ij}$. Now, confidence is smoothly decreased. Will sell orders appear progressively, or will there be a sudden panic where a large fraction of agents want to sell?

Interestingly, one finds that for small enough influence (or strong heterogeneities of agents' anticipations), i.e. for $J \ll \Delta$, the average opinion $O(t) = \sum_i \varphi_i(t)/N$ evolves continuously from $O(t = 0) = +1$ to $O(t \to \infty) = -1$ (see Figure 20.1). The situation changes when imitation is stronger since a discontinuity then appears in $O(t)$ around a

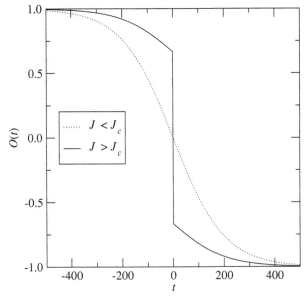

Fig. 20.1. Schematic evolution of the average opinion as a function of external news in the Dahmen-Sethna model. For small enough mutual influence, $J < J_c$, the shift from optimism to pessimism is progressive, whereas for $J > J_c$, it occurs suddenly.

'crash' time t_c, when a *finite* fraction of the population simultaneously change opinion. The gap $O(t_c^-) - O(t_c^+)$ opens continuously as J crosses a critical value $J_c(\Delta)$. For J close to J_c, one finds that the sell orders again organize as avalanches of various sizes, distributed as a power-law with an exponential cut-off. In the fully connected case where $J_{ij} \equiv J/\mathcal{N}$ for all ij, one finds an exponent $\mu = 5/4$. Note that in this case, the value of the exponent μ is to a large extent *universal*, and does not depend, for example, on the detailed shape of the distribution of the h_i's, but only on some global properties of the connectivity matrix J_{ij}.

A slowly oscillating $\bar{h}(t)$ can therefore lead to a succession of bull and bear markets, with a strongly non Gaussian, intermittent behaviour, since most of the activity is concentrated around the crash times t_c.

The above model is very interesting because it may describe several situations where there is a conflict between personal opinions and social pressure. For example, the way mobile phones invaded the French market is qualitatively similar to the plot shown in Fig. 20.1. In this case, $O(t)$ would be the number of persons without mobile phones.

20.3 Models of feedback effects on price fluctuations

20.3.1 Risk-aversion induced crashes

Set up of the model

The above average 'stimulus' $\bar{h}(t)$ may also strongly depend on the past behaviour of the price itself. For example, past positive trends are, for many investors, incentives to buy, and vice-versa. Since the 'true' price of a stock is difficult to estimate, investors are tempted to extract from past price changes themselves some information that other agents might possess. This is a rational decision if indeed some other agents have this information, but is also the mechanism by which bubbles and excess volatilities are created.

Actually, for a given past trend amplitude, price drops tend to feedback more strongly on the investors behaviour than price rises. We have discussed some indirect empirical evidence for this in Chapter 8, when the leverage effect was discussed. Risk-aversion indeed creates an asymmetry between positive and negative price changes. This is also reflected in option markets, where the price of out-of-the-money puts (i.e. insurance against crashes) is anomalously high.

Similarly, past periods of high volatility increase the risk of investing in stocks, and simple portfolio theories (such as discussed in Chapter 12) then suggest that sell orders should follow. A simple mathematical transcription of these effects is to supplement Eq. (20.1) with an dynamical equation for the excess demand Φ_t which encodes the above feedback effects:

$$\Phi_{t+1} - \Phi_t = m + a\,\delta x_t - b\,\delta x_t^2 - a'\,\delta x_t - K(x_t - x_0) + \chi \xi_t, \tag{20.5}$$

where:

- m is the average increase (or decrease) of the overall optimism $\bar{h}(t)$ between t and $t + 1$, that tends to increase (decrease) the excess demand. We will set to m zero in the following.

- $a > 0$ describes trends following effects, that is how past price increments amplifies the average propensity to buy or sell.
- $b > 0$ describes risk aversion effects, and encodes the fact that price drops are given more weight than price increases: the term $-b\,\delta x_t^2$ can be seen as giving to the parameter a a dependence on δx_t, such that a increases when the price goes down. Alternatively, δx_t^2 can be seen as a proxy for the short term volatility; an increase of volatility should decrease the demand for the stock. The most important point is that a term proportional to δx_t^2 breaks the $\delta x_t \rightarrow -\delta x_t$ symmetry.
- $a' > 0$ is a stabilizing term (which has an effect opposite to that of a) that arises from market clearing mechanisms: the very fact that the price moves clears a certain number of orders, and reduces the order imbalance. In fact, because the size of the queue in the order book is on average an increasing function of the distance from the current price, one expects non linear corrections to this market clearing argument, that will add terms of higher order in δx_t (see below).
- $K > 0$ is a mean-reverting force that models the fact that if the price x_t is too far above a reference fundamental price x_0, sell orders will be generated, and vice-versa.
- Finally, ξ is a random variable noise term representing random external news that induces more buy (or sell) orders. The coefficient χ can be seen as the susceptibility to these news. A more refined version of the theory should allow this coefficient itself to depend on past values of δx: an increase of the volatility tends to make agents more nervous and feeds back onto itself, leading to volatility clustering. A simple model of this type will be explored in Section 20.3.2 below.

Before analyzing the properties of the model defined by Eqs. (20.1, 20.5), one should add the following remarks: (i) we consider here a model for absolute price changes, which is suitable for short time scales. On longer time scales, the model should rather be expressed in terms of log prices. This was discussed at length in Chapter 8; (ii) Eq. (20.5) can be thought of as an expansion in powers of the price increment δx_t. One therefore expects further terms, in particular a term of the form $-c\delta x_t^3$ that plays a stabilizing role in some circumstances. Eliminating Φ_t from the equations (20.1) and (20.5), and taking the continuous time limit, one derives a 'Langevin' equation for the price 'velocity' $u \equiv dx/dt$:

$$\frac{du}{dt} = -\alpha u - \beta u^2 - \gamma u^3 - \kappa(x - x_0) + \tilde{\xi}(t) \tag{20.6}$$

where $\alpha = (a' - a)/\lambda$, $\beta = b/\lambda$, $\gamma = c/\lambda$ $\kappa = K/\lambda$ and $\tilde{\xi}(t) = \chi\xi(t)/\lambda$.

This simple equation encapsulates a number of interesting phenomena. In physical terms, equation (20.6) represents the Langevin dynamics of the position u of a fictitious particle in a time dependent 'potential' (see Fig. 20.2):

$$V(u) = \kappa(x - x_0)u + \alpha\frac{u^2}{2} + \beta\frac{u^3}{3} + \gamma\frac{u^4}{4} + \cdots, \tag{20.7}$$

such that

$$\frac{du}{dt} = -\frac{\partial V}{\partial u} + \tilde{\xi}(t). \tag{20.8}$$

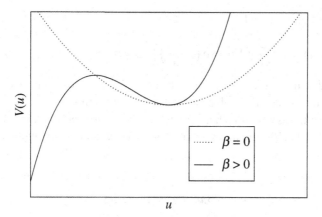

Fig. 20.2. Effective potential $V(u)$ in which the fictitious 'particle' (representing price increments) evolves. Here, we have set $\kappa = \gamma = 0$. For $\beta > 0$, corresponding to risk aversion effects, the potential develops a 'barrier' separating a stable region for $u \sim 0$ from an unstable 'crash' region for $u < u^{*}$.

The linear model

Let us first assume that non linear effects, captured by the coefficients β and γ, are absent, or negligible. The equation giving the evolution of the price then reduces to a forced harmonic oscillator with friction:

$$\frac{d^2 x}{dt^2} = -\alpha \frac{dx}{dt} - \kappa (x - x_0) + \tilde{\xi}(t). \qquad (20.9)$$

For the linear process to be stable, one has to assume that $\alpha > 0$, i.e. that trend following effects are not too strong. On short time scales, mean reversion effects to the fundamental price are expected to be small. The above equation then describes the price changes as an Ornstein-Uhlenbeck process; the price increment correlation function is given by:

$$\langle u(t)u(t + \tau) \rangle = \frac{D}{2\alpha} \mathrm{e}^{-\alpha \tau}, \qquad (20.10)$$

where D is the variance of $\tilde{\xi}(t)$. The volatility $\langle u(t)^2 \rangle$ is equal to $D/2\alpha \propto \chi^2/\lambda(a' - a)$. This means that the volatility is higher when the market participants react strongly to external news (χ large) or when the market is not very deep (λ small), as expected. We also find that the volatility diverges as the trend following effects (a) tend to outweigh the market liquidity (a').

The above result, Eq. (20.10) also shows that for time lags much larger than $1/\alpha$ (but sufficiently small such that κ can be neglected), price increments are uncorrelated and the price behaves diffusively. On shorter time scales, however, some positive correlations appear, that lead to a superdiffusive behaviour.

On very long time scales, on the other hand, the price difference $x - x_0$ becomes large and κ cannot be neglected. The process becomes an Ornstein-Uhlenbeck on the price itself,

and the price variogram reads:

$$\langle (x(t+\tau) - x(t))^2 \rangle = \frac{D}{\kappa\alpha} \left(1 - \exp\left(-\frac{\kappa\tau}{\alpha}\right) \right) \qquad \alpha\tau \gg 1. \tag{20.11}$$

In this model, the price is therefore superdiffusive on short time scales, diffusive on intermediate time scales and subdiffusive on very long time scales. On very long time scales, however, the assumption that the reference price x_0 is constant becomes unjustified, and one should then also take into account the (possibly random) evolution of the reference price itself.

Bubbles and crashes

Suppose now that feedback effects are strong compared to the stabilizing effects: $a > a'$ and $b > 0$, such that $\alpha < 0$ and $\beta > 0$. Suppose that the initial price is $x \approx x_0$ and that $\gamma = 0$ for simplicity. The above potential develops a minimum for a non zero value of u given by $u^* = -\alpha/\beta > 0$. The fictitious 'particle' hovers around this minimum. This means that price returns fluctuate around a *positive* mean u^* that has no economic justification (recall that a 'true' rise of future expectations, modeled by the parameter m, is chosen to be zero here). This growth is entirely sustained because of trend following effects: this is a speculative bubble. In a first approximation, the price (or log price) grows linearly with time, as $u^* t$. Interestingly, when t reaches a time t_c given, within this simple approximation, by $t_c = \alpha^2/4\beta\kappa u^* = -\alpha/4\kappa$, the minimum of the potential ceases to exist and u flows to $-\infty$: the bubble collapses and the price plummets. The price decay is again stopped by the mean reverting 'fundamental' term κ. Not surprisingly, the bubble lifetime t_c is long if trend following effects are strong ($|\alpha|$ large) and/or if the mean reverting fundamental effect is weak (κ small).

If trend following effects are not too strong, α is positive and $V(u)$ has a local minimum for $u = 0$, and a local maximum for $u^* = -\alpha/\beta < 0$, beyond which the potential decreases to $-\infty$. A 'potential barrier' $V^* = \alpha u^{*2}/6$ separates the (meta-)stable region around $u = 0$ from the unstable region $u < u^*$ (see Fig. 20.2). The nature of the motion of u in such a potential is the following: starting at $u = 0$, $x = x_0$, the fictitious particle oscillates in the vicinity of $u = 0$, corresponding to a Brownian-like motion of the price itself. However, occasionally, an 'activated' event (i.e. driven by the news/noise term $\tilde{\xi}(t)$) brings the particle near u^*. Once the potential barrier is crossed, the fictitious particle reaches $u = -\infty$ in finite time. (The stabilizing term γu^3 would actually prevent this divergence and create a secondary minimum for a negative value of u, corresponding to the crash regime.) In financial terms, the regime where u oscillates around $u = 0$ and where β can be neglected, is the 'normal' fluctuation regime. This normal regime can however be interrupted by 'crashes', where the time derivative of the price becomes very large and negative, due to the risk aversion term b which destabilizes the price by amplifying the sell orders. The interesting point is that these two regimes can be clearly separated since the average time t^* needed for such crashes to occur is exponentially large in V^*/D (corresponding to the time needed to overcome the 'barrier' V^*), and can thus appear only very rarely. A very long time scale is therefore naturally generated in this model. Note that in this line of thought,

a crash occurs because of an improbable succession of unfavorable events, and not due to a single large event in particular. However, volatility feedback effects mean that stronger and stronger fluctuations might precede the crash, each large fluctuation increasing D and making the crash more probable.

20.3.2 A simple model with volatility correlations and tails

In this section, we show that a very simple feedback model where past high values of the volatility influence the present market activity does lead to tails in the probability distribution and, by construction, to volatility correlations. The present model is close in spirit to the ARCH models which have been much discussed in this context. The idea is to write, as in Section 7.1.2:

$$\delta x_k = x_{k+1} - x_k = \sigma_k \xi_k, \tag{20.12}$$

where ξ_k is a random variable of unit variance, and to postulate that the present day volatility σ_k depends on how the market feels the past market volatility. If the past price variations happened to be high, the market interprets this as a reason to be more nervous and increases its activity, thereby increasing σ_k. One could therefore consider, as a toy-model of the feedback, the following dynamical equation:[†]

$$\sigma_{k+1} - \sigma_0 = (1 - \epsilon)(\sigma_k - \sigma_0) + g\epsilon |\sigma_k \xi_k|, \quad 0 < \epsilon < 1 \tag{20.13}$$

which means that the volatility would mean return towards an 'equilibrium' value $\sigma = \sigma_0$, but is continuously 'excited' by the observation of the past day activity through the absolute value of the close to close price difference $x_{k+1} - x_k$. The coefficient g measures the strength of the feedback coupling. Now, writing

$$|\sigma_k \xi_k| = \sigma_k(\langle |\xi| \rangle + \tilde{\xi}), \tag{20.14}$$

with $\tilde{\xi}$ a centered noise, and going to a continuous-time formulation, one finds that the volatility probability distribution $P(\sigma, t)$ obeys (for $g \ll 1$) the following Fokker–Planck equation:

$$\frac{\partial P(\sigma, t)}{\partial t} = \epsilon \frac{\partial (\sigma - \sigma_0) P(\sigma, t)}{\partial \sigma} + Dg^2 \epsilon^2 \frac{\partial^2 \sigma^2 P(\sigma, t)}{\partial \sigma^2}, \tag{20.15}$$

where D is the variance of the noise $\tilde{\xi}$. The equilibrium solution of this equation, $P_{\rm e}(\sigma)$, is obtained by setting the left-hand side to zero. One finds:

$$P_{\rm e}(\sigma) = \frac{\exp(-\sigma_0/\sigma)}{\Gamma(\mu)\sigma^{1+\mu}}, \tag{20.16}$$

with $\mu = 1 + (Dg^2 \epsilon)^{-1} > 1$. This is an inverse Gamma distribution, which goes to zero very fast when $\sigma \to 0$, but has a long tail for large volatilities. Qualitatively, the empirical

[†] In the simplest ARCH model, the following equation is rather written in terms of the variance, and second term of the right-hand side is taken to be equal to: $\epsilon(\sigma_k \xi_k)^2$ – see Bollerslev et al. (1994).

distribution of volatility can indeed, as shown in Section 7.2.2, be quite accurately fitted by an inverse Gamma distribution.

For a large class of distributions for the random noise ξ_k, for example Gaussian, one can show, using a saddle-point calculation, that the tails of the distribution of σ are bequeathed to those of δx. The distribution of price changes has thus power-law tails, with the same exponent μ. Interestingly, a short-memory market, corresponding to $\epsilon \approx 1$, has much wilder tails than a long-memory market: in the limit $\epsilon \to 0$, one indeed has $\mu \to \infty$. In other words, over-reactions is a potential cause for power-law tails. Similarly, strong feedback (larger g) decreases the value of μ.

20.3.3 Mechanisms for long ranged volatility correlations

The temporal correlation function of the volatility σ can be exactly calculated within the above model,[†] and is found to be exponentially decaying (as in the Heston model, see Section 8.4), at variance with the slow power-law or logarithmic decay observed empirically.

At this point, the slow decay of the volatility can be ascribed to two rather different mechanisms. One is the existence of traders with many different time horizons. If traders are affected not only by yesterday's price change amplitude $|x_k - x_{k-1}|$, but also by price changes on coarser time scales $|x_k - x_{k-p}|$ ($p > 1$), then the correlation function is expected to be a sum of exponentials with decay rates given by p^{-1}. Interestingly, if the different p's are uniformly distributed on a log scale, the resulting sum of exponentials is to a good approximation decaying as a logarithm. More precisely, the following relation holds:

$$\frac{1}{\log(p_{max}/p_{min})} \int_{p_{min}}^{p_{max}} d(\log p) \exp(-\tau/p) \approx \frac{\log(p_{max}/\tau)}{\log(p_{max}/p_{min})}, \tag{20.17}$$

whenever $p_{min} \ll \tau \ll p_{max}$. This logarithmic behaviour can be compared to the logarithmic form of the (log-)volatility correlation function assumed in multifractal descriptions (see Section 7.3.1). Now, the human time scales are indeed in a natural quasi-geometric progression: hour, day, week, month, trimester, year. This could be a simple explanation – almost trivial – for the quasi-logarithmic (or slow power-law) behaviour of the volatility correlation.[‡]

Yet, some recent numerical simulations of models where traders are allowed to switch between different strategies (active/inactive, or chartist/fundamentalist) also suggest strongly intermittent behaviour, and a slow decay of the volatility correlation function without the explicit existence of logarithmically distributed time scales. Is there a simple, universal mechanism that could explain these ubiquitous long range volatility correlations?

A possibility is that the volume of activity exhibits long range correlations because agents switch between different strategies depending on their relative performance. Imagine for example that each agent has two strategies, one active strategy (say trading every day), and one inactive, or less active strategy. The 'score' of the inactive strategy (i.e. its cumulative profit) is constant in time, or more precisely equal to the long term average interest rate. The

[†] On this point, see: Graham & Schenzle (1982) and Bouchaud & Mézard (2000).

[‡] The fingerprints, in real markets, of these human time scales have recently been unveiled in Zumbach & Lynch (2001).

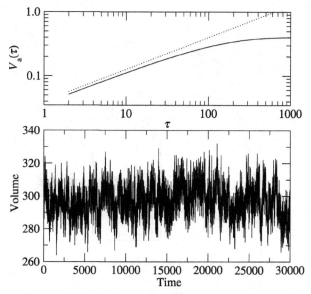

Fig. 20.3. Bottom: Total daily volume of activity (number of players) in the Minority Game with inactive players. Top: Variogram of the activity as a function of the time lag, τ, in log-log coordinates. The dotted line is the square-root singularity predicted by Eq. (20.18).

score of active strategy, on the other hand, fluctuates up and down, due to the fluctuations of the market prices themselves. Since to a good approximation the market prices are not predictable, this means that the score of any active strategy will behave like a random walk, with an average equal to that of the inactive strategy (assuming that transaction costs are small). Therefore, on some occasions the score of the active strategy will happen to be higher than that of the inactive strategy and the agent will be active, before the score of the active strategy crosses that of the inactive strategy. The time during which an agent is active is thus a random variable with the same statistics as the return time to the origin of a random walk (the difference of the scores of the two strategies). Interestingly, the return times t_r of a random walk are well known to be very broadly distributed, asymptotically as $t_r^{-1-\mu}$ with $\mu = 1/2$, such that the *average* return time is infinite. Hence, if one computes the correlation of activity in such a model, one finds long range correlations due to long periods of times where many agents are active (or inactive).[†]

This 'random time strategy shift' scenario can be studied more quantitatively in the framework of simple models, such as the Minority Game model, introduced below. In the version where agents can refrain from playing, fluctuations of the volume of activity $v(t)$, due to the above mechanism, are observed: see Figure 20.3. More precisely, the volume variogram is approximately given by:

$$V(\tau) = \langle (v(t) - v(t + \tau))^2 \rangle = V_\infty(1 - \exp(-\alpha\sqrt{\tau})), \tag{20.18}$$

[†] For more details on this argument, see: Bouchaud et al. (2001).

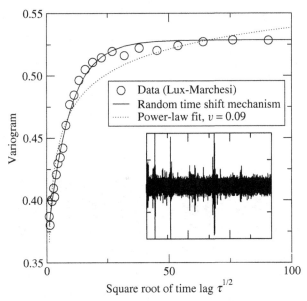

Fig. 20.4. Variogram of the absolute returns of the Lux-Marchesi model as a function of the square root of the time lag, $\sqrt{\tau}$, and fit using Eq. 20.18. Inset: Price returns time series, exhibiting volatility clustering. Data: courtesy of Thomas Lux.

where the small τ square root singularity is related to the power-law decay of the return time of a one dimensional random walk (here, the score of the active strategy). Perhaps surprisingly, this simple model also allows one to reproduce quantitatively the volume of activity correlations observed on the New-York stock exchange market (see Fig. 7.12). This mechanism is very generic and probably also explains why this effect arises in more realistic market models: the only ingredient is the coexistence of different strategies with different levels of activity (active/inactive, high frequency chartists/low frequency fundamentalists, etc.), which switch from one to another at times determined by the return times of a random walk (the relative scores of the strategies, the price itself, etc.). In that respect, it is interesting to note that the same quantitative behaviour is seen in a simple agent based market model introduced by Lux & Marchesi (1999, 2000), where agents can alternatively be chartist and fundamentalist, according to the best performing strategy – see Fig. 20.4. Needless to say, more work is needed to establish this simple scenario as the genuine origin of long range volatility correlations in financial markets.

20.4 The Minority Game

Agent-based market models, where a collection of agents with simple behavioral rules interact and give rise to a fictitious price history, were pioneered in the beginning of the nineties, and are now intensively studied. The **Minority Game** was introduced in Challet and Zhang (1997) as an extremely simplified model where the consequence of agents *heterogeneity* can be studied in details. Although rather remote from real financial markets,

the structure of the model is extremely rich and allows one to investigate several relevant issues.

The 'Minority Game' consists, for each agent, to choose at each time step between two possibilities, say ± 1. An agent wins if he makes the minority choice, i.e if he chooses at time t the option that the minority of his fellow players also choose. The game is interesting because by definition, not everybody can win.

Each agent takes his new decision based on the past history – say on the last M outcomes of the game. A *strategy* maps a string of M results into a decision. The number of strings is 2^M; to each of them can be associated $+1$ or -1. Therefore, the total number of strategies is 2^{2^M}.

Each agent is given from the start a certain number of strategies from which he can try to determine the best one. His 'bag' of strategies is fixed in time – he cannot try new ones or slightly modify the ones he has in order to perform better. The only thing he can do is to try to rank the strategies according to their past performance. So he gives scores to his different strategies, say $+1$ every time a strategy gives the correct result, -1 otherwise. The crucial point here is that he assigns these scores to *all* his strategies depending on the outcome of the game, as if these strategies had been effectively played. Doing this neglects the fact that the outcome of the game is in fact affected by the strategy that the agent is actually playing at time t.

The parameters of this model are: the number of agents N, the number of history strings 2^M, and the number of strategies per agent. The last parameter does not really affect the results of the model, provided it is larger than one. In the limit of large N and M, the only relevant parameter is $\alpha = 2^M/N$. For $\alpha > \alpha_c$, where α_c can be computed exactly (and is found to be equal to $0.33740\ldots$ when the number of strategies is equal to two, see Challet et al. (2000a)), the game is 'predictable' in the sense that conditionally to a given history h, the winning choice w is statistically biased towards $+1$ or -1. One can introduce a degree of predictability q, defined as:[†]

$$q = \frac{1}{2^M} \sum_{h=1}^{2^M} \langle w|h \rangle^2 \tag{20.19}$$

is non zero for $\alpha > \alpha_c$, and goes continuously to zero for $\alpha \to \alpha_c^+$, where the game becomes unpredictable by the agents (however, agents with, say, a longer memory $M' > M$ would still be able to extract information from the history). Correspondingly, for $\alpha > \alpha_c$, the score of the strategies behave as *biased* random walks, with a positive or negative drift. In this predictable (or 'inefficient') phase, some strategies perform systematically better than others. In the unpredictable (or 'efficient') phase $\alpha < \alpha_c$, on the other hand, the score of all strategies behave as unbiased random walks. In particular, the time during which an agent keep playing a given strategy is related to the first passage time problem of a random walk, and is therefore a power-law variable with an exponent $\mu = 1/2$, of infinite mean. This remark is related to the mechanism for long range volatility fluctuations discussed in the previous section. When the game is extended as to allow agents not too play if all their

[†] The definition of q is similar to the definition of the Edwards–Anderson order parameter in spin-glasses, see Mézard et al. (1987). Not taking the square of $\langle w|h \rangle$ would give zero trivially, because of the symmetry of the game between $w = 1$ and $w = -1$.

strategies have negative scores, one finds, as a function of α, an efficient active phase where a finite fraction of agents play. In this phase, the duration of active periods of any given agent follows a power-law distribution with an exponent $\mu = 1/2$, thereby producing non trivial temporal correlations in the total volume of activity (See Fig. 20.3).

The point $\alpha = \alpha_c$ is a critical point corresponding to a second order phase transition, of the kind often encountered in statistical physics. The critical nature of the problem at and around this point leads to interesting properties, such as power-laws distribution and correlations, that may have some relations with the power-laws observed in financial time series. Of course, the Minority Game as such is too simple to be compared to financial markets: there is no price and no exchange, and it is not clear to what extent the minority rule directly applies to financial markets. One can argue that the optimal strategy is to act like the majority for a while, but to pull out of the market as the minority. Notwithstanding, there are several simple extensions of the Minority Game that reproduce, in the vicinity of the efficient/inefficient transition point, some of the stylized facts of financial time series.

However, as already discussed above in the context of the simple percolation model of herding, why should reality spontaneously 'self organize' such as to lie in the immediate vicinity of a critical point?[†] In the context of the Minority Game or its extensions, there is a rather natural answer: large α corresponds to a relatively small number of agents, and the resulting time series is to some extent predictable – and thus exploitable. This is an incentive for new agents to join the game, therefore reducing α, until no more profit can be extracted from the time series itself. This mechanism spontaneously tunes α down to α_c.[‡]

20.5 Summary

- Large fluctuations in financial time series can probably be ascribed to herding effects and/or price feedback mechanisms. Herding (or imitation) effects are captured at the simplest level by a percolation like model, where agents are randomly connected as to create (possibly large) clusters of correlated behaviour. A more sophisticated model also takes into account personal opinions, and leads to avalanches of opinion changes mediated by imitation.
- Price feedback mechanisms, on the other hand, illustrate how bubbles and crashes can emerge from trend following strategies, limited by fundamentalist mean-reverting effects. Volatility feedback effects generically leads to fat tails in the distribution of price changes.
- Simple agent-based models, where agents switch between different strategies, can also shed light on financial price series. In particular, the Minority Game where agents are allowed not to play, leads to a generic mechanism for long range volatility correlations.

[†] This question is of a general nature, and has been discussed in relation with 'Self-Organized Criticality': see Bak (1996).

[‡] A similar mechanism was argued to operate in a more sophisticated, agent-based model of markets. Small levels of investments lead to a very smooth, predictable evolution of the price, whereas the impact of larger levels of investment destabilizes the price dynamics and leads to an unpredictable time series. See the discussion in: Giardina & Bouchaud (2002).

- FURTHER READING

P. Bak, *How Nature Works: The Science of Self-Organized Criticality*, Copernicus, Springer, New York, 1996.

T. Bollerslev, R. F. Engle, D. B. Nelson, *ARCH models*, Handbook of Econometrics, vol 4, ch. 49, R. F. Engle and D. I. McFadden Edts, North-Holland, Amsterdam, 1994.

S. Chandraseckar, *Stochastic problems in physics and astronomy*, Rev. Mod. Phys., **15**, 1 (1943), reprinted in 'Selected papers on noise and stochastic processes', N. Wax Edt., Dover, 1954.

J. D. Farmer, *Physicists attempt to scale the ivory towers of finance*, in Computing in Science and Engineering, November 1999, reprinted in Int. J. Theo. Appl. Fin., **3**, 311 (2000).

J. Feder, *Fractals*, Plenum, New York, 1988.

U. Frisch, *Turbulence: The Legacy of A. Kolmogorov*, Cambridge University Press, 1997.

N. Goldenfeld *Lectures on phase transitions and critical phenomena*, Frontiers in Physics, 1992.

A. Kirman, *Reflections on interactions and markets*, Quantitative Finance, **2**, 322 (2002).

M. Levy, H. Levy, S. Solomon, *Microscopic Simulation of Financial Markets*, Academic Press, San Diego, 2000.

B. B. Mandelbrot, *The Fractal Geometry of Nature*, Freeman, San Francisco, 1982.

M. Mézard, G. Parisi, M. A. Virasoro, *Spin Glasses and Beyond*, World Scientific, 1987.

A. Orléan, *Le Pouvoir de la Finance*, Odile Jacob, Paris, 1999.

E. Samanidou, E. Zschischang, D. Stauffer, T. Lux, in *Microscopic Models for Economic Dynamics*, F. Schweitzer, Lecture Notes in Physics, Springer, Berlin, 2002.

R. Schiller, *Irrational Exuberance*, Princeton University Press, 2000.

A. Schleifer, *Inefficient Markets: An Introduction to Behavioral Finance*, Oxford University Press, Oxford, 2000.

J. P. Sethna, K. Dahmen, C. Myers, *Crackling noise*, Nature, **410**, 242 (2001).

D. Stauffer, A. Aharony, *Introduction to Percolation*, Taylor-Francis, London, 1994.

- SPECIALIZED PAPERS: POWER-LAWS

J. P. Bouchaud, M. Mézard, *Wealth condensation in a simple model of the economy*, Physica A, **282**, 536 (2000).

J. P. Bouchaud, *Power-laws in economics and finance: some ideas from physics*, Quantitative Finance, **1**, 105 (2001).

X. Gabaix, *Zipf's law for cities: an explanation*, Quarterly Journal of Economics, **114**, 739 (1999).

R. Graham, A. Schenzle, *Carleman imbedding of multiplicative stochastic processes*, Phys. Rev., **A 25**, 1731 (1982).

O. Malcai, O. Biham, S. Solomon, *Power-law distributions and Levy-stable intermittent fluctuations in stochastic systems of many autocatalytic elements*, Phys. Rev. E, **60**, 1299 (1999).

S. Solomon, P. Richmond, *Power laws of wealth, market order volumes and market returns*, Physica A, **299**, 188 (2001).

D. Sornette, R. Cont, *Convergent multiplicative processes repelled from zero: power laws and truncated power laws*, J. Phys. I France, **7**, 431 (1996).

D. Sornette, D. Stauffer, H. Takayasu, *Market fluctuations, multiplicative and percolation models, size effects and predictions*, in: The Science of Disaster, A. Bunde, H.-J. Schellnhuber, J. Kropp Edts, Springer-Verlag, 2002.

G. Zumbach, P. Lynch, *Heterogeneous volatility cascade in financial markets*, Physica A, **298**, 521, (2001).

- SPECIALIZED PAPERS: MARKET IMPACT

X. Gabaix, P. Gopikrishnan, V. Plerou, H. E. Stanley, *A Theory of Power-Law Distributions in Financial Markets Fluctuations*, Nature, **423**, 267 (2003).

A. Kempf, O. Korn, *Market depth and order size*, Journal of Financial Markets, **2**, 29 (1999).

F. Lillo, J. D. Farmer, R. N. Mantegna, *Master curve for price impact function*. Nature, **421**, 129 (2003).

V. Plerou, P. Gopikrishnan, X. Gabaix, H. E. Stanley, *Quantifying Stock Price Response to Demand Fluctuations*, Phys. Rev. E, **66**, 027104 (2002).

• SPECIALIZED PAPERS: HERDING AND FEEDBACK

A. Beja, M. B. Goldman, *The dynamic behavior of prices in disequilibrium*, Journal of Finance, **35**, 235 (1980).

J.-P. Bouchaud, R. Cont, *A Langevin approach to stock market fluctuations and crashes*, European Journal of Physics, **B 6**, 543 (1998).

R. Cont, J. P. Bouchaud, *Herd behavior and aggregate fluctuations in financial markets*, Macroeconomics Dynamics, **4**, 170 (2000). See also the numerous references of this paper for other works on herding in economics and finance.

A. Corcos, J.-P. Eckmann, A. Malaspinas, Y. Malevergne, D. Sornette, *Imitation and contrarian behavior: hyperbolic bubbles, crashes and chaos*, Quantitative Finance, **2**, 264 (2002).

K. Dahmen, J. P. Sethna, *Hysteresis, avalanches, and disorder induced critical scaling*, Physical Review B, **53**, 14 872 (1996).

J. D. Farmer, *Market Force, Ecology and Evolution*, e-print adap-org/9812005, Int. J. Theo. Appl. Fin., **3**, 425 (2000).

J. D. Farmer, S. Joshi, *The price dynamics of common trading strategies*, Journal of Economic Behavior and Organization, **49**, 149 (2002).

A. Kirman, *Ants, rationality and recruitment*, Quarterly Journal of Economics, **108**, 137 (1993).

A. Orléan, *Bayesian interactions and collective dynamics of opinion*, Journal of Economic Behaviour and Organization, **28**, 257 (1995).

B. Rosenow, *Fluctuations and Market Friction in Financial Trading*, Int. J. Mod. Phys. C., **13**, 419 (2002).

• SPECIALIZED PAPERS: AGENT BASED MODELS

P. Bak, M. Paczuski, M. Shubik, *Price variations in a stock market with many agents*, Physica A, **246**, 430 (1997).

S. Bornholdt, *Expectation bubbles in a spin model of markets: Intermittency from frustration across scales*, Int. J. Mod. Phys. C **12**, 667, (2001).

J. P. Bouchaud, I. Giardina, M. Mézard, *On a universal mechanism for long ranged volatility correlations*, Quantitative Finance, **1**, 212 (2001).

C. Chiarella, X.-Z. He, *Asset price and wealth dynamics under heterogeneous expectations*, Quantitative Finance, **1**, 509, (2001).

I. Giardina, J. P. Bouchaud, *Bubbles, crashes and intermittency in agent based market models*, European Journal of Physics, **B31**, 421 (2003); Physica A, **299**, 28.

C. Hommes, *Financial markets as nonlinear adaptive evolutionary systems*, Quantitative Finance, **1**, 149, (2001).

G. Iori *A microsimulation of traders activity in the stock market: the role of heterogeneity, agents' interactions and trade frictions*, Int. J. Mod. Phys. C **10**, 149 (1999).

A. Kirman, J. B. Zimmermann, *Economics with Interacting Agents*, Springer-Verlag, Heidelberg, 2001.

A. Krawiecki, J. A. Holyst, D. Helbing, *Volatility clustering and scaling for financial time series due to attractor bubbling*, Phys. Rev. Lett., **89**, 158701 (2002).

B. LeBaron, *A builder's guide to agent-based financial market*, Quantitative Finance, **1**, 254 (2001).

M. Lévy, H. Lévy, S. Solomon, *Microscopic simulation of the stock market*, Journal de Physique, **I5**, 1087 (1995).

T. Lux, M. Marchesi, *Scaling and criticality in a stochastic multiagent model*, Nature, **397**, 498 (1999).

T. Lux, M. Marchesi, *Volatility Clustering in Financial Markets: A Microsimulation of Interacting Agents*, Int. J. Theo. Appl. Fin., **3**, 675 (2000).

R. G. Palmer, W. B. Arthur, J. H. Holland, B. LeBaron, P. Taylor, *Artificial economic life: a simple model of a stock market*, Physica D, **75**, 264 (1994).

M. Raberto, S. Cincotti, S. M. Focardi, M. Marchesi, *Agent-based simulation of a financial market*, Physica A, **299**, 320 (2001).

H. Takayasu, H. Miura, T. Hirabayashi, K. Hamada, *Statistical properties of deterministic threshold elements: the case of market prices*, Physica A, **184**, 127–134 (1992).

L.-H. Tang, G.-S. Tian, *Reaction-diffusion-branching models of stock price fluctuations* Physica A, **264** 543 (1999).

- SPECIALIZED PAPERS: THE MINORITY GAME

D. Challet: See the up to date collection of papers on the following web page: http://www.unifr.ch/econophysics/minority.

D. Challet, Y. C. Zhang, *Emergence of cooperation and organization in an evolutionary game*, Physica A, **246**, 407 (1997).

D. Challet, M. Marsili, R. Zecchina, *Statistical mechanics of systems with heterogeneous agents: Minority Games*, Phys. Rev. Lett., **84** 1824 (2000a).

D. Challet, M. Marsili, Y.-C. Zhang, *Stylized facts of financial markets and market crashes in Minority Games*, Physica A, **276**, 284 (2000b).

D. Challet, A. Chessa, M. Marsili, Y. C. Zhang, *From Minority Games to real markets*, Quantitative Finance, **1**, 168 (2001), and references therein.

P. Jefferies, M. Hart, P. M. Hui, N. F. Johnson, *Traders dynamics in a model market*, Int. J. Theo. Appl. Fin., **3**, 443 (2000).

P. Jefferies, N. F. Johnson, M. Hart, P. M. Hui, *From market games to real-world markets*, Eur. J. Phys., **B20**, 493 (2001).

M. Marsili, R. Mulet, F. Ricci-Tersenghi and R. Zecchina, *Learning to coordinate in a complex and non-stationary world*, Phys. Rev. Lett., **87**, 208701 (2001).

Y. C. Zhang, *Towards a theory of marginally efficient markets*, Physica A, **269** 30 (1999).

- SOME RECENT READS

D. Challet, M. Marsili, Y. C. Zhang, *Minority Games: Interacting agents in Financial Markets*, Oxford University Press, 2004.

M. B. Gordon, J. P. Nadal, D. Phan, V. Semeshenko, *Discrete Choices under Social Influence*, Generic Properties e-print arXiv 0704:2333.

Q. Michard and J.-P. Bouchaud. *Theory of collective opinion shifts: from smooth trends to abrupt swings*. The European Physical Journal B **47**, 151 (2005).

E. Samanidou, E. Zschischang, D. Staüer and T. Lux, *Agent-based Models of Financial Markets*, Rep. Prog. Phys **70**, 409–450 (2007).

V. Semeshenko, M. B. Gordon, J. P. Nadal, *Collective states in social systems with interacting learning agents*, e-print arXiv 0704:2324.

G. Teyssiere, A. Kirman, *Long Memory In Economics*, Springer, 2006.

M. Wyart, J. P. Bouchaud, *Self referential behaviour, overreactions and conventions in financial markets*, Journal of Economic Behavior & Organization, **63**, 1 (2007).

Index of most important symbols

Λ	loss level; Λ^* loss level (or value-at-risk) associated to a given probability \mathcal{P}^*	172
Λ_q	hypercumulant of order q	30
μ	exponent of a power-law, or a Lévy distribution	7
ν	exponent describing the decay of a temporal correlation function	38
π_{ij}	'product' variable of fluctuations $\delta x_i \delta x_j$	191
ρ	interest rate on a unit time interval τ	232
	or correlation coefficient	65
$\rho(\lambda)$	density of eigenvalues of a large matrix	161
$\rho_{ij}^+(\theta)$	exceedance correlation as a function of level θ	189
$\rho t, ij$	tail dependence or covariance	191
σ	volatility	5
σ_1	volatility on a unit time step: $\sigma_1 = \sigma\sqrt{\tau}$	132
σ_p	risk associated with portfolio \mathbf{p}	203
ς	skewness	7
Σ	'implied' volatility	242
τ	elementary time step	88
ϕ_0	common factor in the one factor model	156
ϕ_k^N	quantity of underlying in a portfolio at time k, for an option with maturity N	232
ϕ_k^{N*}	optimal hedge ratio	259
ϕ_M^*	hedge ratio used by the market	260
Φ	order imbalance	356
ψ_k^N	hedge ratio corrected for interest rates: $\psi_k^N = (1+\rho)^{N-k-1}\phi_k^N$	233
ξ	random variable of unit variance	39
Υ	vol of the vol in Heston model	142
Ω	reverting frequency in Ornstein-Uhlenbeck models	142
\equiv	equals by definition	4
\approx	is approximately equal to	17
\propto	is proportional to	24
\simeq	is to leading order equal to	9
\sim	is of the order of, or tends to asymptotically	7
$\mathrm{erfc}(x)$	complementary error function	29
$\log(x)$	natural logarithm	6
$\Gamma(x)$	gamma function: $\Gamma(n+1) = n!$	11

Index

Printed in the United States
By Bookmasters